Molecular Plant Virology

Volume II

Replication and Gene Expression

Editor

Jeffrey W. Davies
Head
Virus Research Department
John Innes Institute
Norwich
United Kingdom

CRC Press
Taylor & Francis Group
Boca Raton London New York

CRC Press is an imprint of the
Taylor & Francis Group, an **informa** business

First published 1985 by CRC Press
Taylor & Francis Group
6000 Broken Sound Parkway NW, Suite 300
Boca Raton, FL 33487-2742

Reissued 2018 by CRC Press

Library of Congress Cataloging in Publication Data
(Revised for vol. 2)
Main entry under title:

Molecular plant virology.

Includes bibliographies and index.
Contents: v. 1. Virus structure and assembly and nucleic-acid protein interactions — v. 2. Replication and gene expression.
1. Plant viruses—Collected works. 2. Molecular biology—Collected works. I. Davies, Jeffrey W.
QR351.M64 1985 576'.6483 84-9469
ISBN 0-8493-6290-3 (v. 1)
ISBN 0-8493-6291-1 (v. 2)

A Library of Congress record exists under LC control number: 84009469

ISBN 13: 978-1-315-89568-0 (hbk)
ISBN 13: 978-1-351-07478-0 (ebk)

Visit the Taylor & Francis Web site at http://www.taylorandfrancis.com and the
CRC Press Web site at http://www.crcpress.com

PREFACE

In calling this series *Molecular Plant Virology,* I had in mind aspects of plant virology of interest to biochemists, molecular geneticists, biophysicists, genetic engineers, or, collectively, "molecular biologists". At the same time, the intention was to provide up-to-date reviews, by expert contributors, on current research topics in plant virology of interest and referential use to virologists and plant biologists. The selected topics are pitched mainly at a research level, but with sufficient introduction and cross-referencing to enable graduate students to enter this fascinating field and, hopefully, not get lost.

Although the volumes are primarily for reference, I encouraged authors to provide readily readable accounts, with, in some cases, interesting historical notes or at times a personal touch. A certain amount of criticism or questioning adds to the flavor.

Molecular biology and virology have run a related course for many years, developing together, merging at times, and gaining from each other. This is not surprising when one considers the nature of virus particles, what we are now learning about their functions, and what molecular biology is.

At this point a sinking feeling told me I might have to define molecular biology. I decided to go back to the origin and roots of the term. Warren Weaver[1] states that the first mention, in the 1938 Annual Report of the Rockefeller Foundation, describes "— a relatively new field, which may be called molecular biology, in which delicate modern techniques are being used to investigate ever more minute details of certain life processes". Later,[2] in a Harvey Lecture, Astbury stated that "Molecular Biology implies searching below the large scale manifestations of classical biology for the corresponding molecular plan". He went on to say that it was "concerned particularly with the forms of biological molecules and with the evolution, exploitation and ramification of these forms. . . . " He further stated that "Molecular Biology is predominantly 3-dimensional and structural" — but explained that "It must, of necessity, enquire at the same time into genesis and function".

Volume I of *Molecular Plant Virology* reviews aspects of the structure of plant viruses, and development and application to physical techniques such as electron microscopy and X-ray crystallography. The structure and assembly of rod-shaped, spherical, and bacilliform virus particles are considered. In the latter case, nucleic acid-protein interaction is emphasized which takes us a step nearer to function.

In Volume II, function and molecular plan (strategy) are the key words but still in relation to structure, albeit nucleic acid sequence rather than particle structure. Here again we see new and different techniques having remarkable influence. Early events in infection are discussed, followed by gene expression and replication of RNA and DNA viruses. These two "structure and function" volumes are being published almost concurrently.

In the near future we will, hopefully, see the "exploitation", in the form of genetic manipulation with plant viruses and the use of plant virus sequences in the construction of gene vectors. The "ramifications" of the study of structure, function, and modification of viruses should lead to a molecular analysis of virus-host relationships, host range and symptom expression, and other plant responses including resistance. This molecular pathology will require, however, a better understanding of the molecular biology of plants, both healthy and infected, a subject which is now rapidly progressing.[3,4,7,8]

Should the reader not be familiar with plant virology, I must recommend as one excellent comprehensive text book, Matthew's *Plant Virology,*[3] and for companion reading with this series, recent reports from research meetings, such as those of Cold Spring Harbor[4] and EMBO or NATO Workshops.[5-7]

Jeffrey W. Davies
Norwich, January 1984

REFERENCES

1. Weaver, W., Molecular biology: origin of the term, *Science,* 170, 581, 1970.
2. Astbury, W. T., Adventures in Molecular Biology, *Harvey Lecture Series,* 46, 3, 1950.
3. Hall, T. C. and Davies, J. W., *Nucleic Acids in Plants,* Vols. I and II, CRC Press, Boca Raton, Florida, 1979.
4. Grierson, D. and Covey, S., *Plant Molecular Biology,* Blackie Press, Glasgow, 1984.
5. Matthews, R. E. F., *Plant Virology,* 2nd ed., Academic Press, New York, 1982.
6. Robertson, H. D., Howell, S. H., Zaitlin, M., and Malmberg, R. L., Plant infectious agents: viruses, viroids, virusoids and satellites, *Current Comm. Molec. Biol.,* Cold Spring Harbor Laboratory, New York, 1983.
7. Ciferri, O. and Dure, III, L., Eds., *Structure and Function of Plant Genomes,* NATO ASI Series, Plenum Press, New York, 1983.
8. van Vloten-Doting, L., Groot, G. S. P., and Hall, T. C., *Molecular Form and Function of the Plant Genome,* NATO ASI Series, Plenum Press, New York, 1985.
9. Hirth, L., Richards, K., and Fritig, B., Plant Viruses, Satellites and Viroids: Genome Structure, Expression and Replication, EMBO Workshop, Strasbourg, 1984.

THE EDITOR

Jeffrey W. Davies, Ph.D., F.L.S., received a B.Sc. (1963) and Ph.D. (1966) at the University of Nottingham where he began his research career with E.C. Cocking on plant protoplasts and plant protein synthesis in cell-free extracts. From 1966 to 1970 he was a faculty member at the University of Edinburgh, in the then new Department of Molecular Biology, headed by M.R. Pollock and W. Hayes. During this time, his research concerned the control of enzyme synthesis and the structure and translation of bacterial RNA. This interest extended to bacteriophages and plant viruses after an EMBO-Fellowship (with L. Bosch) in Leiden, The Netherlands (1970), when he was appointed Assistant Scientist in Paul Kaesberg's Biophysics Laboratory, at the University of Wisconsin, Madison. His research there concerned translation and replication of RNA bacteriophages and plant viruses, especially brome mosaic virus. In 1976, Dr. Davies joined A. van Kammen's Department of Molecular Biology in Wageningen, the Netherlands, where he continued research on the molecular biology of plant viruses.

In 1979, he was appointed Head of the Virus Research Department (formerly the ARC Virus Research Unit, Cambridge) now at the John Innes Institute in Norwich. Dr. Davies' current interests include the development of plant viruses as plant gene vectors, and the molecular pathology of plant virus diseases.

CONTRIBUTORS

S. N. Covey, Ph.D.
Senior Scientific Officer
Virus Research Department
John Innes Institute
Norwich
United Kingdom

J. W. Davies, Ph.D.
Head
Virus Research Department
John Innes Institute
Norwich
United Kingdom

William G. Dougherty, Ph.D.
Assistant Professor
Department of Plant Pathology
North Carolina State University
Raleigh, North Carolina

Rob Goldbach, Ph.D.
Associate Professor
Department of Molecular Biology
Agricultural University
Wageningen
Netherlands

Ernest Hiebert, Ph.D.
Professor
Department of Plant Pathology
University of Florida
Gainesville, Florida

A. J. Maule, Ph.D.
Senior Scientific Officer
Virus Research Department
John Innes Institute
Norwich
United Kingdom

John G. Shaw, Ph.D.
Professor
Department of Plant Pathology
University of Kentucky
Lexington, Kentucky

John Stanley, Ph.D.
Senior Scientific Officer
Virus Research Department
John Innes Institute
Norwich
United Kingdom

Albert van Kammen, Ph.D.
Professor
Department of Molecular Biology
Agricultural University
Wageningen
Netherlands

TABLE OF CONTENTS

Volume II

REPLICATION AND GENE EXPRESSION

Chapter 1

EARLY EVENTS IN PLANT VIRUS INFECTIONS

John G. Shaw*

TABLE OF CONTENTS

* This review was completed in September, 1983.

I. INTRODUCTION

Of all the steps in the development and expression of a plant virus infection, undoubtedly the least understood are those which come first, i.e., the so-called early events. Our knowledge of the events in infected cells, during which active viral gene expression, replication, and progeny virion accumulation occur, is considerably more detailed than our understanding of what precedes these events. This gap continues to increase with time and, in recent years, there have been successively fewer publications dealing with the subject of this chapter. While it is surely not necessary to point out the vital importance of the first stages in virus infections, it is puzzling that so little effort has been devoted to events without which the dynamic activities involved in the production of the new virus particles and in the development of some very serious diseases will not occur.

It is the purpose of this article to bring together some of the information that bears upon the early events in the establishment of virus infections of plants. This will not result in the offering of new concepts, profound insights, or inspired suggestions because, at least to this reviewer, the subject is so complex and so poorly understood that few such expressions of wisdom are readily apparent. I should, in fact, warn the reader at the outset of what may seem a decidedly gloomy treatment of the subject. This mood, if apparent, should be regarded as an expression of the necessity to emphasize those aspects of an important subject about which far too little is known and, yet, about which rather dogmatic but clearly unsubstantiated statements are sometimes made. It is also hoped, of course, that this account will remind readers of the pressing need for more energetic and inspired investigation of the subject.

II. SCOPE OF THIS CHAPTER

For probably any virus-host system, the early events in the establishment of infection can be thought of as those essential activities which occur during the period from the first contact between virion and host cell to the point where the parental viral nucleic acid can begin to function in translation, transcription or integration into the host genome. This definition can become complicated in referring to infections resulting from cell-to-cell spread of virus in tissues. In such cases, the early events may be quite different from those in cells that are inoculated directly. Thus, this article will focus upon the early events in cells, single or in tissue, which are the first to become infected after inoculation.

Questions concerning the mechanisms and activities involved in the earliest stages of plant virus infections will be seen to remain largely unresolved. It may thus be appropriate to list here some of the more apparent reasons for this lack of information:

1. The early events in infection of crop plants, in their natural settings, occur on a scale which cannot at present be effectively investigated. Somewhat artificial experimental systems and procedures must therefore be used for the study of these events.
2. The requirement for large numbers of virus particles to ensure the infection of significant numbers of cells leads to what is the greatest problem concerning these types of investigations — the difficulty of distinguishing the particles and their constituents which play an active role in the establishment of infection from those which do not.
3. The vast majority of plant viruses are of small size and simple structure. This makes them very difficult to detect and describe by electron microscopy in the early stages of infection.
4. The apparent requirement for wounding of cells or tissues in order to initiate infections results in a much less definitive interaction of plant virus particles and cells than does the adsorption of animal and bacterial virus attachment structures to cell receptors.
5. A lack of investigative effort is the major contributing factor to our poor understanding

of the early events in plant virus infection. While other aspects of plant virology receive intensive investigation and generate great excitement, the initial stages in the viral life cycle continue to be studied on an almost casual basis with a paucity of fresh approaches. In addition, only one plant virus has been used to any extent in such investigations.

It is hopefully obvious that most of the important virus infections of plants do not occur in laboratories and greenhouses. Rather, they occur in fields, pastures, orchards, and forests usually with very little role played by human beings. They occur, in many cases, because an insect, in the course of seeking to nourish itself, has acquired virus particles from an infected plant and has subsequently introduced them into a different plant. They also occur after transmission by other invertebrates and by fungi and pollen.

There is virtually no information on the early events in such natural situations. For rather obvious reasons, it has been necessary, in order to obtain even the most meager insights into the processes involved in these events, to utilize systems and procedures which are undoubtedly quite different from those which occur under natural conditions. This review will deal with information that has come from the use of these experimental systems. At the same time, it is hoped that some appreciation of the probable artificiality inherent in such studies will be maintained.

Most of this chapter will focus upon investigations of the early events in virus infections of leaves and protoplasts and of the interaction of virus particles with subcellular fractions and membranous substances. This will be preceded by brief descriptions of the various stages involved in the early events and some general discussion of the manner in which they have been investigated in plants.

A fairly recent review of early events in plant virus infections has been presented by de Zoeten.[1]

III. STAGES IN THE EARLY EVENTS

The early events in virus infections have often been considered as involving three steps: adsorption (or attachment, retention, binding, fixation), penetration (or entry), and uncoating (or disassembly). Some of these terms suggest activities or mechanisms which, for many host-virus systems, have not been adequately demonstrated. Perhaps more worrisome is that the use of these terms frequently suggests a series of discrete stages, often in a given order, in cases where there is inadequate evidence of such stages or where two or more of the activities appear to occur simultaneously or in a different order. Nevertheless, the terms do have wide acceptance and some of them will be used in this chapter.

The initial contact between virus particle and host cell is referred to as adsorption or attachment. The process during which the virion or its nucleic acid and, in some cases, other constituents, passes into the cytoplasm of the cell has been described as entry or penetration. Uncoating is the removal of various components of the mature virus particle and the consequent release of the viral genome and other constituents which play a part in the establishment of infection. For the reasons mentioned above, and from what will follow, it will be apparent that these terms can be used in a rather careless manner in referring to early events in virus infections of plants.

It may be worth mentioning here two other terms which, while not used to refer to events, have often been featured in discussions of plant virus infections. "Infectible sites" are said to be created on the surfaces of leaves as a result of abrasion during the mechanical inoculation procedure. These sites may be converted to "infective centers" if virus particles are present at an appropriate time. Such sites and centers, which are defined in an "operational" sense, obviously exist, but the use of the terms seems inappropriate here and may in fact promote casual, superficial consideration of the locations and mechanisms actually involved during the early events of infection. These terms will not be used in this chapter.

The early events in virus infection may be said to have been completed when the nucleic acid (and, in some cases, other virion constituents) of parental virus particles becomes available for the first reactions leading either to the synthesis of new viral protein and nucleic acid molecules or to its integration into the host genome. For the vast majority of plant viruses, the first activities after the early events are thought to involve translation of at least part of the single-stranded, positive-sense, parental viral RNA (see Chapter 3, this volume). Other types of viruses will have different requirements at this stage. These reactions, and those that follow as the infection proceeds, will be treated in other chapters in this series.

IV. OTHER VIRUS-HOST SYSTEMS

Since much more effort has been devoted to the study of the early events in animal and bacterial virus infections than to those in plant virus infections, it may be appropriate to provide a brief review of some of the systems where useful information is available.

A. Animal Viruses

Members of nearly every group of the animal viruses have been investigated with respect to the early stages of infection. There have been several recent reviews of this subject.[2-6]

1. Adsorption

The interaction between proteins or glycoproteins on the exterior of virus particles and receptors on cell surfaces is the initial step leading to infection by most animal viruses. Viral attachment proteins have been most extensively investigated in enveloped viruses that contain spikes (e.g., in certain myxoviruses, togaviruses, and rhabdoviruses) and in the nonenveloped adenoviruses and reoviruses. Much less information about virion structures involved in attachment is available for those viruses which do not contain spikes and, in particular, for the small icosahedral viruses which are similar in structure to so many plant viruses. Capsid proteins are presumably involved in attachment of viruses such as the picornaviruses to host cell receptors.

The receptors to which virus particles become attached are generally thought to be cell surface glycoproteins or lipoproteins, although, in most cases, their chemical structure has not been established. There are estimated to be about 10^5 receptor units on a typical human cell surface.[7] Receptors are specific for the attachment of certain viruses and thus are important determinants of susceptibility and tissue tropism. In some cases, it has been shown that the interaction between virion and receptor is, at first, a rather tenuous one but that, within a short time, it becomes more firm (and irreversible) as the result of the involvement of additional receptors. Enhanced membrane fluidity may be involved in these events. The lateral movement of complexes formed by the attachment of Mengovirus to sites in cell membranes results in the formation of patches and then the polar distribution (capping) of the virus-receptor complexes.[8]

2. Entry

Once the interaction between virion and cell receptors has been established, the next event is the passage of part or all of the virus particle into the cell. For some years, there has been considerable controversy as to which of two mechanisms, adsorptive endocytosis (viropexis) or fusion, is of primary importance in the entry process.[4] There have been numerous electron microscope studies in which viruses (especially the enveloped viruses) have been shown to have entered animal cells by adsorptive endocytosis.[2] In this process, virus particles are trapped within invaginations of the cell surface which eventually pinch off to form intracellular vesicles. Such particles, though now inside the plasma membrane, are still separated from the cytoplasm by the vesicle membrane.

In the fusion mechanism, the viral envelope fuses with the plasma membrane and internal portions of the virion pass directly into the cell (thus accomplishing entry and, at least, initial uncoating at one time). Some nonenveloped viruses may also be capable of direct entry after the interaction of plasma membrane and surface components of the virion.

There are recent reports in which it is claimed that both the endocytotic and fusion processes are involved in the establishment of some infections.[9,10] Some enveloped viruses appear to associate with coated pits in the plasma membrane, to be taken into coated vesicles by endocytosis and then translocated via large, uncoated vesicles into lysosomes. Fusion of viral envelope with lysosomal membrane then releases the nucleocapsid directly into the cytoplasm without exposure of it to the contents of the lysososme. The fusion stage seems to be dependent upon the low pH in the lysosome. It has been concluded that most enveloped animal viruses engage a fusion mechanism at some stage of the entry process for passage of virus particle constituents across a membrane barrier that, initially, separates the virus from its ultimate intracellular location.[4]

With nonenveloped viruses, both viropexis and fusion mechanisms of entry have been reported but more emphasis has been given to changes in virion structure as a result of attachment than to the details by which the virus gains access to particular intracellular sites. In some of these cases, entry and uncoating seem to be the result of a related set of events rather than discrete stages in the establishment of an infection.

3. Uncoating

This remains one of the most poorly understood events in the viral life cycle. In a few cases, there is reasonably good evidence of the subcellular location at which some uncoating reactions occur, but in almost no case is there a detailed explanation of the mechanisms involved in these reactions. Uncoating of many animal viruses does not occur in a single step but rather, especially with the more complex viruses, as a series of reactions leading to the eventual release of the minimal constitutents required for infection. The process may be regarded as completed when the viral nucleic acid is present inside the cell in a form suitable for transcription or translation.

With the enveloped viruses, uncoating begins with removal of the lipoprotein envelope. As noted above, this may occur at the time of initial contact of virus and plasma membrane or after endocytosis. Subsequent uncoating reactions result in the release of various structural proteins. In some cases, at least the final step may occur in the nucleus or at the nuclear membrane.

In the case of adenovirus, a subviral particle, from which some of the surface proteins have been removed and which is nuclease sensitive, appears in the cell after entry. In subsequent reactions, the capsid is degraded and the viral DNA appears in the nucleus.[11] Polyoma virus is thought to be transported as intact virions in small vesicles to the host cell nucleus and to be uncoated via intermediate structures after entering the nucleus.[12] Reovirus particles are reported to be converted within phagolysosomes into intermediate subviral particles.[13] A second uncoating reaction then results in the appearance of subviral particles with activated transcriptase activity in the cytoplasm.[14]

Various aspects of the uncoating of picornaviruses have been studied for many years.[5] After attachment of virus particles to the surfaces of susceptible cells, many of the particles are structurally altered, i.e., they lose one (VP4) of the four capsid proteins and become sensitive to proteolytic and detergent degradation. This early modification is considered to be essential for the establishment of infection though some of the altered particles may be eluted from the cells. Subsequent dissociation of pentameric VP2 proteins then apparently leads to the release of viral RNA. It has been suggested that complexes of lipid, carbohydrate, and protein materials of the plasma membrane are active in bringing about the stepwise uncoating of poliovirus and in preventing thermal degradation of subviral particles or virion RNA.[15] It is still not clear whether the final uncoating reactions occur in association with the cell membrane or after passage of modified particles into the cytoplasm. The final product is messenger-sense, single-stranded 35S viral RNA.

B. Bacterial Viruses

The attachment of bacterial viruses to host cells and subsequent entry of viral nucleic acid has been recently reviewed.[16]

The most common types of bacterial viruses, the tailed phages, possess some morphological features not found in any other types of viruses. These phages utilize elaborate structures, including contractile or noncontractile tail sheaths, collars, baseplates, fibers, and spikes to achieve attachment to receptors in the cell surface and injection of the viral genome and small amounts of other materials into the cell. In some cases, lipopolysaccharides in the outer cell envelope are thought to function as receptors. There is no example of a virus infecting higher plants whose structure or likely mode of interaction with the host cell even remotely resembles this situation.

The icosahedral, single-stranded DNA phages (e.g., φX174) become attached by means of spike proteins to the lipopolysaccharide of the cell surface. The capsid proteins undergo a rearrangement and DNA and spike proteins enter the cell. In the case of the filamentous, single-stranded DNA phages (e.g., f1, M13), the DNA and pilot protein (present in two to four molecules per particle) enter the cell and the major capsid protein becomes embedded in the cell membrane. It was previously thought that the DNA of these viruses enters the tips of the F pili of male cells but this now seems uncertain. Isometric, single-stranded RNA phage particles (e.g., MS2) contain one molecule of a protein (gpA) which is required for attachment of the virus to F pili. The particles migrate to the base of the pili and RNA and a fragment of gpA enter the cell. All the small phages seem to possess mechanisms for overcoming the barrier posed by the cell wall. Therefore, even though they are somewhat similar in structure to many of the plant viruses, it seems unlikely that the adsorption and entry processes they employ will be shown to be involved in plant virus infections.

V. EARLY EVENTS IN PLANT VIRUS INFECTIONS

For virtually all studies of the molecular biology of plant virus infection, it has been necessary to employ somewhat artificial systems and procedures. With respect to the later stages of infection, such systems probably support activities which are quite similar to those occurring under natural conditions. Thus, it is reasonable to suppose that the synthesis of viral proteins and nucleic acids and the assembly and accumulation of progeny virions in an infected protoplast occur in much the same manner as they do in the cell of a field-grown plant into which virus particles have been placed by a vector. It is with the study of the earlier events, however, that considerable deviations from the natural situations are introduced.

A. Experimental Systems

For many years, virtually all studies of early events in virus infections of plants were conducted with mechanically inoculated leaves. In more recent years, protoplasts have offered a more effective means of obtaining information concerning some of the activities of viruses in plant cells. In addition, there have been several studies of the interaction of virus particles with various subcellular fractions (especially those containing membranous materials) and with various reagents and synthetic materials thought to be similar to membrane constituents. A review of some of the investigations with these various experimental systems will be presented below.

The vast majority of viruses which infect plants are nonenveloped and contain positive-sense, single-stranded RNA. It is therefore quite disturbing that nearly all the studies pertinent to this chapter have dealt with only one such virus, tobacco mosaic virus (TMV). To the author's knowledge, there have been no investigations with members of the phytoreoviruses, the enveloped viruses (rhabdoviruses, tomato spotted wilt virus), or the DNA-containing viruses (caulimoviruses, geminiviruses). The need to extend such studies to include other plant viruses is obvious.

B. Experimental Procedures

Probably the most obvious difference between natural and experimental plant virus infections lies in the procedures involved in inoculation. Few infections in nature are likely to have arisen because an agent sprinkled an abrasive substance on a leaf and then proceeded to rub the leaf with a suspension of virus particles. Nor is it likely that the external membranes of the cells in a field grown plant were bathed in a near-lethal dose of a polycationic substance at the time of inoculation. In comparison with the hundred or fewer virus particles required for successful inoculation of a healthy plant via a single aphid,[17] much larger amounts of virus are used for the production of a small number of primary infections (local lesions) in a mechanically inoculated leaf. As will be mentioned later, massive amounts of virus have been used in the inoculum in most studies of the early events.

The main reason, of course, for the use of such methods has been to overcome the problems of detection of the few virus particles and the few cells in the plant with which they interact. The usual procedure has been to manually establish initial contact between virus particles and plant tissues, cells, or protoplasts under conditions which should promote infection and to attempt to determine the fate of the virus particles and the immediate responses of the host. Several methods of analysis of these events have been used. These include determination of the numbers of infections and time of appearance of symptoms in the host tissue or cells, measurement of the extent of attachment to or retention by the host of the virus particles, changes in the structure and infectivity of the retained virus particles, and electron microscopic observations of the virus-host complex.

Over a period of several decades, there have been many reports of the effects of various treatments before, during, and after inoculation on the production of signs or symptoms of disease (usually numbers of local lesions). Some of these have purported to provide information on the early events in infection. Most of this literature will not be discussed here since it has been reviewed on several occasions.[18-20] It is difficult to judge the significance of some of these types of studies because the analysis of an effect on an initial activity is based upon detection of the end result of a series of complex intervening reactions. In some cases, there is a real question about the actual stage in the infection process (i.e., an early or one of many later events) that has been affected by the treatment.

C. Mechanically Inoculated Leaves

For well over one-half century, persons working with plant viruses have spent many hours applying virus-containing fluids to the surfaces of leaves. For just as long a period, questions concerning the means by which this procedure leads to the production of infections have been pondered. What is the effect of mechanical inoculation on the surface and underlying structures of the leaf? What is the location of those virus particles which will take an active part in the development of infection? What is the physical structure of these particles and their protein and nucleic acid components? What constituents and activities of host cells are required for the establishment of infection? How do answers to these questions change with time after inoculation?

1. Attachment and Entry

For virus particles, or at the very least, viral nucleic acid molecules, to enter epidermal cells during or after mechanical inoculation of a leaf, a considerable barrier must be overcome. Attached to the outer surface of the cellulosic epidermal cell wall is the cuticle, made up of cutin, a polyester of hydroxy and epoxy fatty acids. In many leaves, pectin is found between or within the cell wall and cuticle layers. The outer surface of the leaf is covered with a layer, often of a distinctive pattern, of epicuticular wax.

In the mechanical inoculation of a leaf, an abrasive substance such as carborundum or Celite is usually used. Rubbing the leaf surface in the presence of such a substance produces varying kinds and degrees of local wounds and structural perturbations in the surface materials of the leaf. These are regarded by some as being quite superficial[21,22] and by others as involving extensive damage to the epidermal cells themselves.[23]

Microscopic observations of leaf surface structure after mechanical inoculation have led to several suggestions of possible portals and mechanisms of entry of virus particles:

1. Direct entry, i.e., local penetration of all surface layers and of the wall and plasmalemma by abrasive particles with simultaneous entry of virus particles into the epidermal cells and subsequent repair of the plasmalemma.[24]
2. Deposition of virus particles on or in the plasmalemma during abrasion of surface layers followed by a pinocytotic uptake of virus particles;[18] it has been argued that pinocytosis does not (and cannot) occur in intact, turgid plant cells.[25]
3. Entry via subcuticular, electron-lucid areas ("blebs") formed in the cell wall as a result of superficial abrasion;[26] some of these areas adjoin the plasmalemma and may be temporarily hydrated as a result of the extrusion and reabsorption of cellular fluids.[21]
4. Entry via ectodesmata (spaces or channels in outer epidermal cell walls) which have been suggested as being involved in the transport of virus particles or viral nucleic acid through the cell wall;[27,28] it is still felt by some that "ectodesmata are . . . best regarded as artefacts of a specialized fixation procedure".[29]

These and other possible avenues of entry of virus particles during or after mechanical inoculation have been discussed elsewhere.[1,20] None of them is supported by direct evidence. Such evidence would require an ability to detect the location and activities of those few particles in the inoculum that actually cause an infection, a feat that seems well beyond our means at this time. There have been several electron microscope investigations in which virus particles from the inoculum have been shown in various locations in or on leaf surfaces after mechanical inoculation.[26,28,30,31] There is no reason, however, to believe that these particles were, or would have been, involved in the establishment of infections. What for now seems a plausible, though hardly illuminating, conclusion is that a few of the particles are placed, during the inoculation process, either in contact with or inside the plasmalemma. Some of these can undergo the subsequent reactions that lead to infection, and the rest remain attached to or embedded in the various surface materials of the leaf. A postinoculation rinse of the leaf removes many but by no means all of these excess or misdirected particles.

It seems rather unlikely that most plant viruses contain specific attachment proteins or that their coat proteins are involved in a specialized cell recognition function. The vast majority of these viruses are nonenveloped particles and contain only one type of coat protein molecule. In contrast, few, if any, animal or bacterial viruses contain fewer than two types of capsid proteins, many contain spikes or other attachment structures, and most are thought to interact with cell receptors by means of specialized virion components. By analogy, the complex plant viruses (the phytoreoviruses, rhabdoviruses, and tomato spotted wilt virus) probably do contain attachment proteins. These structures may be utilized during interaction with insect cells since most of these viruses are thought to multiply in their vectors. It might, in fact, be expedient to consider them primarily as viruses of invertebrate animals rather than of plants.

There is no direct evidence of the involvement of specific cell receptors in the attachment or adsorption of plant virus particles. In an effort to identify substances in leaves that may have the characteristics of such receptors, Kiho and colleagues[32] have examined the ability of membrane fractions and constituents to bind TMV. In the presence of poly-L-ornithine, a particulate, presumably membranous, fraction from tobacco leaves was reported to bind the virus and polymerized viral coat protein. Binding activity was found in preparations from other plant species including some thought not to be hosts of TMV. Treatment of the membrane fraction with salt, detergent, and pronase yielded soluble material which bound TMV in the absence of poly-L-ornithine.[33] It was claimed that the main component of the active material was lipid.[34] Binding of TMV was favored by low pH and low ionic strength conditions but was inhibited in 1 *M* NaCl. The authors suggested that binding of virus particles to lipid and protein constituents of membranes is a necessary early step in the development of an infection. Whether the materials they have isolated can be regarded as receptors for virus particles remains to be determined.

2. *Uncoating*

During or immediately after inoculation, some virus particles in the inoculum will be situated in as yet unidentified locations which permit the establishment of infection. At this point, it becomes essential for the virus particle to be uncoated.

The requirement for uncoating became evident with the realization that the nucleic acid constituents of virus particles were the essential elements for infectivity.[35-38] This was soon followed by a number of reports describing the kinetics and extent of local lesion development after inoculation of leaves with intact virus or viral nucleic acid and often after treatment of the leaves with various inactivating agents.[20] These experiments provided evidence for a stage shortly after mechanical inoculation during which the separation of viral nucleic acid from coat protein occurred.

Attempts to investigate the uncoating process have also been made by examining the physical state and subcellular location of virus particles and their constituents at various times after mechanical inoculation of leaves.[39-54] TMV has been the most frequently used virus, though studies with brome mosaic virus (BMV) and turnip yellow mosaic virus (TYMV) have also been reported. In this type of experiment, inocula consisted of purified virus particles labeled with radioisotopes in their protein or nucleic acid moieties. At various times after mechanical inoculation and rinsing of the leaves, the amount and physical state of radioactive material in various fractions of leaf extracts were determined. The use of labeled virus ensured that only particles from the inoculum (parental virions), and not progeny virus particles, were analyzed.

Most of these studies have produced generally similar results. Depending upon the host tissue and the concentration of virus in the inoculum, some 1 to 10% of the TMV particles and up to 20% of the TYMV particles applied to leaves was retained after a postinoculation rinse. Within a few minutes after inoculation, substantial amounts of coat protein had been dissociated from the retained virus particles. With TMV and BMV, the released proteins were detected as individual molecules or small aggregates but, interestingly, in the case of TYMV, a large number of empty coat protein shells was found.[42] It is generally assumed that the coat proteins removed from the parental virus particles have no further role, in which case they may eventually be broken down in autophagic vacuoles or by some other mechanism.[55]

In nearly all cases, the amounts of dissociated protein increased over a period of several hours after inoculation. Within an hour after inoculation with TMV, some of the parental viral RNA was found to be free of coat protein.[45,49,50] Most of this RNA was degraded but a very small amount of it was shown to be of full length. The intact, uncoated TMV RNA was found in the particulate fractions of leaf extracts and could only be solubilized by treatment with deoxycholate.[50] Considerable amounts of what appeared to be incompletely uncoated TMV particles were found at all periods after inoculation in many of these experiments.

In further studies, it was reported that the rate and extent of uncoating of TYMV was considerably increased when the inoculum was high in ionic strength.[42] Pretreatment of leaves with high concentrations of poly-L-ornithine resulted in increased retention and uncoating of TMV.[56,57] It was suggested that uncoating of TMV might be a two-stage process since the initial removal of some of the coat protein from retained virus particles was not affected by low temperature or cycloheximide but further removal was inhibited by these treatments.[44] The significance of this observation is questionable since it was not shown that those particles actually involved in the infection process were affected by the treatments.

Plant virus particles seem to undergo at least some degree of uncoating in plants that are not thought to be hosts.[41,44,58] In fact, the author is not aware of any reports of failure to detect virus particle dissociation after mechanical inoculation of any plant.

The main problem in attempting, from the results of these types of experiments, to describe the process of uncoating of plant viruses concerns the extremely large amounts of radioactive virus that have been applied to leaves. Concentrations of milligrams of virus per milliliter (in one case, 15 mg/mℓ) have been routinely used with the result that, after a postinoculation rinse, a typical tobacco leaf has retained an amount of virus equivalent to 10^5 to 10^6 particles per epidermal cell! Under such conditions, it seems obvious that the vast majority of these particles will have little to do with the actual infection process and that their presence will most certainly obscure the activities of the particles of interest. It has been shown, for instance, that the amount of intact, completely uncoated, parental TMV RNA detected in inoculated tobacco leaves is only 10^{-3} to 10^{-5} of the RNA present in the particles in an inoculum of high virus concentration.[45,49,50] Until such time as this type of experiment is performed with much lower concentrations of virus, the observations reported to date must remain of questionable significance.

There have been several studies aimed at defining the site at which uncoating of plant virus particles occurs during or after mechanical inoculation. It has been suggested that uncoating of potato virus X (PVX) may occur on the cuticle,[28] but this conclusion appeared to be based upon failure to detect virus particles in the epidermal cell walls or ectodesmata of inoculated leaves. Tobacco rattle virus particles which had been introduced by infiltration into the intercellular spaces of leaves were found to be attached at their ends to cell walls and to become shorter with time.[30] It was concluded that uncoating occurs outside the cell with only the viral RNA entering the cell proper.[1,30] Rapid uncoating of TMV particles in intracellular spaces or on leaf surfaces has also been observed.[59] There seems to be little doubt that virus particles which remain in contact with leaf tissue after inoculation become degraded.[60] However, to equate this phenomenon with the uncoating reaction that yields the viral RNA molecules actually involved in infection is, at this time, simply not justified.

In other attempts to identify potential uncoating sites, the ability of certain subcellular fractions to bring about changes in virion stability has been investigated. Particles of several plant viruses were observed attached to membranes in extracts of host and nonhost leaves which had been rinsed and frozen immediately after inoculation.[61] These particles were degraded by treatment of the extracts with SDS or Triton® X-100; unattached particles were not altered by such treatments. A lipid fraction extracted with chloroform and isopropanol from tobacco leaf membrane preparations was found capable, in the presence of poly-L-ornithine, of altering the particle stability of TMV.[62,63] After a period of about 1 hr, the RNA of treated virus became RNase sensitive, and, upon addition of Triton® X-100, coat protein was released. Similar results were obtained when lecithin was used in place of the membrane lipid material. Reports of this kind may provide some experimental support for the suggestion made many years ago by Caspar[64] that, at least with respect to TMV, "a plausible place to look for virus disruption in the cell is in lipid-containing structures, since such an environment could affect these stabilizing interactions in a way similar to nonpolar solvents and detergents.'' Membranes, perhaps the plasma membrane itself, would seem to be appealing candidates for such structures.

The requirement of wounds in inoculated tissues for the establishment of infection has usually been considered from the point of view of the provision of portals for the entry of virus particles or viral nucleic acid. It should be kept in mind, however, that some of the responses of cells to the kinds of wounds produced during mechanical inoculation of leaves may be important in the uncoating of virus particles and in subsequent events in infection.

Wounding results in profound changes in damaged cells and in those adjacent to them.[65] The plasmalemma and internal membranes may be instantaneously torn, disordered, or depolarized and membrane components are degraded by enzymes released from vacuoles. The osmotic balance within such cells is rapidly altered. Within minutes, many biosynthetic pathways are activated and there is a dramatic increase in RNA synthesis, polyribosome formation, and protein synthesis. Vigorous production of lipids and phospholipids for the reconstruction of membranes is rapidly initiated. These or other events may well play an important role in the uncoating of plant viruses at the surface of or inside cells undergoing wound response and wound healing reactions.

Attempts have been made to determine the location on TMV particles at which uncoating begins after mechanical inoculation of leaves.[66,67] In these studies, TMV particles which contained labeled coat protein molecules at various locations along the particles were prepared by partial disassembly (with alkali and SDS) and reconstitution procedures. It has since been shown that both in vitro disassembly[68,69] and assembly[70,71] are initiated at different sites than was previously thought (Chapter 2, Volume 1, this series). Thus, the particles of Kiho[66] which were assumed to be labeled at the 5′-ends were probably labeled internally and those thought labeled at the 3′-ends were likely 5′-labeled. Similarly, the two types of end-labeled particles of Shaw[67] were likely labeled at the opposite ends from those stated. Kiho's experiments[66] were based upon the absence of labeled coat protein in partially uncoated TMV particles which were associated with ribosomes 3 hr after inoculation. The results would suggest that the first part of the RNA to be uncoated, and thus available for ribosome attachment, was the 5′-terminus (instead of the 3′-terminus as originally suggested). Such a polar uncoating process may draw some support from the observation that TMV particles from which 2 to 3% of the coat protein had been removed by mild alkali treatment were, under some conditions, more infectious than were intact particles.[72] Shaw's study,[67] in which the proportion of labeled protein released from and remaining with virus particles a few minutes after inoculation was determined, suggested that coat protein molecules were initially released from many positions along the length of the particle. Analysis of the sensitivity of partially uncoated TMV particles to RNase and SDS led Hayashi[52] to suggest that coat protein molecules had been removed from internal regions.

A polar removal of coat protein molecules beginning at the 5′-end of TMV RNA is certainly the more appealing of these alternatives since it would result in the initial exposure of the part of the RNA where translation begins. In addition, it would mimic the manner in which protein molecules are removed by treatment with alkali, detergent, and other reagents. The suggestion by Shaw[67] that protein is first removed from two to many internal regions is much less attractive and perhaps less significant since, unlike the study of Kiho,[66] it was not based upon parental virus particles that could be shown to be involved in a potentially useful role in infection.

Probably the most meaningful information dealing with the fate of virus particles after mechanical inoculation of leaves has been that of Kiho[66,73-75] on the appearance of uncoated parental TMV RNA in polysomes. This approach is important, at the very least, in identifying the numbers of particles or RNA molecules which have the real potential for a role in the development of infection since translation of at least part of the viral genome is an early requirement. Kiho[73,74] inoculated tobacco leaves with [^{32}P] TMV and detected radioactivity in the polysome region of sucrose gradients layered with leaf extracts prepared at various times after inoculation. Even with inocula containing 5 mg TMV per milliliter, 5 to 8% of the radioactivity retained by the leaves was found in the polysome region during the period between 1 and 2 hr after inoculation. It was suggested that these "parental polysomes" were membrane bound since deoxycholate treatment was required for their extraction. Ribosomes appeared to be associated primarily with the RNA in incompletely uncoated virus particles.[66,75]

Kiho et al.[58] also analyzed extracts of plants in which multiplication of TMV is not detectable. Shortly after inoculation, parental TMV RNA was found associated with ribosomes. This constitutes evidence against the presence of virus-specific receptors in plants and suggests that some degree of viral gene expression can occur in almost any plant cell.

It would be desirable to have further studies of the type performed by Kiho with much lower concentrations of virus in the inocula and with other plant viruses. In addition, the involvement of parental virus RNA in double-stranded RNA in infected cells should be examined as a link between the earliest events and those in which initial transcription occurs.

D. Protoplasts

In the last 15 years, a good deal of information about various aspects of plant virus infections has come from the use of protoplasts. In particular, those stages involving the synthesis of viral nucleic acid and proteins and the appearance of progeny virions have been very effectively investigated in infected protoplasts. There have been several reviews of plant virus-protoplast systems.[76,77] Rather surprisingly, there have been relatively few reports dealing with the early events in the infection of protoplasts. It might have been supposed that, with the development of a system in which virtually all the cells in a sample can be infected at a given time, a fairly detailed description of events over the first few hours after inoculation would have been achieved. Even with this system, however, there still remain the problems of detecting virus particles that actually initiate infections and of identifying those parts of the protoplast with which they interact.

1. Attachment and Entry

There have been several electron microscopy investigations of the manner in which plant virus particles gain entry into protoplasts. These have been adequately reviewed elsewhere[76] and will only be mentioned briefly here. In studies with TMV and tomato fruit protoplasts, Cocking and Pojnar[78] concluded that virus particles enter by means of a pinocytotic mechanism. Similar conclusions were reached after examination of tobacco protoplasts inoculated with poly-L-ornithine-treated TMV.[79,80] Takebe[81] stated that several viruses appear to enter protoplasts by an endocytosis-like process but added the cautionary note that proof of the involvement in infection of such a mechanism was not available. In similar investigations, Burgess et al.[82,83] reached rather different conclusions. Virus particles were observed associated with complex surface lesions caused by the poly-L-ornithine in the inoculum mixture, and it was proposed that these damaged areas in the plasmalemma provided a means for direct entry of virus particles. The possibility of such a means of entry has also been suggested by others.[84,85] Zhuravlev et al.[86] argued against the stimulation by poly-L-ornithine of a pinocytotic uptake of TMV by tobacco protoplasts since low temperature and certain metabolic inhibitors did not affect attachment of virus particles. In a later study,[87] it was claimed that the lack of uptake of radioactive sucrose during inoculation of protoplasts with poly-L-ornithine-treated-TMV also argued against entry by pinocytosis. An "electroosmotic mechanism of virus entry" in cases where poly-L-ornithine was not used in inoculation of protoplasts was proposed.

The results of electron microscopy studies of inoculated protoplasts have been regarded by some as giving rise to a controversy between those investigators favoring either endocytotic or direct entry (through wounds) mechanisms. This seems to be an unnecessary view, since in these same reports, it has been carefully stressed that transfer of virus particles or viral RNA to the cytoplasm has not been detected. There is, in fact, every reason to believe that parental virus particles do end up in small vesicles and that wounds in the plasmalemma are caused by poly-L-ornithine. The main problem with assuming that a controversy exists is that it suggests that one view is correct and that the other is not. It would seem just as logical to assume that virus particles can initiate infections after entry by either route or, on the other hand, to speculate that neither is the essential means of entry.

While for many virus-protoplast systems the use of a polycationic substance (usually poly-L-ornithine) is essential for efficient infections,[88] the presence of such a substance introduces a complicating factor to attempts to describe the early events. The mere requirement for a third component in an inoculation mixture indicates that a simple virion-receptor combination will not suffice for the initiation of infection. It also suggests that somewhat different processes are involved in inoculation of leaf cells, where poly-L-ornithine has little effect on the number of detectable infections,[81] and in inoculation of protoplasts. Attempts to examine the distribution of radioactive poly-L-ornithine in subcellular fractions of inoculated protoplasts have proven difficult because of the tendency of the material to adhere to virtually any surface with which it comes in contact.[89]

The formation of aggregates of virus particles by poly-L-ornithine is thought to be important for successful infection of protoplasts[84,90] though there is some uncertainty about the essential characteristics of such complexes.[91] It has been frequently proposed[81] that poly-L-ornithine promotes a decrease in the negative surface charges of virus particles and/or protoplasts[92] such that more favorable interactions may be established. The amount of TMV retained by protoplasts is considerably higher in the presence than in the absence of poly-L-ornithine.[93,94] Systems where virus particles or protoplast surfaces may be less negatively charged do not appear to require poly-L-ornithine for infection.[76,95] The suggested involvement of poly-L-ornithine in stimulating endocytosis or small wounds in protoplasts has been mentioned above. The continued uncertainty of the actual role of poly-L-ornithine has been a major factor in the failure to obtain more information about the early events in inoculated protoplasts.

An alternative method for inoculating protoplasts involves the use of polyethylene glycol.[96] It is claimed that virus particles situated between fusing protoplasts initiate infection and that vesicles are not detected during these events. Large amounts of virus appear to be required for efficient infection. Whether this system will be any more useful than others in providing significant information about the early events in infection is not known.

As with mechanically inoculated leaf cells, the numbers of virus particles used in most protoplast inoculations are so large as to make it difficult to examine the fate of those actually involved in infection. After inoculation and washing, hundreds or thousands of particles may be retained by each protoplast that ultimately becomes infected.[79,84,93,94,97-99] Takebe[76] has estimated that an average of 1.7 TMV particles were involved in the infection of a single tobacco protoplast when the concentration of virus in the inoculum was much lower (2.5 ng/mℓ) than normally used. With higher concentrations of virus, it is likely that most of the retained particles remain in aggregates with poly-L-ornithine and do not take part in infection.[93]

2. Uncoating

The uncoating of the virus particles which have been retained by protoplasts has been only very superficially examined. Thirty minutes after mixing of poly-L-ornithine-treated, radioactive TMV and tobacco protoplasts, Wyatt and Shaw[93] found that only very small amounts (3 to 5%) of protein and RNA had been released from parental virus particles and that these amounts did not increase with time. Some virus particles had been partially uncoated but many sedimented in sucrose gradients to the same extent as intact TMV and most particles appeared to be bound by the poly-L-ornithine. The amounts of intact, fully uncoated, parental viral RNA molecules in extracts of the inoculated protoplasts was not determined. Zhuravlev et al.[100] have reported quite different results even though the conditions for inoculation of tobacco protoplasts with ^{32}P-labeled TMV were virtually the same. After 1 hr, over 20% of the label in protoplast extracts was found to be soluble in trichloroacetic acid after RNase treatment and this rose to 40 to 70% a few hours later. Since the RNase-sensitive, acid-soluble label in control samples (protoplast extracts to which purified [^{32}P] TMV had been added directly) was quite high (16%), the above estimates of the degree of uncoating may be somewhat exaggerated. Nevertheless, the results of the two studies were sufficiently different to raise the question of the significance of either and to give emphasis to the view[76] that such experiments should be conducted with much lower concentrations of virus in the inoculum.

E. Other Approaches

It seems obvious that studies of the type mentioned in the preceding sections will not lead to an adequate description of the early events in virus infection. However, there seems no reason to doubt that new approaches can and will be utilized to aid in overcoming our current lack of understanding.

In recent years, several components of eukaryotic cells, including the plasmalemma, clathrin-coated vesicles, the Golgi apparatus, the endoplasmic reticulum, and lysosomes or vacuoles, have been recognized as being involved in complex and dynamic systems of membrane "traffic" pathways.[101] These systems are responsible for the movement of various materials into and out of cells and between various compartments and organelles within cells, and for the recycling of membranes or membrane constituents. So remarkable is the extent of this activity that an encounter between parental virus particles and some part of the array of membranes seems inevitable. In order to define more clearly some of the initial stages in virus infection, it would seem essential that there be a much greater appreciation of the complex subcellular structures and activities with which the virus is confronted.

There have been a few studies of the interaction of virus particles or coat proteins with artificially generated membranes. It was reported that TMV particles passed through thin phospholipid membranes in the presence of magnesium ions.[102] Others have claimed that TMV protein induced destabilizing and fusogenic effects in lipid bilayers and suggested that the ends of TMV particles may be regions for attachment of virions to host membranes.[103,104] Whether such observations have any real relationship to attachment and entry processes in TMV infections is presently quite uncertain. These types of study need to be more clearly coordinated with information on the structure and mechanical properties of the plasmalemma of plant protoplasts.[105]

There is a very large literature on the in vitro disassembly of plant virus particles. In most cases, the agents or conditions used to bring about the dissociation of viral coat proteins and nucleic acids are quite harsh and not likely to be encountered by virus particles at the locations at which in vivo uncoating occurs. It is clearly beyond the scope of this article to discuss in vitro disassembly, but it should be mentioned that there is a clear role to be played in the investigation of the early events in infection by those engaged in the study of virus structure and stability.

A noteworthy example of the potential application of this approach may be seen in the work of Durham and colleagues.[106-115] In a series of investigations involving hydrogen ion titration of several viruses in the presence and absence of divalent cations, it was proposed that most, if not all, plant viruses contain various calcium ion binding sites.[106-112] In the presence of sufficient calcium ions, the net charge on TMV and papaya mosaic virus particles was approximately zero over a wide pH range.[110,111] This observation led Durham and

Bancroft[110] to suggest that in a medium containing divalent càtions (e.g., "extracellular fluid"), a virus particle "could approach a membrane, preparatory to disassembly, without being electrically repelled."

It was stated that the removal of bound cations is a necessary part of the uncoating reaction.[108,113] According to the hypothesis of Durham[114] a virus particle is disassembled after contact with or during penetration of a membrane. The process is driven by the release of calcium ions as the virion enters the cell and encounters the low calcium ion concentration (about 10^{-7} *M*) therein. The free energy of calcium ion dilution, in combination with the hydrophobic environment of the membrane, brings about the dissociation of the parental virus particle. Dissociated coat protein molecules may or may not be released into the membrane lipid bilayer.

Whether or not Durham's hypothesis ultimately proves valid, it offered a fresh and stimulating approach to considerations of the early events in virus infection. As pointed out,[114] cation-mediated uncoating may be difficult to demonstrate in vivo though there is some evidence that virus particles may promote the passage of calcium ions through liposome membranes.[115] It would seem that it should be possible to demonstrate infection of protoplasts with viruses that normally require poly-L-ornithine in the absence of this substance if the divalent ion contents of the materials and solutions used are carefully controlled and if the plasmalemma plays the role assumed by Durham.[114]

Other useful applications of investigations of virus particle structure to the mechanisms involved in the uncoating process may be close at hand with the simple, isometric plant viruses. There are a number of detailed studies of the structure of such viruses and of the switching of their capsids, under relatively mild conditions, from rigid to swollen, flexible configurations. Uncoating of these viruses is undoubtedly initiated by some rearrangement or disturbance of the forces and units which maintain the stability of the intact virion. It should be possible to determine whether in vitro structural alterations and in vivo uncoating are controlled and characterized by similar mechanisms.

The tymoviruses present an intriguing case where observations of in vitro and in vivo disassembly may have some relationship. It has been reported that exposure of intact TYMV to high pH results in the release of a few protein subunits (possibly a pentamer or hexamer cluster) and of the entire RNA contents from the particle.[116] The remainder of the viral protein remains as an empty shell. Freezing and thawing of TYMV at pH 7.5 produces similar results and the capsid from which the RNA has been released has been shown by analytical ultracentrifugation, X-ray scattering, and X-ray diffraction to contain a hole formed by the loss of five to nine adjacent coat protein molecules.[117] Empty capsids are also produced when suspensions of particles of another tymovirus, belladonna mottle virus (which, unlike TYMV, does not contain polyamines), are adjusted from pH 5.5 to neutrality or above.[118] It would be quite interesting to determine whether the empty shells observed in leaves very shortly after inoculation with TYMV[42] contain "holes". It is also worth noting that one proposed scheme for the uncoating of picornaviruses involves the loss of as few as one of the pentamers from the capsid and the subsequent release of the viral RNA.[5]

Some recent and surprising observations by Wilson[119] may have a most important bearing upon the question of uncoating. When intact TMV particles were incubated in a rabbit reticulocyte lysate translation system, characteristic TMV RNA translation products were synthesized. Furthermore, some of the particles were found to have been partially uncoated and to be associated with clusters of ribosomes. An attractive model to account for some of these observations has been offered and it has been proposed that "co-translational disassembly" may be involved in in vivo uncoating of TMV.[119]

VI. CONCLUDING REMARKS

It is somewhat troubling to conclude this chapter by emphasizing how little we understand of the events in plant virus-host cell interactions which lead to the establishment of infection. Little comfort is derived from the statement of Dales[120] that, with regard to animal viruses, "there has been much less effort expended than scientifically desirable on investigating the early events during the normal, infectious process." Some of the more obvious reasons for the deficiency of understanding and effort have been mentioned in this chapter. What seems quite apparent is the urgent need to rectify this situation if we are ever to have adequate knowledge of the viral life cycle.

There would seem to be considerable differences in some of the early events of the infection processes between most plant viruses and those that infect bacterial and animal cells. The simple plant viruses probably do not contain specific attachment proteins and the surfaces of plant cells do not appear to contain special receptors for the attachment of most viruses. It thus seems unlikely that the question of whether or not an infection will occur is determined at the moment when virus particles first contact plant cells.

One aspect of this subject that has been investigated on enough occasions to generate some firmly held views is that concerning the entry process. However, the central question remains: where are the particles which will be active in generating an infection located during and after the process of inoculation? In most cases, it can be argued (1) that the act of inoculation produces a wound or opening in the plasmalemma of a plant cell sufficient for direct entry of virus particles, or (2) that inoculation produces less severe wounds with the result that virus particles are placed in, on, or near the plasmalemma. The latter situation would then require some mechanism for the passage of either the intact virion or, at least, its nucleic acid into the interior of the cell.

There are arguments for, but as yet insufficient evidence of, each of these alternative entry processes. A direct entry mechanism has been traditionally assumed because nearly all plant viruses seem to require some kind of disturbance or wounding (whether by humans, various invertebrates, or fungi) of cells or tissues in order to be able to produce infections. It can be further argued that the requirement, in most cases, of poly-L-ornithine for the infection of protoplasts but not for mechanical inoculation of leaf cells indicates a wound-producing (hence portal-providing) role for the polycation. The absence of demonstrable interactions between specific viral coat proteins and cell surface receptors may also favor a direct entry process. This alternative, however, leaves us with the predicament of exposing our lack of progress since such a process came to be commonly accepted half-a-century ago.

Perhaps the main argument in support of a direct involvement of the plasmalemma in the entry process is the clear role of this membrane in attachment and entry of animal and bacterial viruses. This mechanism would not require the repair of holes in the plasmalemma. It would also accommodate those situations where a polycationic or fusogenic substance is not necessary for the infection of protoplasts.

Little is to be gained by suggesting, at this time, that one or the other of these entry mechanisms is the more favored. As with arguments concerning the viropexis and fusion mechanisms of entry of animal viruses, a good deal of detailed information is required before such an issue can begin to be settled.

Uncoating of plant viruses is also a poorly understood part of the early events. The problem of differentiating those particles which take place in infection from excess particles which do not and from particles which may undergo some but not all of the vital events has been mentioned above. Many of the TMV particles that remain associated with cells or tissue after inoculation appear to undergo only partial dissociation; whether these represent aborted particles or whether the exposed RNA in them can be translated in vivo is not known. Some intact, completely uncoated RNA is detectable though how much of it is involved in infection is not yet established. The very few studies with isometric viruses present a slightly less confusing picture in that an array of incompletely uncoated particles is not detected. It is of interest that some degree of uncoating of any virus seems to be detected in any plant to which it is applied. This observation, if meaningful, would support the idea that the success or failure of infection and the host range are determined by later events.

What seems apparent is a process based upon a sort of mass production strategy. Plant viruses do not possess the means to bring about their own dispersal or, in most cases, to initiate infections in the absence of wound-producing agents. However, particles of many of them are produced in very large quantities in infected cells. This alone may ensure that, whenever opportunities arise for transmission of virus to other plants, large reservoirs of potential inoculum will be available.

ACKNOWLEDGMENTS

The author is very grateful to his colleagues S. A. Ghabrial, R. J. Jacob, J. A. Lesnaw, M. L. Perdue, T. P. Pirone, and H. E. Wheeler for their helpful suggestions and criticisms.

REFERENCES

1. **de Zoeten, G. A.,** Early events in plant virus infection, in *Plant Diseases and Vectors*, Maramorosch, K. and Harris, K. F., Eds., Academic Press, New York, 1981, chap. 7.
2. **Howe, C., Coward, J. E., and Fenger, T. W.,** Viral invasion: morphological, biochemical and biophysical aspects, in *Comprehensive Virology*, Vol. 16, Fraenkel-Conrat, H. and Wagner, R. R., Eds. Plenum Press, New York, 1980, 1.
3. **Lonberg-Holm, K. and Philipson, L.,** *Virus Receptors*, Part 2, *Animal Viruses*, Chapman and Hall, London, 1981.
4. **Bukrinskaya, A. G.,** Penetration of viral genetic material into host cell, *Adv. Virus Res.*, 27, 141, 1982.
5. **Crowell, R. L. and Landau, B. J.,** Receptors in the initiation of picornavirus infections, in *Comprehensive Virology*, Vol. 18, Fraenkel-Conrat, H. and Wagner, R. R., Eds., Plenum Press, New York, 1983, 1.
6. **Dimmock, N. J.,** Initial stages in infection with animal viruses, *J. Gen. Virol.*, 59, 1, 1982.
7. **Lonberg-Holm, K., Crowell, R. L., and Philipson, L.,** Unrelated animal viruses share receptors, *Nature (London)*, 259, 679, 1976.
8. **Gschwender, H. H. and Traub, P.,** Mengovirus-induced capping of virus receptors on the plasma membrane of Ehrlich ascites tumour cells, *J. Gen. Virol.*, 42, 439, 1979.
9. **Lenard, J. and Miller, D. K.,** Uncoating of enveloped viruses, *Cell*, 28, 5, 1982.
10. **Matlin, K. S., Reggio, H., Helenius, A., and Simons, K.,** Pathway of vesicular stomatitis virus entry leading to infection, *J. Mol. Biol.*, 156, 609, 1982.
11. **Lyon, M., Chardonnet, Y., and Dales, S.,** Early events in the interaction of adenoviruses with Hela cells. V. Polypeptides associated with the penetrating inoculum, *Virology*, 87, 81, 1978.
12. **Mackay, R. L. and Consigli, R. A.,** Early events in polyoma virus infection: attachment, penetration and nuclear entry, *J. Virol.*, 19, 620, 1976.
13. **Borsa, J., Morash, B. D., Sargent, M. D., Copps, T. P., Lievaart, P. A., and Szekely, J. G.,** Two modes of entry of reovirus particles into L cells, *J. Gen. Virol.*, 45, 161, 1979.
14. **Borsa, J., Sargent, M. D., Lievaart, P. A., and Copps, T. P.,** Reovirus: evidence for a second step in the intracellular uncoating and transcriptase activation process, *Virology*, 111, 191, 1981.
15. **De Sena, J., Heifner, W. R., and Stolov, N. S.,** Studies on the *in vitro* uncoating of poliovirus. IV. Characteristics of solubilized membrane-modifying and -stabilizing factors, *Virology*, 128, 354, 1983.
16. **Goldberg, E.,** Bacteriophage nucleic acid penetration, in *Virus Receptors*, Part 1, *Bacterial Viruses*, Randall, L. L. and Philipson, L., Eds., Chapman and Hall, London, 1980, 115.
17. **Pirone, T. P. and Raccah, B.,** personal communication, 1983.
18. **Mundry, K. W.,** Plant virus-host cell relations, *Annu. Rev. Phytopathol.*, 1, 173, 1963.
19. **Siegel, A. and Zaitlin, M.,** Infection process in plant virus diseases, *Annu. Rev. Phytopathol.*, 2, 179, 1964.
20. **Matthews, R. E. F.,** *Plant Virology*, Academic Press, New York, 1970, chap. 7.

21. **Favali, M. A., Conti, G. G., and Bassi, M.,** Some observations on virus-induced local lesions by transmission and scanning electron microscopy, *Acta Phytopathol. Acad. Sci. Hung.*, 12, 141, 1977.
22. **Dreyer, S. A., Seymour, V., and Cleland, R. E.,** Low proton conductance of plant cuticles and its relevance to the acid-growth theory, *Plant Physiol.*, 68, 664, 1981.
23. **Jones, A. M. and Vanderhoef, L. N.,** Effect of abrading the cuticle using emery, *J. Exp. Bot.*, 32, 405, 1981.
24. **Rawlins, T. E. and Tompkins, C. M.,** Studies on the effect of carborundum as an abrasive in plant virus inoculations, *Phytopathology*, 26, 578, 1936.
25. **Cram, W. J.,** Pinocytosis in plants, *New Phytol.*, 84, 1, 1980.
26. **Gerola, F. M., Bassi, M., Favali, M. A., and Betto, E.,** An electron microscopy study of the penetration of tobacco mosaic virus into leaves following experimental inoculation, *Virology*, 38, 380, 1969.
27. **Brandts, D. H.,** Relation between ectodesmata and infection of leaves by C^{14}-labeled tobacco mosaic virus, *Virology*, 26, 554, 1965.
28. **Merkens, W. S. W., de Zoeten, G. A., and Gaard, G.,** Observations on ectodesmata and the virus infection process, *J. Ultrastruct. Res.*, 41, 397, 1972.
29. **Hallam, N. D.,** Fine structure of the leaf cuticle and the origin of leaf waxes, in *The Plant Cuticle*, Cutler, D. F., Alvin, K. L., and Price, C. E., Eds., Academic Press, London, 1982, 197.
30. **Gaard, G. and de Zoeten, G. A.,** Plant virus uncoating as a result of virus-cell wall interactions, *Virology*, 96, 21, 1979.
31. **Golinowski, W., Kupidlowska, E., and Skrzeczkowski, L.,** Degradation of potato virus X (PVX) in the intracellular space of potato mesophyll, *Phytopathol. Z.*, 101, 136, 1981.
32. **Kiho, Y.,** Model system for the adsorption of tobacco mosaic virus, *Jpn. J. Microbiol.*, 18, 313, 1974.
33. **Kiho, Y. and Shimomura, T.,** Binding of tobacco mosaic virus to membrane material isolated from tobacco leaves, *Jpn. J. Microbiol.*, 20, 537, 1976.
34. **Kiho, Y., Shimomura, T., Abe, T., and Nozu, Y.,** Infectivity suppressing and virus-binding activities of a membrane material isolated from tobacco leaves, *Microbiol. Immunol.*, 23, 735, 1979.
35. **Hershey, A. D. and Chase, M.,** Independent functions of viral protein and nucleic acid in growth of bacteriophage, *J. Gen. Physiol.*, 36, 39, 1952.
36. **Gierer, A. and Schramm, G.,** Infectivity of ribonucleic acid from tobacco mosaic virus, *Nature (London)*, 177, 702, 1956.
37. **Fraenkel-Conrat, H. and Williams, R. C.,** Reconstitution of active tobacco mosaic virus from the inactive protein and nucleic acid components, *Proc. Natl. Acad. Sci. U.S.A.*, 41, 690, 1955.
38. **Fraenkel-Conrat, H.,** The role of nucleic acid in the reconstitution of active tobacco mosaic virus, *J. Am. Chem. Soc.*, 78, 882, 1956.
39. **Reddi, K. K.,** Studies on the formation of tobacco mosaic virus ribonucleic acid. VII. Fate of tobacco mosaic virus after entering the host cell, *Proc. Natl. Acad. Sci. U.S.A.*, 55, 593, 1966.
40. **Reddi, K. K.,** Studies on the formation of tobacco mosaic virus ribonucleic acid. VIII. Fate of parental viral RNA in the host cell, *Proc. Natl. Acad. Sci. U.S.A.*, 62, 604, 1969.
41. **Kurtz-Fritsch, C. and Hirth, L.,** Premiers evenments de l'infection virale chez les plantes, *C. R. Acad. Sci. Paris, Ser. D*, 265, 220, 1967.
42. **Kurtz-Fritsch, C. and Hirth, L.,** Uncoating of two spherical plant viruses, *Virology*, 47, 385, 1972.
43. **Shaw, J. G.,** *In vivo* removal of protein from tobacco mosaic virus after inoculation of tobacco leaves, *Virology*, 31, 665, 1967.
44. **Shaw, J. G.,** *In vivo* removal of protein from tobacco mosaic virus after inoculation of tobacco leaves. II. Some characteristics of the reaction, *Virology*, 37, 109, 1969.
45. **Shaw, J. G.,** Uncoating of tobacco mosaic virus ribonucleic acid after inoculation of tobacco leaves, *Virology*, 42, 41, 1970.
46. **Hirashima, A. and Hirai, T.,** The fate of tobacco mosaic virus introduced into tobacco leaves, *Ann. Phytopathol. Soc. Jpn.*, 35, 21, 1969.
47. **Hirashima, A. and Hirai, T.,** Dissociation of tobacco mosaic virus particles immediately after infection of tobacco leaves, *Ann. Phytopathol. Soc. Jpn.*, 35, 25, 1969.
48. **Hayashi, T., Machida, H., Abe, T., and Kiho, Y.,** In vivo uncoating of tobacco mosaic virus irradiated with ultraviolet light, *Jpn. J. Microbiol.*, 13, 386, 1969.
49. **Machida, H., Hayashi, T., Abe, T., and Kiho, Y.,** In vivo uncoating of tobacco mosaic virus after infection of tobacco leaves, *Jpn. J. Microbiol.*, 13, 351, 1969.
50. **Machida, H. and Kiho, Y.,** In vivo uncoating of tobacco mosaic virus. II. Complete uncoating to TMV-RNA, *Jpn. J. Microbiol.*, 14, 441, 1970.
51. **Onda, H. and Hirai, T.,** Degradation of RNA of partially uncoated tobacco mosaic virus during extraction from tobacco leaves, *Jpn. J. Microbiol.*, 14, 221, 1970.
52. **Hayashi, T.,** Fate of tobacco mosaic virus after entering the host cell, *Jpn. J. Microbiol.*, 18, 279, 1974.
53. **Hayashi, T.,** Fate of tobacco mosaic virus after entering the host cell. II. Electron microscopy of virus particles recovered from inoculated leaves, *Jpn. J. Microbiol.*, 19, 399, 1975.

54. **Hayashi, T.,** Fate of tobacco mosaic virus after entering the host cell. III. Partial uncoating, *Microbiol. Immunol.*, 21, 317, 1977.
55. **Hershko, A. and Ciechanover, A.,** Mechanisms of intracellular protein breakdown, *Annu. Rev. Biochem.*, 51, 335, 1982.
56. **Shaw, J. G.,** Effect of poly-L-ornithine on the attachment of tobacco mosaic virus to tobacco leaves and on the uncoating of viral RNA, *Virology*, 48, 380, 1972.
57. **Zhuravlev, Y. N., Yudakova, Z. S., Shumilova, L. A., and Seletskaya, L. D.,** Uptake of ^{14}C-TMV by tobacco leaves. Effect of poly-L-ornithine and metabolic inhibitors, *Acta Phytopathol. Acad. Sci. Hung.*, 9, 201, 1974.
58. **Kiho, Y., Machida, H., and Oshima, N.,** Mechanisms determining the host specificity of tobacco mosaic virus. I. Formation of polysomes containing infecting viral genome in various plants, *Jpn. J. Microbiol.*, 16, 451, 1972.
59. **Kassanis, B. and Kenten, R. H.,** Inactivation and uncoating of TMV on the surface and in the intercellular spaces of leaves, *Phytopathol. Z.*, 91, 329, 1978.
60. **Kohler, E.,** Studien zur Kinetik der in situ erfolgenden Inaktivierung des in die Blattflache auf genommenen Impfvirus (Residualvirus), *Phytopathol. Z.*, 74, 21, 1972.
61. **Niblett, C. W.,** Attachment of plant viruses to host and non-host plants, *Proc. Am. Phytopathol. Soc.*, 2, 87, 1975.
62. **Kiho, Y., Abe, T., and Ohashi, Y.,** Disassembly of tobacco mosaic virus by membrane lipid isolated from tobacco leaves and polyornithine, *Microbiol. Immunol.*, 23, 1067, 1979.
63. **Kiho, Y. and Abe, T.,** Modification of tobacco mosaic virus by polyornithine and lecithin, *Microbiol. Immunol.*, 24, 617, 1980.
64. **Caspar, D. L. D.,** Assembly and stability of the tobacco mosaic virus particle, *Adv. Protein Chem.*, 18, 37, 1963.
65. **Kahl, G.,** Molecular biology of wound healing: the conditioning phenomenon, in *Molecular Biology of Plant Tumors*, Kahl, G. and Schell, J. S., Eds., Academic Press, New York, 1982, chap. 7.
66. **Kiho, Y.,** Polycistronic translation of plant viral ribonucleic acid, *Jpn. J. Microbiol.*, 16, 259, 1972.
67. **Shaw, J. G.,** *In vivo* removal of protein from tobacco mosaic virus after inoculation of tobacco leaves. III. Studies on the location on virus particles for the initial removal of protein, *Virology*, 53, 337, 1973.
68. **Perham, R. N. and Wilson, T. M. A.,** The polarity of stripping of coat protein subunits from the RNA in tobacco mosaic virus under alkaline conditions, *FEBS Lett.*, 62, 11, 1976.
69. **Wilson, T. M. A., Perham, R. N., Finch, J. T., and Butler, P. J. G.,** Polarity of the RNA in the tobacco mosaic virus particle and the direction of protein stripping in sodium dodecyl sulfate, *FEBS Lett.*, 64, 285, 1976.
70. **Butler, P. J. G., Finch, J. T., and Zimmern, D.,** Configuration of tobacco mosaic virus RNA during virus assembly, *Nature (London)*, 265, 217, 1977.
71. **Lebeurier, G., Nicolaieff, A., and Richards, K. E.,** Inside-out model for self-assembly of tobacco mosaic virus, *Proc. Natl. Acad. Sci. U.S.A.*, 74, 149, 1977.
72. **Helms, K. and Zaitlin, M.,** Enhancement of infectivity of tobacco mosaic virus particles partially uncoated by alkali, *Virology*, 41, 549, 1970.
73. **Kiho, Y.,** Interaction between viral RNA and cellular components in tobacco leaves infected with tobacco mosaic virus, *Jpn. J. Microbiol.*, 13, 312, 1969.
74. **Kiho, Y.,** Polysomes containing infecting viral genome in tobacco leaves infected with tobacco mosaic virus, *Jpn. J. Microbiol.*, 14, 291, 1970.
75. **Kiho, Y.,** Association of infecting tobacco mosaic virus ribonucleic acid with polysome, *Jpn. J. Microbiol.*, 16, 152, 1972.
76. **Takebe, I.,** Protoplasts in the study of plant virus replication, in *Comprehensive Virology*, Vol. 11, Fraenkel-Conrat, H. and Wagner, R. R., Eds., Plenum Press, New York, 1977, 237.
77. **Mühlbach, H. P.,** Plant cell cultures and protoplasts in plant virus research, *Curr. Topics Microbiol. Immunol.*, 99, 81, 1982.
78. **Cocking, E. C. and Pojnar, E.,** An electron microscopic study of the infection of isolated tomato fruit protoplasts by tobacco mosaic virus, *J. Gen. Virol.*, 4, 305, 1969.
79. **Otsuki, Y., Takebe, I., Honda, Y., and Matsui, C.,** Ultrastructure of infection of tobacco mesophyll protoplasts by tobacco mosaic virus, *Virology*, 49, 188, 1972.
80. **Hibi, T. and Yora, K.,** Electron microscopy of tobacco mosaic virus infection in tobacco mesophyll protoplasts, *Ann. Phytopathol. Soc. Jpn.*, 38, 350, 1972.
81. **Takebe, I.,** The use of protoplasts in plant virology, *Annu. Rev. Phytopathol.*, 13, 105, 1975.
82. **Burgess, J., Motoyoshi, F., and Fleming, E. N.,** Effect of poly-L-ornithine on isolated tobacco mesophyll protoplasts: evidence against stimulated pinocytosis, *Planta*, 111, 199, 1973.
83. **Burgess, J., Motoyoshi, F., and Fleming, E. N.,** The mechanism of infection of plant protoplasts by viruses, *Planta*, 112, 323, 1973.

84. **Kubo, S., Robinson, D. J., Harrison, B. D., and Hutcheson, A. M.,** Uptake of tobacco rattle virus by tobacco protoplasts, and the effect of phosphate on infection, *J. Gen. Virol.*, 30, 287, 1976.
85. **Kassanis, B., White, R. F., Turner, R. H., and Woods, R. D.,** The mechanism of virus entry during infection of tobacco protoplasts with TMV, *Phytopathol. Z.*, 88, 215, 1977.
86. **Zhuravlev, Y. N., Pisetskaya, N. F., Yudakova, Z. S., and Reifman, V. G.,** Uptake of labelled tobacco mosaic virus by tobacco protoplasts in the presence of metabolic inhibitors and at low temperature, *Acta Virol.*, 20, 435, 1976.
87. **Zhuravlev, Y. N., Yuakova, Z. S., and Pisetskaya, N. F.,** Infection of tobacco protoplasts with TMV in the absence of poly-L-ornithine and the electroosmotic mechanism of virus entry, *Phytopathol. Z.*, 98, 296, 1980.
88. **Takebe, I. and Otsuki, Y.,** Infection of tobacco mesophyll protoplasts by tobacco mosaic virus, *Proc. Natl. Acad. Sci. U.S.A.*, 64, 843, 1969.
89. **Shaw, J. G. and Fannin, F. F.,** unpublished observations, 1981.
90. **Mayo, M. A. and Roberts, I. M.,** Inoculation of tobacco protoplasts with tobacco rattle virus: experiments with aggregrates formed in inocula, *J. Gen. Virol.*, 40, 19, 1978.
91. **Mayo, M. A. and Roberts, I. M.,** Some effects of buffers on the infectivity and appearance of virus inocula used for tobacco protoplasts, *J. Gen. Virol.*, 44, 691, 1979.
92. **Nagata, T. and Melchers, G.,** Surface charge of protoplasts and their significance in cell-cell interaction, *Planta*, 142, 235, 1978.
93. **Wyatt, S. D. and Shaw, J. G.,** Retention and dissociation of tobacco mosaic virus by tobacco protoplasts, *Virology*, 63, 459, 1975.
94. **Zhuravlev, Y. N., Pisetskaya, N. F., Shumilova, L. A., Musorek, T. I., and Reifman, V. G.,** Attachment of labelled TMV to tobacco mesophyll protoplasts, *Virology*, 64, 43, 1975.
95. **Okuno, T. and Furusawa, I.,** Modes of infection of barley protoplasts with brome mosaic virus, *J. Gen. Virol.*, 38, 409, 1978.
96. **Cassells, A. C. and Cocker, F. M.,** TMV inoculation of tobacco protoplasts in the presence of protoplast fusion agents, *Z. Naturforsch.*, 35c, 1057, 1980.
97. **Motoyoshi, F., Bancroft, J. B., and Watts, J. W.,** A direct estimate of the number of cowpea chlorotic mottle virus particles absorbed by tobacco protoplasts that become infected, *J. Gen. Virol.*, 21, 159, 1973.
98. **Hibi, T., Rezelman, G., and Van Kammen, A.,** Infection of cowpea mesophyll protoplasts with cowpea mosaic virus, *Virology*, 64, 308, 1975.
99. **Shaw, J. G.,** The influence of protoplast concentration on retention of tobacco mosaic virus by tobacco protoplasts, *Virology*, 84, 544, 1978.
100. **Zhuravlev, Y. N., Reifman, V. G., Shumilova, L. A., Yudakova, Z. S., and Pisetskaya, N. F.,** Absorption of P^{32}-labelled tobacco mosaic virus by isolated tobacco protoplasts and deproteinization of the virus in them, *Sov. Plant Physiol.*, 22, 941, 1975.
101. **Farquhar, M. G.,** Multiple pathways of exocytosis, endocytosis, and membrane recycling: validation of a Golgi route, *Fed. Proc.*, 42, 2407, 1983.
102. **Petkau, A. and Chelack, W. S.,** Permeability of a thin phospholipid membrane to ions and tobacco mosaic virus, *Biochim. Biophys. Acta*, 135, 812, 1967.
103. **Banerjee, S., Vanderbranden, M., and Ruysschaert, J. M.,** Interaction of tobacco mosaic virus protein with lipid membrane systems, *FEBS Lett.*, 133, 221, 1981.
104. **Banerjee, S., Vandenbranden, M., and Ruysschaert, J. M.,** Tobacco mosaic virus protein induces fusion of liposome membranes, *Biochim. Biophys. Acta*, 646, 360, 1981.
105. **Wolfe, J. and Steponkus, P. L.,** Mechanical properties of the plasma membrane of isolated plant protoplasts, *Plant Physiol.*, 71, 276, 1983.
106. **Durham, A. C. H. and Hendry, D. A.,** Cation binding by tobacco mosaic virus, *Virology*, 77, 510, 1977.
107. **Durham, A. C. H. and Abou Haidar, M.,** Cation binding by tobacco rattle virus, *Virology*, 77, 520, 1977.
108. **Durham, A. C. H., Hendry, D. A., and Von Wechmar, M. B.,** Does calcium ion binding control plant virus disassembly?, *Virology*, 77, 524, 1977.
109. **Pfeiffer, P. and Durham, A. C. H.,** The cation binding associated with structural transitions in bromegrass mosaic virus, *Virology*, 81, 419, 1977.
110. **Durham, A. C. H. and Bancroft, J. B.,** Cation binding by papaya mosaic virus and its protein, *Virology*, 93, 246, 1979.
111. **Hendry, D. A. and Durham, A. C. H.,** Titration behavior of three strains of tobacco mosaic virus, *Virology*, 100, 65, 1980.
112. **Durham, A. C. H., Bancroft, J. B., and Borque, D. P.,** The generality of cation-binding sites in rod-shaped viruses, *Biosci. Rep.*, 1, 547, 1981.
113. **Durham, A. C. H. and Butler, P. J. G.,** A prediction of the structure of tobacco mosaic virus protein, *Eur. J. Biochem.*, 53, 397, 1975.

114. **Durham, A. C. H.,** The roles of small ions, especially calcium, in virus disassembly, takeover, and transformation, *Biomedicine*, 28, 307, 1978.
115. **Durham, A. C. H.,** Viruses make holes in membranes, presented at 4 Int. Congr. Virology, The Hague, August 30 to September 6, 1978.
116. **Keeling, J. and Matthews, R. E. F.,** Mechanism for release of RNA from turnip yellow mosaic virus at high pH, *Virology*, 119, 214, 1982.
117. **Katouzian, Safadi, M. and Berthet-Colominas, C.,** Evidence for the presence of a hole in the capsid of turnip yellow mosaic virus after RNA release by freezing and thawing. Decapsidation of TVMV in vitro, *Eur. J. Biochem.*, 137, 47, 1983.
118. **Virudachalam, R., Sitaraman, K., Heuss, K. L., Markley, J. L., Argos, P.,** Evidence for pH-induced release of RNA from belladonna mottle virus and the stabilizing effect of polyamines and cations, *Virology*, 130, 351, 1983.
119. **Wilson, T. M. A.,** Cotranslational disassembly of tobacco mosaic virus in vitro, *Virology*, 137, 255, 1984.
120. **Dales, S.,** Penetration of animal viruses into cells, in *Transport of Macromolecules in Cellular Systems*, Silverstein, S. C., Ed., Abakon-Verlags, Berlin, 1978, 47.

Chapter 2

GENOME STRUCTURE AND GENE EXPRESSION OF PLANT RNA VIRUSES*

William G. Dougherty and Ernest Hiebert

TABLE OF CONTENTS

* The literature review for this chapter was complete up to September 1983.

I. INTRODUCTION

Slightly more than 10 years ago, viral RNA was translated in a wheat germ cell-free translation system.[1-3] Since then, a tremendous amount of information has been generated in the quest to understand how plant viral genomes are expressed and regulated, and how these "simple" pathogens interact with their plant host cells. This information has been acquired by a variety of different experimental approaches, each one aimed at elucidating a particular aspect of the viral replicative cycle. Although these studies have increased our knowledge of plant viruses, the exact mechanisms involved in viral genome expression and regulation remain to be determined. Further technological advances in molecular biology should make possible the resolution of the important structural features which control viral gene expression and hence viral phenotype.

In this review of gene expression of RNA plant viruses, we have followed established viral classification groupings.[4] We begin the discussion of each virus group by briefly describing the structural organization of the virion and of the viral genomic RNA(s). Then, the expression of the viral RNA in cell-free systems and in the host cells is presented. Finally, we have attempted to integrate in vivo and in vitro observations and present an overview of the genomic expression of each virus group. Many of the proposed strategies and product relationships that are illustrated are speculative and represent our own interpretation and predictions.

A number of other reviews dealing with related topics have been published and the reader is referred to them for additional general information.[5-10]

II. VIRUSES WITH MONOPARTITE GENOMES

The nine virus groups discussed in this section have a single-stranded, ribonucleic acid component which is infectious. The initial five virus groups have members possessing a particle morphology which is spherical. The remaining four virus groups have members which are anisometric in morphology.

A. Tymovirus Group

Turnip yellow mosaic virus (TYMV) is the type member of the tymovirus group which includes 15 other members.[4] TYMV has spherical particles 30 nm in diameter, with capsid protein monomers (22,000 M_r) arranged in a T = 3 icosahedral structure with pentamer clustering.[11-12] Purified preparations of TYMV analyzed by equilibrium density-gradient centrifugation, reveal a number of minor components with densities less than and greater than the major infectious virions. These minor components vary from empty virus particle shells to virions containing genomic RNA and the discrete subgenomic mRNA for the capsid protein. Additionally, less than full-length RNA and the mRNA for the capsid protein are found in certain minor components.[13-15] Similar types of heterogeneous minor particle distributions have been reported for other tymoviruses.[16]

1. Viral Genome Structure

The genomic TYMV RNA and the monocistronic mRNA for the capsid protein have been analyzed. The genomic RNA is infectious and has a molecular weight of 2.0×10^6.[17] The 5′ terminus has a m⁷Gppp cap structure and the following 110 nucleotide sequence has been determined.[18] Two AUG initiation codons are present in this sequence and it appears that the second AUG initiation codon, which is 95 to 97 nucleotides from the 5′ terminus, is used in translation.[18] The entire 695-nucleotide sequence for the capsid protein mRNA has been determined[19] and is identical to the 3′ terminal sequence of the genomic RNA. The capsid protein mRNA also has a 5′ terminal m⁷Gppp cap structure and the initiation codon lies 19 nucleotides downstream from this cap structure. The capsid protein coding region (567 nucleotides) is followed by a 3′ untranslated region of 109 nucleotides.[19] This untranslated region has a three-dimensional structure which shares a number of structural characteristics of a tRNA molecule.[20-22] The 3′ terminal nucleotide of TYMV genomic RNA and capsid protein mRNA can be aminoacylated both in vitro and in vivo.[23,24] In vitro, valine can be added onto TYMV RNA and eggplant mosaic virus (EMV) RNA.[25] This valine is neither used in cell-free protein synthesis nor does it interfere with the translation of tymovirus RNA. Studies using *Xenopus laevis* oocytes or chinese cabbage leaves have shown that aminoacylation occurs in vivo.[24,26] TYMV RNA is first adenylated and valine is then added to this 3′ terminal adenosine. The valylated RNA appears to be processed to a small 3′ terminal fragment (4 to 5S) in oocytes and possibly in the chinese cabbage system.[24,26] It is presently unclear what role this valylation reaction plays in the replicative cycle of TYMV. However, it has been proposed that this valylated tRNA-like fragment may be involved in replication.[24] Encapsidated TYMV RNA is not valylated.[22-26]

2. Cell-Free Translation Studies

TYMV RNA initially was translated in a wheat germ cell-free extract into a number of proteins ranging from 65,000 to 10,000 M_r.[17] Translation of a small RNA (2.5×10^5 M_r) present in TYMV particles resulted in the synthesis of TYMV capsid protein.[17,27-31] These initial studies have been expanded by cell-free translation analysis of TYMV RNA in rabbit reticulocyte lysates, yeast S-23 extracts, and ascites cell extracts. In general, after optimizing the cell-free translation systems, the genomic RNA produces two major high-molecular-weight products, 195,000 and 150,000, which have the same N terminus and common amino acid sequences.[29-33] A number of lower molecular weight products also are observed. Analysis of these products has revealed that they are translated from the same initiation codon as the 190,000 and 150,000 M_r proteins and represent a nested set of proteins varying only in the amino acids present at the C terminus.[33] Whether these proteins are authentic viral products, remains to be confirmed by in vivo analysis. The addition of a yeast suppressor tRNA [amber tRNA(UAG)] does not enhance the synthesis of the 195,000 M_r protein; rather, a higher molecular weight protein of 210,000 is synthesized.[34] Capsid protein is not synthesized by translation of the full-length genomic RNA. Similar results have been reported for the translations of EMV RNA[35-37] and two other tymoviruses RNAs (wild cucumber mosaic virus and okra mosaic virus).[16]

Recent studies have implicated posttranslational proteolytic processing in the formation of viral gene products.[38,39] In the rabbit reticulocyte lysate, the 195,000 M_r protein is processed proteolytically into a N-terminal protein fragment with a molecular weight of 120,000 and a C-terminal fragment with a molecular weight of 78,000[38,39] The cleavage of the 195,000 M_r protein appears to be both a dilution-sensitive and dilution-insensitive process.[39] A dilution-insensitive process would infer an intramolecular virus-mediated event. Since the 150,000 M_r protein, which represents the N terminus 75% of the 195,000 M_r protein, is not processed, it has been proposed that the processing function may be associated with the C-terminal protein sequences of the 195,000 M_r protein.[39] The processing of the TYMV 195,000 M_r protein also has been observed in the wheat germ and ascites cell extracts, suggesting that the process is indeed virus encoded.[38-40] The protease inhibitor TLCK, an inhibitor of trypsin-like proteases, prevents the processing of the 195,000 M_r protein.[39]

3. In Vivo Observations

The replicative forms of three different tymoviruses have been analyzed by polyacrylamide gel electrophoresis. Each double-stranded RNA species has a molecular weight greater than 2.45×10^6.[41] This probably represents a replicative form of the genomic RNA. No other double-stranded RNA has been detected. A number of research groups have shown that the replication of TYMV is associated with the outer chloroplast membrane.[42,43] Cytochemical studies demonstrate that the replicative intermediate involved in TYMV replication is predominantly a single-stranded structure.[44] It is suggested that TYMV may synthesize its genomic RNA and capsid protein mRNA from a complementary RNA species of the genomic RNA.[44]

TYMV-induced proteins in chinese cabbage cells have been detected. Ten virus-associated proteins, which range in molecular weight from 190,000 to 10,000, have been observed.[45] Presently, it is unclear if these proteins are all virus coded and how they are synthesized.

A possible strategy for TYMV genome expression is presented below (Figure 1). The strategy that has been proposed is similar to a number of other viruses. TYMV genome expression resembles the tobacco mosaic virus (TMV) genome expression (see Section II.F) in the synthesis of two large proteins (TYMV = 195,000 and 150,000 M_r; TMV = 183,000 and 126,000 M_r) which are initiated at the same codon in the genomic RNA and share amino acid sequences. However, these two viruses differ in that the use of a supressor tRNA does not increase the synthesis of TYMV 195,000 M_r product. Apparently the TYMV 195,000 M_r protein is not a read-through product of the 150,000 M_r protein termination site such as is found in the TMV genomic RNA expresssion. Both viruses are very similar in the manner in which they express the capsid protein gene. A discrete subgenomic mRNA for the capsid protein is synthesized. This strategy may enable the virus to amplify and to regulate the expression of this specific gene.

The translational strategy of TYMV genome is similar to that which has been proposed for turnip rosette virus (see Section II.B) (TRoSV). Translation of the genomic RNA of this member of the sobemovirus group also produces a high-molecular-weight protein which is processed posttranslationally into lower molecular weight products. TRoSV capsid protein, as in the case of TYMV, is translated from a discrete subgenomic mRNA.

B. Sobemovirus Group

Southern bean mosaic virus (SBMV), the type member of the sobemoviruses, is a small spherical plant virus which has a single-stranded genomic RNA.[4] The genomic RNA is encapsidated by 180 protein monomers each with a molecular weight of approximately

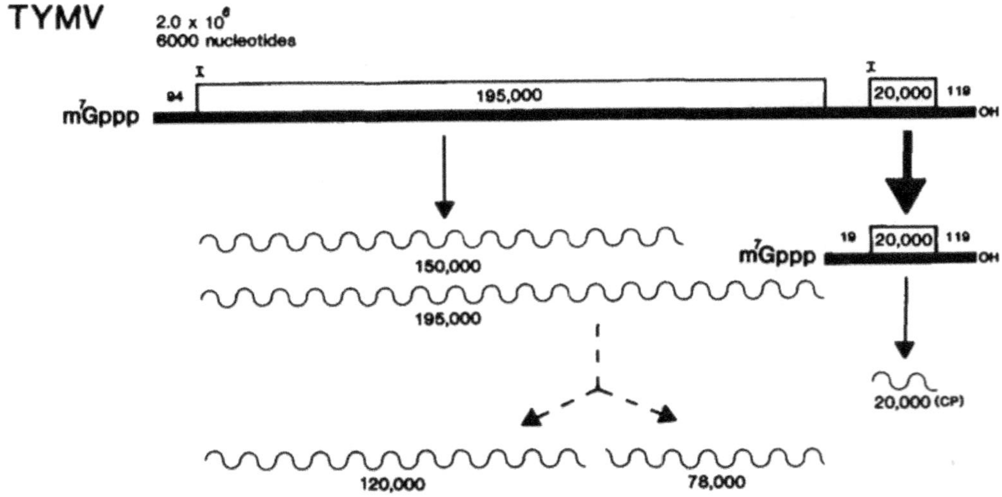

FIGURE 1. A scheme of genomic organization and expression strategy of the RNA of turnip yellow mosaic virus (TYMV). The molecular weight and approximate number of nucleotides which make up the genomic RNA are given. An explanation of the symbols and abbreviations used in this figure and the other figures throughout this chapter is as follows: RNA, ▬; Protein, ∿. Additional information about the viral RNA may be depicted and includes; m⁷Gppp, 5′ terminal cap blocking structure; VpG, genome-linked protein; pp or ppp, 5′ terminal di- or triphosphate; (poly A)$_n$, polyadenylate region; OH, lacking a polyadenylate region; I, translation initiation site. Small numbers above RNA indicate the number of nucleotides comprising the 5′ or 3′ untranslated regions. Possible open reading frames are represented by boxes above and below the RNA. Additional information about the viral proteins include: estimated molecular weights presented below the protein; cp, capsid protein. A transcription event, ➡, translation event, ⟶, and a proteolytic processing event, - - ➤, are illustrated in the figure. Structural elements which have not been analyzed or replication cycle events which are speculative are indicated by ?.

29,000.[46,47] The capsid proteins are arranged in a T = 3 icosahedral structure with pentamer clustering and with a diameter of 28 to 30 nm.[47] (See M. Rossman, Volume 1, Chapter 4 of this series.)

Three members of the sobemovirus group have been studied in detail; they are the cowpea strain of SBMV (SBMV-C), the bean strain of SMBV (SBMV-B), and turnip rosette virus (TRosV).

1. Viral Genome Structure

The entire nucleotide sequence of the genomic RNA of SBMV-C has been determined and has an apparent molecular weight of ~1.4 × 10⁶ and contains at least three open reading frames (Figure 2).[48,48a] A low-molecular-weight protein (VpG) is covalently attached to the 5′ terminal nucleotide and the 3′ terminus lacks a polyadenylate region.[49] The 5′ terminal VpG of the SBMV-C RNA has a molecular weight of ~12,000.[49] Secondary structure of the genomic RNA and the 5′ terminal VpG of the genomic RNA affect infectivity.[50,51] Removal of the VpG with proteinase K treatment renders the RNA noninfectious.[51] In addition to the genomic RNA, less-than-full-length RNA species are present in RNA preparations isolated from purified virions.[52,53] A discrete subgenomic RNA molecule, which contains the gene for the capsid protein, also is encapsidated and has a molecular weight of 0.38 × 10⁶.[54] This subgenomic RNA species has a small-molecular-weight protein, similar to the VpG found on the genomic RNA, covalently attached to the 5′ terminus.[55]

The RNA of TRosV also has been analyzed. TRosV genomic RNA has a molecular weight of 1.4 × 10⁶ and has a small protein (12,700 M_r) at the 5′ terminal nucleotide.[56,57] A subgenomic RNA species (0.5 × 10⁶ M_r) codes for the capsid protein. A polyadenylate region at the 3′ end has not been detected.[58]

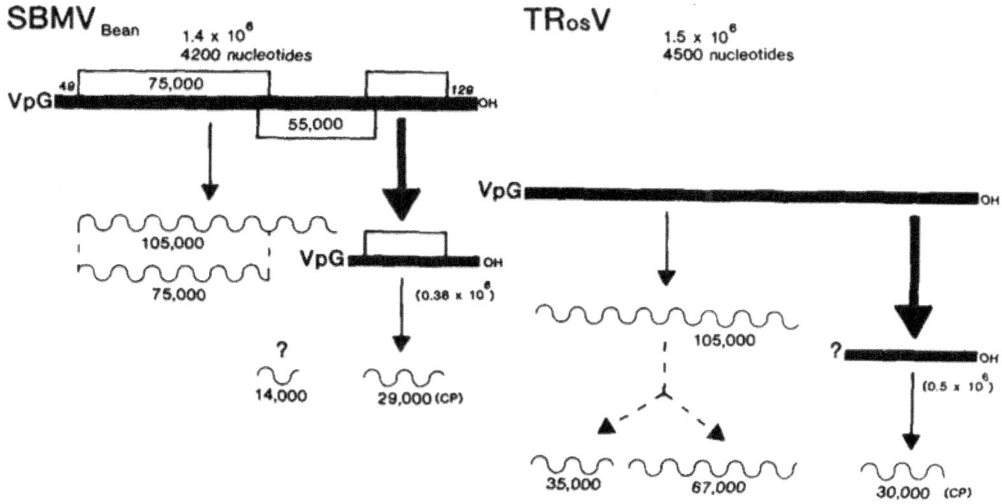

FIGURE 2. A scheme of the genomic structure and expression strategy of the sobemovirus RNA of southern bean mosaic (SBMV) and turnip rosette virus (TRosV). See Figure 1 for an explanation of the figure.

2. Cell-Free Translation Studies

RNA, isolated from SBMV-B and SBMV-C, has been translated in a rabbit reticulocyte lysate and wheat germ extract with similar results.[52,53] Translation of SBMV RNA, separated by sucrose, rate-zonal centrifugation, results in the synthesis of four major products. Translation of SBMV-B RNA produces proteins with estimated molecular weights of 105,000, 75,000, 30,000, and 14,000.[52] The 105,000 and 75,000 M_r proteins are synthesized exclusively from the genomic RNA.[52,54] These products have been shown to share amino acid sequences as all of the tryptic peptides of the 75,000 M_r protein are contained in the 105,000 M_r protein.[52] The capsid protein (30,000 M_r) is synthesized from a small subgenomic RNA which has a molecular weight of 0.38×10^6.[52-54] The 14,000 M_r protein is synthesized by SBMV-B RNA present in all fractions of the sucrose gradient.[52] This low-molecular-weight protein is unrelated to any of the other proteins synthesized in the cell-free systems by SBMV-B RNA.[52]

Similar results have been obtained in the translation of SBMV-C RNA in the wheat germ and rabbit reticulocyte cell-free systems.[53] Two major products of SBMV-C translation have molecular weight of 100,000 and 70,000. A 30,000 M_r product of translation is identical to the SBMV-C capsid protein. This protein is made exclusively from a small subgenomic mRNA. Additionally, a family of low-molecular-weight proteins (~20,000 M_r) is synthesized by RNA isolated from throughout a separating sucrose gradient. Similar to the translational products of SBMV-B RNA, the 100,000 and 70,000 M_r proteins share amino acid sequences while the 30,000 and 20,000 M_r proteins are unique. Translation of the SBMV-C RNA in wheat germ extracts produces a number of additional small-molecular-weight proteins, many of which share amino acid sequences with the 20,000 M_r product.[53] Proteolytic processing of SBMV translation products has not been reported.

Translation of TRosV RNA in the rabbit reticulocyte system also produces four major products.[58] These products have molecular weight of 105,000, 67,000, 35,000, and 30,000. The 30,000 M_r protein, identical to the capsid protein of TRosV, is synthesized almost exclusively from an encapsidated subgenomic RNA with a molecular weight 0.5×10^6.[58] Capsid protein, however, also may be synthesized in minor amounts from the full-length genomic RNA. The three major high-molecular-weight translation products of the full-length, genomic RNA have been shown to be related by peptide mapping. Time course studies have shown that the 67,000 M_r protein most likely is processed proteolytically from the 105,000 M_r protein by posttranslational cleavage.[58] This cleavage can be prevented by the addition of *n*-ethylmaleimide, a thiol protease inhibitor. Additionally, incubation of the translation mix at 37°C also prevents the formation of the 67,000 M_r protein while the synthesis of the 105,000 M_r protein remains unaffected.[58]

TRosV products with similar molecular weights have been observed in wheat germ translation studies.[59]

3. In Vivo Observations

The infection of soybean protoplasts or soybean callus tissue with SBMV-B[60,61] or turnip protoplasts with TRosV[62] has been described. These systems, however, have not been used to examine the synthesis of viral proteins or RNAs in a temporal fashion.

Two double-stranded RNAs have been isolated from SBMV-C infected plant tissue.[63] Although molecular size has not been estimated, these two double-stranded RNA structures may represent the replicative forms of the genomic RNA and of the subgenomic, capsid protein mRNA.

Two strategies for genome expression appear to be used by different members of the sobemovirus group and they are presented below (Figure 2).

SBMV-B and SBMV-C appear to have a similar cell-free translation strategy, differing only in the molecular weights of the products. TRosV genome expression shares part of the translation strategy used by SBMV; however, posttranslational proteolytic processing may play a role in product formation. TRosV genomic expression, therefore, appears similar to TYMV gene expression in the possible proteolytic processing of a polyprotein whose translation is initiated at the 5′ terminus of the genomic RNA. Use of a discrete subgenomic RNA for the expression of the capsid protein gene has been observed for sobemoviruses and other unicomponent virus systems (TMV, TYMV).

C. Tombusvirus Group

Tomato bushy stunt virus (TBSV) is the type member of the tombusvirus group. The virion structure is comprised of 180 repeating capsid monomers (41,000 M_r) in a T = 3 icosahedral organization with prominent dimer clustering.[64-66] Extensive studies have been carried out on the interaction of the capsid protein monomer and the viral RNA.[67,68] Turnip crinkle virus (TCV), a tentative member of this group, is morphologically similar and will be discussed in this section. (See also M. Rossman in Volume 1, Chapter 4 of this series.)

1. Viral Genome Structure

Nucleic acid isolated from purified TBSV consists of a major single-stranded RNA species (28S) and minor less-than-full-length RNA species (13 to 24S).[69,70] The molecular weight of the infectious TBSV RNA is 1.5×10^6. RNA isolated from TCV has a similar heterogeneous distribution[71] and additionally may contain a satellite RNA species (0.5×10^6 M_r) which can affect the severity of viral-induced symptoms.[71,72] Presently, it is unknown if the population of less-than-full-length RNA species are authentic mRNAs or if they represent fragmented RNAs produced during purification.[71] The 5′ and 3′ terminal structures of TBSV and TCV RNAs have not been examined in any detail. However, neither viral RNA binds to an oligodeoxythymidylate cellulose column under conditions which bind polyadenylated RNAs.[73]

2. Cell-Free Translation Studies

Unfractionated TBSV RNA, translated in a wheat germ cell-free extract, produces three major products (38,000, 35,000, and 32,000 M_r) and one minor product (87,000 M_r).[70,74] The same RNA stimulates the synthesis of two major products with molecular weights of

38,000 and 35,000 when translated in a rabbit reticulocyte lysate. The 38,000 M_r protein appears to be identical to the capsid protein by serological and biochemical analyses. The 32,000 M_r protein is serologically related to the capsid protein in immunoprecipitation studies. TBSV RNA has been fractionated by sucrose, rate-zonal centrifugation, and translated in the cell-free systems. The 38,000 and the 32,000 M_r proteins are synthesized predominantly from less-than-full-length RNA species. Full-length genomic RNA programs the synthesis of the 35,000 M_r protein in the cell-free extracts.[70,74]

TCV RNA has been translated in a rabbit reticulocyte lysate and a wheat germ extract.[71,75] Translation of unfractionated TCV RNA produces three major proteins with molecular weights of 38,000, 36,000, and 25,000. Other minor products also are synthesized and have molecular weights of 80,000, 56,000, 35,000, and 21,000. The 38,000 M_r product has been shown to be identical to the TCV capsid protein. Additionally, a number of small-molecular-weight products, serologically related to the capsid protein, are synthesized in the wheat germ system. TCV RNA, fractionated by sucrose, rate-zonal centrifugation, has been translated in both cell-free systems. Each of the major cell-free translation products is translated from a particular size RNA. The 25,000 M_r protein is synthesized primarily by full-length genomic RNA species. TCV 38,000 M_r capsid protein is synthesized predominately by a population of small RNAs. The 36,000 M_r protein is the product of an intermediate-sized RNA.Although a particular size class of RNA is primarily responsible for the synthesis of a single protein, most RNA fractions synthesize all the viral products to a varying extent. Tryptic peptides of the 25,000 and 56,000 M_r proteins are unique and are contained in the tryptic peptides of the 80,000 M_r protein. All of the tryptic peptides of the 36,000 M_r protein are contained in the tryptic peptide of the 56,000 M_r protein.[71]

3. In Vivo Observations

TBSV- and TCV-infected tissues have been studied cytologically.[76,77] Infection with TBSV results in the formation of multivesicular bodies which apparently are derived from peroxisomes. Double-stranded RNA has been detected in these vesicular bodies.[76] Infection with TCV also results in the formation of multivesicular bodies; however, it appears that the membrane structures are derived from mitochondria.[77]

Replicative forms of both TBSV- and TCV-genomic RNAs have been isolated from infected plant tissue and analyzed by gel electrophoresis.[69] Two replicative forms have been isolated from TBSV infected plant tissue. One is the replicative form of the genomic RNA and the other is a double-stranded RNA molecule $\sim 1.5 \times 10^6$ M_r.[69] This smaller replicative form may be involved in the synthesis of a subgenomic mRNA ($\sim 0.75 \times 10^6$). Only a single replicative form (3.0×10^6), twice the size of the genomic RNA, has been isolated from TCV infected plant tissue.[69]

Polysomes from TBSV RNA also have been isolated and analyzed by cell-free translation in a wheat germ extract. Two proteins, with molecular weights of 35,000 and 38,000, are translated from TBSV-infected plant polysomal RNA and are not synthesized by polysomal RNA isolated from healthy plant tissue.[70]

The genetic expression strategy of these two viruses is unclear. However, the interrelationship of the viral products and a possible genetic map of the TBSV and TCV genome is presented below (Figure 3).

The one curious aspect of the cell-free translation products of these two viruses is that they apparently have 5′ proximal gene which codes for a protein with a low-molecular-weight (TBSV, 35,000 M_r; TCV, 25,000 M_r). Other unicomponent viruses usually code for a protein which is much larger in molecular weight. The significance of this is unclear at the present time.

D. Tobacco Necrosis Virus and Tobacco Necrosis Satellite Virus

Tobacco necrosis virus (TNV) and its satellite virus (STNV) were two of the original

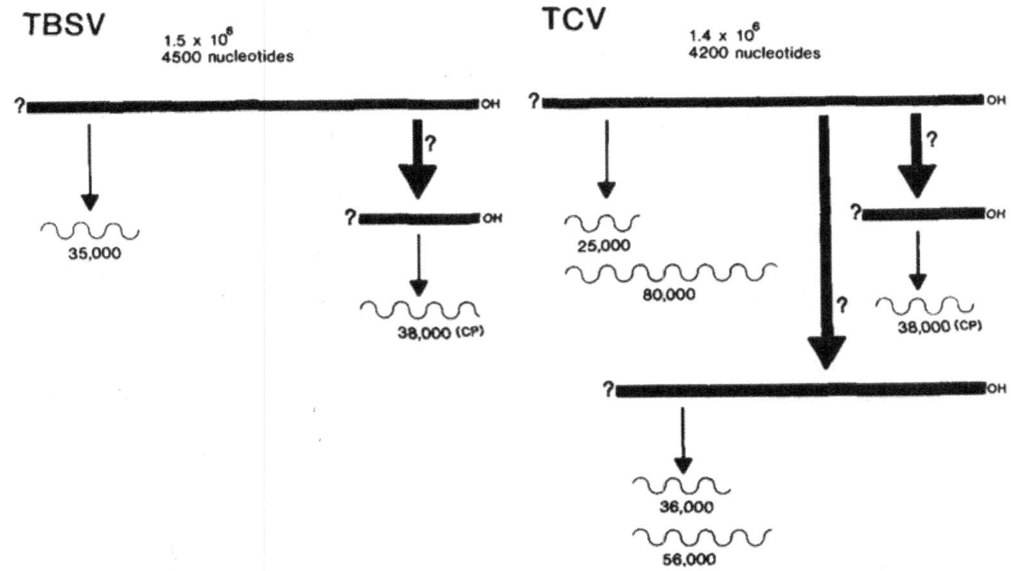

FIGURE 3. A scheme of the genomic organization and expression strategy of the tombusvirus RNA of tomato bushy stunt virus (TBSV) and turnip crinkle virus (TCV). See Figure 1 for an explanation of the figure.

plant viral RNAs to be analyzed in a cell-free translation system.[78,79] TNV is a spherical virus, 30 nm in diameter, with a capsid protein monomer which has a molecular weight of 30,000.[80] STNV is a spherical virus, 18 nm in diameter. The capsid protein (21,600 M_r) amino acid sequence has been determined.[81] The capsid monomers for STNV are arranged in a T = 1 icosahedral structure which accounts for the small diameter of the virion.[82,83]

1. Viral Genome Structure

The encapsidated RNA of TNV has been analyzed. Genomic TNV RNA, 1.4×10^6 M_r, and less-than-full-length RNAs are isolated from virions.[84] The genomic RNA, which is infectious, has a 5′ terminus which ends in a nucleotide diphosphate (ppA). There is no indication that a polyadenylate region is present at the 3′ terminus.[85,86]

The nucleotide sequence of the RNA of STNV has been determined.[87,88] The entire genome is 1239 nucleotides long and the 5′ terminus ends in a nucleotide di- or triphosphate (ppA, pppA). There is a 30-nucleotide untranslated leader sequence before the initiation codon (AUG) of the only open reading frame (587 nucleotides long) on the genome. A 3′-terminal untranslated region of 622 nucleotides completes the genomic RNA.[88] STNV RNA shows no nucleic acid homology with TNV RNA, and it is completely dependent on TNV RNA for replication.[89] Sequence heterogeneity of STNV RNA has been found to be common after successive passage in host tissue.[90]

2. Cell-Free Translation Studies

TNV RNA has been translated in a wheat germ cell-free system.[84] A protein similar to the TNV capsid protein (30,000 M_r) is the major translation product. Additionally, proteins with molecular weights of 63,000, 43,000, and 26,000 are synthesized as minor translation products. Less-than-full-length TNV RNAs have no messenger activity in the wheat germ cell-free system. Full-length genomic RNA appears to be the template for the four products described above. These results have been rationalized by proposing that the translation of TNV RNA involves multiple internal initiation sites, thus the synthesis of individual gene products.[84]

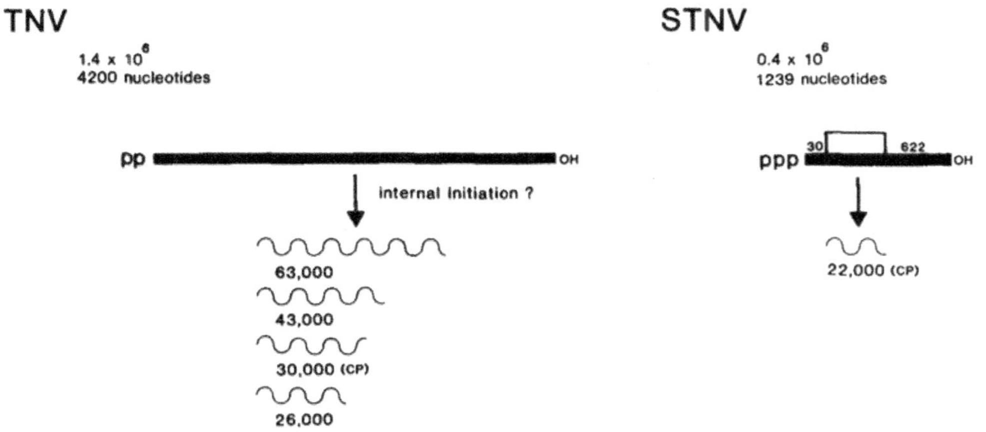

FIGURE 4. A scheme of the genomic organization and expression strategy of tobacco necrosis virus (TNV) RNA and its satellite virus (STNV) RNA. See Figure 1 for an explanation of the figure. The gene order and initiation site of the TNV products is unknown.

STNV RNA translation in the wheat germ system produces primarily STNV capsid protein (22,000 M_r).[84] Additionally, two other proteins (~18,000 and 35,000 M_r) have been identified. The origin of these proteins and the relationship of these proteins to the STNV capsid protein have not been determined.

3. In Vivo Observations

The expression of TNV or STNV RNA in plant cells has received limited study. Analysis of TNV-infected plant tissue has revealed three double-stranded viral RNA species with molecular weights of 2.6×10^6, 1.0×10^6, and 0.95×10^6.[91] Nucleic acid analysis of these RNA species reveals a nested set of RNAs which are representative of varying portions of the 3′-terminal nucleotide sequence. These double-stranded RNAs have been melted and the single-stranded forms have been translated in a wheat germ cell-free system. Single-stranded RNA from the smallest replicative form produces capsid protein during translation in the wheat germ system.[91]

A proposed translation expression strategy for TNV RNA may involved internal genomic initiation sites for protein synthesis (Figure 4). A similar strategy has been proposed for the RNA bacteriophages.[92,93] However, this hypothesis is not in agreement with the identification of discrete less-than-full-length replicative forms. This observation might indicate that subgenomic mRNAs are synthesized during the replicative cycle. Further work is required to answer this question. Only one other plant virus RNA, that of carnation mottle virus, has been proposed to use internal initiation sites in expressing its genome.[94] The technology for producing sensitive nucleic acid probes should be used to reinvestigate the hypothesized genome expression mode of these two viruses.

The RNA of STNV is a monocistronic messenger for the capsid protein (Figure 4). It is unclear why this viral genome has such a long untranslated 3′-terminal region. However, it might be important in the viral replicative cycle as it is conserved. It is also unclear why these two viral RNAs do not have blocking structures at their 5′ termini.

E. Luteovirus Group

Barley yellow dwarf virus (BYDV) is the type member of the luteovirus group.[4] Luteoviruses are unicomponent and have isometrical particles. Members of this group are approximately 25 nm in diameter and have a sedimentation coefficient of 115 to 120S.[95,96]

Luteoviruses, in general, are phloem-limited viruses, reach low titer in the plant, are not mechanically transmitted, and can be transmitted by aphids in a persistent manner.[97,98]

1. Viral Genome Structure

Luteoviruses appear to have a single-stranded RNA genome,[99] although there are reports in the literature claiming erroneously that potato leaf roll virus (PLRV) has a double-stranded DNA genome.[100,101] The major RNA species of PLRV is approximately 2.0×10^6 M_r as determined by sucrose density-gradient centrifugation and by gel electrophoresis under non-denaturing conditions.[99] The genomic RNA has a small protein covalently attached (7,000 M_r), does not contain a polyadenylate sequence at the 3′ terminus, and is infectious when used to inoculate tobacco mesophyll protoplasts.[99] Low-molecular-weight, minor RNA species also are detected when the RNA is isolated from purified BYDV[102] or tobacco necrotic dwarf virus (TNDV) virions.[103] These minor RNAs are heterogeneous in size (2.0×10^6 to 0.1×10^6 M_r) except in the case of TNDV, which has two minor discrete species ($\sim 0.1 \times 10^6$ M_r).[103] Presently, it is unknown if these minor RNA species represent encapsidated subgenomic mRNAs or degradation products.

2. Cell-Free Translation Studies

There is only one report on the translational activity of luteoviruses.[99] PLRV RNA has been translated in a rabbit reticulocyte lysate. Products are synthesized which range in molecular weight from 125,000 to 20,000 with a major translation product of 71,000 M_r. PLRV capsid protein (24,000 M_r) is not synthesized in the cell-free system.

3. In Vivo Observations

Little is known about luteovirus RNA replication and expression in plant tissue. Recently, it has been demonstrated that PLRV, PLRV RNA, and TNDV can be used to infect tobacco protoplasts.[99,104] PLRV RNA, which has been treated with proteinase K, still infects tobacco protoplasts.[99] Use of the tobacco protoplast system may now provide a means to carry out detailed expression studies.

F. Tobamovirus Group

Tobamoviruses are unicomponent plant viruses with a rigid rod morphology, 300 nm in length, and 18 nm wide.[4,105] The prototype member of the group, tobacco mosaic virus (TMV), is one of the best characterized plant virus from a molecular, biological, and physiochemical viewpoint. For a recent review, see the article by Hirth and Richards,[106] and the chapter by Lomonossoff and Wilson in Volume 1 of this series.

1. Viral Genome Structure

TMV has an infectious nucleic acid which is a single-stranded RNA molecule with a molecular weight of 2.0×10^6.[4,106] The entire nucleotide sequence (6395 nucleotides) of the genomic RNA of the vulgare strain of TMV has been determined.[107] The genomic RNA has a m^7Gppp "cap" structure covalently attached to the first nucleotide (guanidine) at the 5′ terminus[108] followed by an untranslated "leader" sequence of 69 nucleotides before the first AUG initiation codon.[107] This AUG initiates an open reading frame which is 3350 nucleotides in length[107] and is translated into a protein with a molecular size of 125,941 (126,000).[109,110] The termination codon of this gene is a "leaky" amber terminator (UGA) and with the appropriate tRNA,[111] ribosomes can translate through this stop codon to another termination codon located 1500 nucleotides downstream toward the 3′ terminus.[107] This read-through protein has a molecular size of 183,253 (183,000).[109-111] A third open reading frame initiates at nucleotide 4903 and continues to a termination codon ending at nucleotide 5709.[107] This gene codes for a protein with a molecular size of 29,987 (30,000).[112] Additionally, the nucleic acid sequence responsible for the initiation of virion assembly in the vulgare strain of TMV is located in this sequence.[113-115] The fourth open reading frame is located proximal to the 3′ terminus and codes for the capsid protein.[110,112,116] The reading frame initiates at nucleotide 5712, terminates at nucleotide 6191, and codes for a 17,604 M_r (17,600) protein.[107] A 204 nucleotide untranslated 3′-terminal sequence is present and begins after the termination codon of the 17,600 M_r protein and ends at the 3′-terminal adenosine.[107,116] The terminal adenosine of the vulgare strain of TMV can be aminoacetylated in vitro with histidine[117] (or valine with the RNA of the cowpea strain of TMV).[110] The significance of this aminoacetylation is not clear; however, it has been proposed that it might be involved in replicase recognition.[117]

The nucleic acid sequences of the 3′-terminal region of two other TMV strains (OM and cowpea)[118-120] and the tobamovirus cucumber green mottle mosaic virus (CGMMV)[121] have been determined. The genetic organization of the TMV OM strain is similar to the TMV vulgare strain, while the TMV cowpea strain and the tobamovirus CGMMV both differ by (1) having the coding sequence of the 30,000 M_r protein and the capsid protein overlap by 10 to 20 nucleotides and (2) by having the nucleotide sequence for the initiation of virion assembly in the capsid protein gene.

Subgenomic mRNAs are used to express the genes coding for the 30,000[112,122,123] and 17,600 M_r[109,112,124-127] TMV proteins. The mRNAs for the 30,000 M_r proteins are encapsidated and are found in the intermediate (I) length particles which are present in purified virion preparations of all TMV strains.[112,122,123] The capsid protein mRNA is not encapsidated by most TMV strains; however, the cowpea strain of TMV and CGMMV encapsidate this subgenomic mRNA in short (S) particles.[112,124-130]

2. Cell-Free Translation Studies

TMV RNA has been translated in a number of different cell-free translation systems. Translation of genomic RNA produces two major proteins, 126,000 and 183,000 M_r, and a number of minor polypeptides ranging in molecular weight from 110,000 to 46,000.[111,112,124-126,132,134,135] The two major proteins share the same initiation codon, and the 183,000 M_r protein is synthesized by a translational read-through of the amber termination codon (UAG) of the 126,000 M_r protein gene.[111] The synthesis of the 183,000 M_r protein can be enhanced in the cell-free translation system by supplementing the translation mix with a yeast amber suppressor tRNA[111] or by using an amber suppressor tRNA from *Drosophila melanogaster*.[131]

The genes coding for the 30,000 and 17,600 M_r proteins are present in the genomic RNA but are not expressed from the full-length viral-RNA in the eukaryotic cell-free systems. However, the 17,600 M_r protein gene, and possibly the 30,000 M_r protein gene, are expressed in a prokaryotic *E. coli* cell-free system.[134] For expression of these genes in a eukaryotic cell-free system, less-than-full-length RNA transcripts are required for translation.[109,110,112,122-127,132,133]

The gene coding for the 30,000 M_r protein apparently is transcribed into three major subgenomic RNAs which are 3′ coterminal and form a nested set of RNAs that are 1900, 1850, and 1500 nucleotides in length.[122,123] These RNAs do not have a 5′-terminal "cap" structure and appear to terminate in nucleotide 5′-tri- and 5′-diphosphates.[122] Because these RNAs contain the nucleic acid sequence involved in initiating virion assembly, they are encapsidated and make up the "I" particles found in TMV preparations.[112,122,123,129,130] Translation of these RNAs produces multiple proteins with molecular weights of 30,000, 29,000, and 23,000.[122] Tryptic peptide mapping reveals that these proteins form a nested set with staggered N termini and identical C termini.[122] The subgenomic mRNAs which are translated into the 30,000 M_r family of proteins contain the sequence for the capsid protein. However, the capsid protein gene is not expressed from these RNAs.[110,112]

The capsid protein gene is expressed in the cell-free systems only from a monocistronic, subgenomic mRNA.[109,112,126,127] This RNA can be isolated from TMV-infected plant tissue[126,127] and is encapsidated in the cowpea strain of TMV to generate 'S' particles approximately 60 nm in length.[112,124,125] This subgenomic RNA in the vulgare strain of TMV has a molecular weight of 250,000, is 693 nucleotides long, and contains a m^{7}Gppp 5′ terminus unlike the mRNAs for the 30,000 M_r family of proteins.[116] The initiation codon for the capsid protein gene lies 9 nucleotides downstream from the cap structure and begins a coding sequence of 480 nucleotides. The coding region is terminated by a UGA stop codon which is followed by a 204 nucleotide untranslated 3 ′-terminal sequence.[107,116] This untranslated region also is present in the mRNA for the 30,000 M_r protein and the genomic RNA.[110]

A number of other TMV-specific translational proteins are observed in addition to the four major cell-free products discussed above. Analysis of the multiple products isolated from size-fractionated RNA from purified TMV and from RNA isolated from alkali-stable TMV particles indicates that these products begin translation from the same initiation codon used for the 126,000 and 183,000 M_r proteins.[133] The multiple products synthesized most likely are the result of (1) specific endonucleolytic cleavage of the TMV genomic RNA, (2) misreading of sense codons as termination codons, or (3) RNA secondary structure impedence of ribosome translocation.[133] Translational analysis of TMV RNA isolated from TMV-infected tobacco tissue also produces a number of products which range in molecular weight from 95,000 to 45,000.[135] The RNA species have been shown to be a nested set of RNAs all containing 3′-coterminal sequences. However, in contrast to the results of RNA isolated from virions,[133] translation of these RNAs results in the synthesis of a nested set of proteins sharing C-terminal amino acid sequences and having staggered N-terminal amino acids.[135] These proteins perhaps could arise from the initiation of protein synthesis on different initiation codons. Further experiments will determine the origin of these proteins and the roles they might play in the TMV replicative cycle.

3. In Vivo Observations

The synthesis of TMV specific proteins has been examined in protoplasts and in whole leaf tissue.[123,132,136] Viral protein synthesis has been shown to occur despite the fact that host protein synthesis is not suppressed during TMV replication.[136] However, the detection of TMV specific proteins has been difficult, because of the high levels of host protein synthesis.

The major cell-free translation products of TMV genomic RNA, the 126,000 and 183,000 M_r products, are detected soon after TMV infection of protoplasts[132,136] or leaf tissue[123] and reach a maximum level 30 to 40 hr after infection. After this time, the ability to detect these proteins decreases. The 126,000 M_r protein, labeled in vivo, has been isolated and compared to the 126,000 M_r protein synthesized in a wheat germ cell-free translation system.[138] The proteins share cyanogen-bromide cleavage patterns and are similar in their amino acid sequence. This protein appears to be closely associated with the plasma membrane and has been implicated as either the TMV-specific replicase or part of the replicase.[138]

The synthesis of the 30,000 M_r protein in protoplasts and in whole leaf tissue has been analyzed. Only a single polypeptide (31,000 M_r in tobacco protoplasts[132] and 30,000 M_r in tobacco leaf tissue)[123] can be detected. This is in contrast to the multiple 30,000-M_r-related proteins synthesized when I-RNAs are translated in cell-free systems.[122] A peptide map comparison of the protein detected in vivo with the protein synthesized in the cell-free systems has not been reported. The temporal analysis of the 30,000 M_r protein reveals that this protein is present early in the TMV replicative cycle and reaches a maximum level 20 hr after infection.[123] After this time the ability to detect this protein decreases until it can no longer be detected 48 to 72 hr postinfection. Relative to the 126,000 M_r protein, very little 30,000 M_r protein is detected. Because of the early appearance of this protein after infection, the 30,000 M_r protein has been implicated to be involved in a function before or during viral RNA replication.[123] Recently, an analysis of a temperature-sensitive mutant of TMV, which infects but does not spread systemically under nonpermissive temperatures, has led to the speculation that this protein is somehow involved in cell-to-cell movement of TMV.[137]

The initial cell-free translation studies with TMV genomic RNA detected little or no TMV capsid protein synthesis.[139] Analysis of RNA from TMV-infected tobacco tissue revealed that a small RNA [low-molecular-weight component (LMC)] which had a sedimentation value between 6 and 18S, was the mRNA for TMV capsid protein.[126,127] The structural characteristics of this RNA have been described in the preceding section. A temporal analysis of the synthesis of this RNA has not been carried out, but the synthesis of capsid protein has been assayed in infected tobacco leaf tissue.[123,132,136] Synthesis of capsid protein initiates approximately 10 hr after infection and reaches a maximum level 30 to 40 hr after infection.

The strategy for the generation of TMV subgenomic mRNAs remains to be determined, although a number of studies have analyzed the double-stranded RNA replicative forms found in TMV-infected tissue.[140,142,142a] A double-stranded replicative form and a replicative intermediate of the genomic RNA have been detected in tobacco tissue.[140] Additionally, three other TMV-specific double-stranded RNAs have been isolated from infected protoplasts and plant leaf tissue. These double-stranded RNAs have molecular weights of 2.25×10^6, 1.1×10^6, and 0.23×10^6.[142] All the identified double-stranded RNAs contain the sequences of the 3′-terminal portion of the genomic RNA and form an overlapping set of RNAs. In addition to these double-stranded RNAs, other workers have identified eight other TMV specific double-stranded RNAs.[142a] Presently, the role these double-stranded RNAs play in the TMV replication cycle is unclear.

In a study designed to analyze the plus- and minus-strand TMV RNA species in infected plant tissue, eight plus-stranded RNA species have been detected with an estimated size of 6400 (genomic RNA), 3500, 3000, 1800, 1500, 1300, 1100, and 700 (capsid protein mRNA) nucleotides.[135] Similar negative-stranded RNA species have been detected and therefore the TMV RNA species may be synthesized by incomplete transcription of the genomic RNA into less-than-full-length negative-stranded RNA species. These RNAs subsequently may be transcribed into mRNA species.[135]

Further work is necessary to ascertain if the minor RNA species and the corresponding proteins are authentic viral products and whether they play an integral role in the TMV replicative cycle.

A proposed strategy of tobamovirus gene expression is presented schematically in Figure 5.

G. Potexvirus Group

Potato virus X (PVX) is the type member of the potexvirus group.[5] Members of this group reach high titers in plant tissue, are mechanically transmitted, and possess a flexuous rod morphology.[143,144] Generally, the members of this group are 480 to 580 nm in length and ~12 nm in diameter (see chapter by AbouHaidar and Erickson in Volume 1 of this series).

1. Viral Genome Structure

Potexviruses have a single-stranded genomic RNA molecule (2.0 to 2.5×10^6 M_r) which is messenger sense.[145] The RNA of PVX and clover yellow mosaic virus (CLMV) has a m^7Gppp structure at the 5′ terminus.[145-149] The 3′ terminus of the potexvirus genome has been examined for three different members.[149,150] Two of the members (papaya mosaic virus [PMV] and CLMV) have a polyadenylate region at the 3′ terminus, which is approximately 25 to 100 adenosine residues long.[150] No detectable polyadenylate terminal sequence is found in the type member, PVX.[149] The nucleotide sequence adjacent to the 5′ terminus of PMV

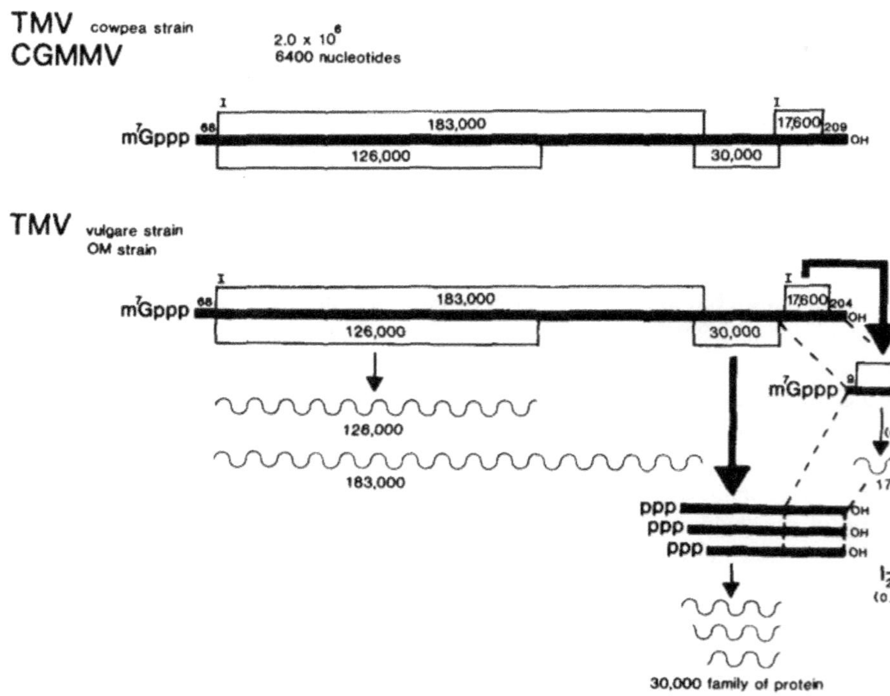

FIGURE 5. A scheme of the genomic organization and expression strategy of the tobamovirus RNA of tobacco mosaic virus (TMV) and cucumber green mottle mosaic virus (CGMMV). See Figure 1 for an explanation of the figure. The discrete low-molecular-weight component (LMC) RNA and the heterologous intermediate-RNAs (I-RNAs) are presented.

has been determined.[151] The sequences GCAA and AAA are repeated several times in the 100 nucleotides proximal to the 5′ terminus. As this is the region involved in the initiation of virion assembly, these sequences may play an important role in encapsidation.[151] The encapsidation of subgenomic mRNA species has been reported for only one member of the potexvirus group, namely, narcissus mosaic virus (NMV).[151a]

2. Cell-Free Translation Studies

The initial investigation of PVX RNA translation used wheat germ extracts.[145] A large number of proteins were synthesized; however, none of the proteins could be immunoprecipitated using antibodies to PVX capsid protein.[145] This work has been expanded by translating PVX RNA in a rabbit reticulocyte lysate and a S-27 extract of yeast.[146,147] Two major products with molecular weights of 180,000 and 145,000 are synthesized. Minor small-molecular-weight products also are synthesized, especially in the wheat germ and yeast extracts. No product comigrating with PVX capsid protein and serologically reactive with antiserum to PVX capsid has been detected. The two high-molecular-weight proteins have been analyzed by tryptic peptide analysis and preliminary evidence indicates that they share amino acid sequences.[147] Cap inhibition studies indicate that synthesis of the two proteins initiates near the 5′ terminus.[147]

The viral RNA of PMV has been translated in the wheat germ and rabbit reticulocyte cell-free systems.[152,152a] Products ranging in molecular weight from 155,000 to 22,000 are synthesized in both cell-free systems. Three major products are synthesized in both systems and have molecular weights of 155,000, 73,000, and 22,000. The number of minor products increases when the wheat germ extract is used. The 22,000 M_r product has been demonstrated to be identical to the PMV capsid protein by serological and biochemical analyses.[152] The 155,000 and 73,000 M_r products share amino acid sequences and represent a nested set of proteins. Preliminary evidence suggests that these products do not share the same initiation codon.[152]

NMV RNA has been analyzed and translated in a wheat germ extract.[152b] Translation of the genomic RNA of NMV results in the synthesis of three products with estimated molecular weights of 116,000, 66,000, and 44,000.[152b] NMV appears to be unique among the potexviruses that have been examined to date in the synthesis and encapsidation of a subgenomic messenger RNA coding for the capsid protein (22,800 M_r). This capsid protein messenger RNA is approximately 840 nucleotides in length (0.27×10^6 M_r) and is encapsidated by NMV capsid protein to form rod-shaped particles, 100 nm in length.

Some interesting experiments have been carried out using partially encapsidated PMV RNA.[152] Previously, it has been demonstrated that potexviruses initiate encapsidation at or near the 5′ terminus.[153] Partially encapsidated virions with an average particle length of 180 nm have been formed and added to a cell-free translation system.[152] Partially encapsidated virions do not stimulate the synthesis of the 155,000 or 73,000 M_r products efficiently. This infers a 5′-terminal location on the viral genome for these product genes. However, capsid protein (22,000 M_r) synthesis is stimulated at a greater rate by this partially encapsidated virion template than the other two products. It has been proposed from this experiment that the capsid protein gene may be located near the 3′ terminus of the viral genome and that the initiation codon for the other two genes must be proximal to the 5′ terminus.[152] It remains unresolved if the capsid protein gene is translated from the genome-length RNA or from a less-than-full-length RNA species which is an authentic subgenomic mRNA or fragmented RNA. The use of partially encapsidated viral RNAs to map the genome of anisometric viruses, however, is an interesting and novel approach.

3. In Vivo Observations

There have been a limited number of reports dealing with potexvirus-infected plant tissue. PVX-infected tobacco leaf tissue has been examined for viral protein synthesis.[154,155] A protein comigrating with PVX capsid protein and some smaller proteins have been detected. The relationships between these in vivo results and the cell-free translation data have not been established.

Members of the potexvirus group appear to express their genomes using two different strategies. NMV and PVX genomic expression may be similar to TMV or TYMV RNA. That is, a high-molecular-weight protein is synthesized from the genomic RNA. Whether this protein is posttranslationally processed by proteases has not been determined. The capsid protein and possibly other genes might be expressed via the synthesis and translation of subgenomic mRNAs. This only has been documented for NMV. PMV may have an expression strategy which initiates translation at internal sites. High-molecular-weight proteins and capsid protein may be synthesized from the genomic RNA via the binding of ribosomes and initiation of translation near the 5′ terminus and at internal sites. Other potexvirus RNAs have been examined in a preliminary fashion. CLMV RNA translation results are similar to those obtained with PMV RNA while foxtail mosaic virus RNA, barrel cactus virus RNA, and viola mottle virus RNA are similar to PVX RNA in their cell-free translation.[152a] The genomic organization and possible expression scheme of PVX RNA and PMV RNA is presented below (Figure 6). NMV RNA appears to be similar to PVX RNA with an additional subgenomic transcript (capsid protein mRNA) being generated from 3′-terminal sequences. Further work is required to determine the strategy(ies) used by members of the potexvirus group.

H. Potyvirus Group

Potato virus Y is the type member of the potyvirus group.[4] This is the largest group of

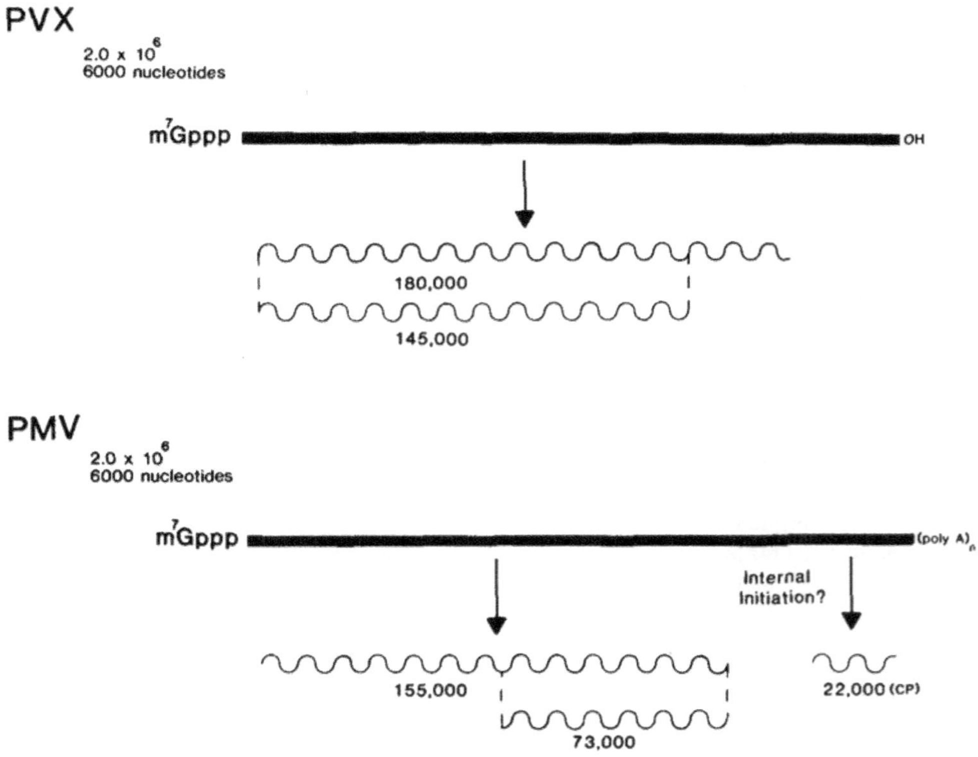

FIGURE 6. A scheme of the genomic organization and expression strategies of the potexvirus RNA of potato virus X(PVX) and papaya mosaic virus(PMV). See Figure 1 for an explanation of the figure.

RNA plant viruses with 48 members and 68 tentative members. The single-stranded genomic RNA is encapsidated by capsid protein monomers (30,000 to 35,000 M_r) to form a virion which has a filamentous-rod morphology 700 to 800 nm in length and 12 to 15 nm in diameter.[156] Little information is available on the structure or assembly of potyviruses.[157,158]

A unique aspect of this group is the formation of cytoplasmic, cylindrical (pinwheel) inclusion bodies in the cytoplasm of potyvirus-infected plant cells.[159] This inclusion body is an organized aggregation of viral protein monomers with a molecular weight of ~70,000 and is serologically virus specific and unrelated to the viral capsid protein and to host proteins.[160] In addition to the cytoplasmic inclusion body, a limited number of potyviruses form a nuclear inclusion body in infected plant cells. Only the nuclear inclusion body of tobacco etch virus (TEV) has been analyzed. It is an organized aggregate of two distinct, virus-specific proteins, with molecular weights of 49,000 and 54,000, in approximately equal molar amounts.[161]

Potyviruses are transmitted by aphids in a nonpersistent manner.[156] A virus-specific, and presumably virus-coded protein is required for aphid transmission.[162] This protein is referred to as a "helper component".

1. Viral Genome Structure

The genome of potato virus Y consists of a single-stranded, messenger-sense RNA molecule with a molecular weight of 3.2×10^6.[163] The 5'- and 3'-terminal structures have been examined only for TEV RNA. The 5' terminus has a small protein (6,000 M_r) covalently attached.[164] This protein can be removed without affecting infectivity or the cell free translation properties of the viral RNA. The 3′ terminus contains a polyadenylate region which is heterogeneous in length.[164,165] Encapsidated subgenomic RNAs have not been detected.[168]

2. Cell-Free Translation Studies

Potyvirus RNAs have been translated rabbit reticulocyte[166-170] and wheat germ[171,172] cell-free systems. TEV RNA, isolated from purified virions, produces major unique products with molecular weights of 87,000, 54,000, 50,000, 49,000, and 30,000. Proteins with the molecular weights of 85,000, 120,000, and 150,000 also are synthesized along with a number of minor products with molecular weights less than 85,000. Immunoprecipitation analysis and peptide mapping studies using *Staphylococcus aureus* V-8 protease have revealed that the 54,000 and 49,000 M_r proteins are similar to the TEV nuclear inclusion proteins. The 30,000 M_r product is similar to the TEV capsid protein. A 70,000 M_r product, similar to the cytoplasmic inclusion protein, usually is not detected. Instead, a 120,000 M_r product appears to contain the amino acid sequence of the 70,000 M_r inclusion protein as determined by immunological studies. It is unclear if this product represents a read-through of two genes (50,000 and 70,000 M_r) by the ribosomes of the cell-free system or an authentic polyprotein. The 85,000 M_r product is a similar type of read-through product or polyprotein of the 54,000 M_r nuclear inclusion protein gene and the 30,000 M_r capsid protein gene. This has been confirmed by immunological studies and peptide mapping. The proposed order of the TEV genome, based on the immunoprecipitation analyses of the cell-free translation products is 5′ — 87,000 — 49,000 — 50,000 — 70,000 — 54,000 — 30,000 — 3′. The genetic map accounts for approximately 95% of the coding capacity of a 3.2×10^6 M_r RNA. A similar cell-free product relationship has been established for pepper mottle virus (PeMV) RNA and 15 other potyviral RNAs.[168,178] PeMV, and most of the other potyviruses analyzed, does not form nuclear inclusion bodies during infection. However, two nuclear inclusion-like proteins, with molecular weight around 54,000 and 49,000, are synthesized in the rabbit reticulocyte cell-free translation system and react with the antisera raised against the two TEV nuclear inclusion proteins.

A similar type of analysis has been carried out with tobacco vein mottling virus (TVMV) RNA.[169,170] Additionally, antiserum to a partially purified "helper component" has been used in the immunoprecipitation analysis. The gene product (TVMV = 75,000; TEV = 87,000 M_r), which has been mapped proximal to the 5′ terminus, reacts specifically with this antiserum.[170]

The molecular weight estimated for the predominant translation product (mapped near the 5′ end) of 17 potyviruses tested ranges from 78,000 for PeMV to 110,000 for watermelon mosaic virus 1 (WMV-1).[178] Antiserum prepared to "helper component" protein (~52,000 M_r) of TVMV reacts with the primary translation product of most of the 17 potyviruses translated to date, with varying degrees of efficiency.[170-172] Antisera prepared to the major constituent protein (~52,000 M_r) of the amorphous inclusions associated with PeMV and WMV-1 infections also reacts efficiently with the respective primary translation product of PeMV and WMV-1.[173] Whether the "helper component" protein and the protein of the amorphous inclusions are identical for PeMV and WMV-1 has not been tested. The difference in molecular weights for the in vivo products compared with the in vitro product suggests that proteolytic processing may be necessary for the formation of active "helper component" protein and of constituent protein for amorphous inclusions.

Potyviral RNA has been translated in a wheat germ cell-free translation system. TEV RNA has been reported only to produce a single polypeptide with a molecular weight of 41,000.[174] We have translated TEV RNA and a number of other potyviral RNAs in a wheat germ extract.[175] TEV cell-free products similar to the proteins synthesized in the rabbit reticulocyte lysate are formed with two exceptions. A major product with a molecular weight around 36,000 does not react with any of the TEV-specific antisera and is not observed in the rabbit reticulocyte lysate. Products which comigrate with proteins observed during infection, and which are immunoreactive with TEV-specific antisera, are not synthesized as they are in the rabbit reticulocyte cell-free translation system. Instead, only the high-molecular-weight read-through proteins or polyproteins are evident in immunoprecipitations with the TEV antisera. Differences in RNA or in the wheat extracts used in these two studies may account for the discrepancies.

3. In Vivo Observations

There are only a limited number of reports regarding the in vivo expression of the potyviral genome. Presumed subgenomic TEV RNA species have been detected in RNA isolated from TEV-infected plant tissue.[176] These single-stranded RNAs, detected using randomly primed complementary DNA to TEV RNA are polyadenylated and have molecular weights of 3.2×10^6, 1.14×10^6, 1.01×10^6, 0.84×10^6, 0.62×10^6, 0.55×10^6, 0.46×10^6, and 0.34×10^6. We have analyzed RNA from infected plant tissue in a similar manner using randomly primed complementary DNA and recombinant DNA molecules containing inserts of complementary DNA of varying lengths to TEV RNA.[177] Similar RNA species are detected regardless of the hybridization probe used in the experiment. Various reconstitution experiments with TEV RNA and RNA isolated from healthy host plant have detected similar RNA species. These RNA zones appear to be artifacts caused by plant ribosomal RNAs and ribosomal RNA breakdown products interfering with the electrophoresis and/or hybridization of fragmented potyviral RNA. Analysis of the polyadenylated RNA fraction from TEV-infected plant tissue reveals a single RNA species with a molecular weights of 3.2×10^6.[177]

The polyadenylated RNA fraction from TEV-infected plant tissue has been translated in a rabbit reticulocyte lysate cell-free system and the products have been analyzed by immunoprecipitation.[177] Four virus-distinct products with molecular weights of 120,000, 87,000, 85,000, and 49,000 can be identified. The 120,000 M_r product is immunoprecipitated with antiserum to TEV cytoplasmic inclusion protein. The 87,000 M_r product is unreactive with the antisera prepared to four, virus-specific proteins. The 85,000 M_r product is immunoprecipitated by antisera to the TEV 54,000 M_r nuclear inclusion protein and the TEV capsid protein. The 49,000 M_r product is immunoprecipitated by antiserum to the 49,000 M_r nuclear inclusion protein.

Thus potyviruses may express their genome from a full-length RNA by (1) synthesis of specific products by initiating translation at internal genome sites analogous to the expression strategy proposed for TNV, (2) synthesis of polyproteins which are co- or posttranslationally processed, or (3) a strategy which combines limited internal initiation and the synthesis of the polyproteins. The consistent observation of high-molecular-weight products (160,000, 120,000, and 85,000 M_r) tends to support the latter two strategies. A possible strategy of expression is presented in Figure 7.

I. Closterovirus Group

Sugar beet yellows virus is the type member of the closterovirus group which contains at least 14 additional members.[4,179] The RNA is encapsidated by protein to form flexuous rod particles with a diameter of 12 nm and ranging in length from 600 to 2000 nm, depending upon the RNA size. The capsid protein monomers have a molecular weight of 23,000 to 27,000.[179]

1. Viral Genome Structure

The closterovirus genome consists of a messenger-sense, single-stranded RNA with a molecular weight which, depending on the virus, may be 2.2 to 6.5×10^6.[180] Information is not available on the terminal structures of closteroviral RNA.

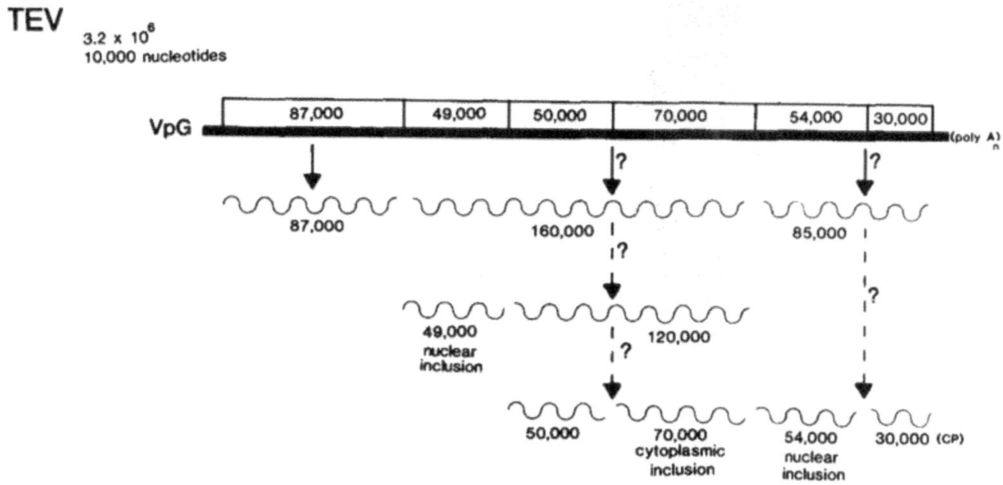

FIGURE 7. A scheme of the genomic organization and expression strategy of the potyvirus tobacco etch virus (TEV) RNA. See Figure 1 for an explanation of the figure.

2. *Cell-Free Translation Studies*

Citrus tristaza virus (CTV) RNA ($6.5 \times 10^6 M_r$) has been translated in a rabbit reticulocyte lysate system.[181] It is difficult to obtain intact virions during CTV purification.[182] As a result, translations have been limited to RNA fractionated by sucrose rate-zonal centrifugation and sedimenting at an estimated range of 30 to 60S. Only a limited amount of RNA was available and therefore optimal translation conditions have not been resolved. A major translation product with a molecular weight of 50,000 and others of 65,000, 33,000, and 26,000 have been detected. The 26,000 M_r product reacts with antiserum specific for CTV capsid protein. The low template activity of the CTV RNA suggests that the translations involve fragmented RNAs rather than subgenomic mRNAs.[183]

3. *In Vivo Observations*

Information about genomic expression of closteroviruses is very limited. Four to nine species of double-stranded RNAs have been detected in closterovirus-infected tissue and these may represent replicative forms of subgenomic messengers.[184] Nine double-stranded RNAs in addition to the double-stranded genomic length RNA have been detected in CTV infections. These RNA species could represent up to nine subgenomic messengers, but the total molecular weight of the double-stranded RNAs exceeds the molecular weight of genomic double-stranded RNA. More information is required to elucidate genome expression strategy.

III. VIRUSES WITH BIPARTITE GENOMES

The five virus groups discussed in this section possess a genome consisting of two RNA species. Both RNA species are required for infection. The initial four groups discussed have spherical virion morphology while tobraviruses have a rigid, rod virion morphology.

A. Nepovirus Group

Tobacco ringspot virus (TRSV) is the type member of the nepovirus group which has at least 25 additional members.[4] The infectious virus genome comprises two unique RNA species. RNAs 1 and 2 are encapsidated separately into isometric particles 28 nm in diameter.[185] The capsid protein monomer for most members has a molecular weight between

55,000 to 60,000. Strawberry latent ringspot virus (SLRV), a possible member of the nepovirus group, has a protein capsid made of two different subunits with molecular weights of 44,000 and 29,000.[186]

1. Viral Genome Structure

The nepovirus genome consists of two messenger-sense, single-stranded, RNAs with molecular weights of 2.8×10^6 (RNA 1) and 1.3 to 2.4×10^6 (RNA 2).[185] A satellite RNA molecule (0.1 to 0.5×10^6 M_r) is associated with some nepovirus strains.[185] The nepovirus RNAs have a polyadenylate sequence at the 3′ terminus.[187] A protein is covalently linked to the genomic RNAs and this protein is required for infectivity.[188-191] No sequence homology has been detected between RNAs 1 and 2 of TBRV in hybridization experiments.[192] The satellite RNA associated with tomato black ring virus (TBRV) has a protein, probably linked to the 5′ terminus and a polyadenylate sequence at the 3′ terminus.[193,194] A similar structure has been found with SLRV satellite RNA. The satellite RNA associated with TRSV does not contain covalently linked protein and lacks a detectable polyadenylate region at the 3′ terminus.[195]

2. Cell-Free Translation Studies

The genomic RNAs of TBRV have been translated in a wheat germ extract and rabbit reticulocyte lysate.[196] Translation products of RNA 1 and 2 have molecular weights of 220,000 and 160,000, respectively. These products represent about 80 and 90% of the genetic potential of the respective RNAs. The 160,000 M_r product reacts with antiserum to TBRV capsid protein while RNA 1 translation products do not react. The TBRV satellite RNA (0.48×10^6 M_r) directs the synthesis of a 48,000 M_r product in wheat germ extracts.[197] This product does not react with antiserum to TBRV capsid protein. Peptide mapping of this satellite RNA product reveals differences when compared with a satellite RNA product from another strain of TBRV. Product differences are not detected when TBRV genome and satellite RNAs are translated in the wheat germ system before and after treatment of the RNAs with protease.

Unfractionated TRSV RNAs have been translated in a wheat germ system before and after treatment with pronase.[191] The polypeptides synthesized by either RNA preparation are indistinguishable. It has been concluded that the genome-associated protein of TRSV is not necessary for cell-free translation. The satellite RNA (0.1×10^6 M_r) associated with TRSV does not direct the synthesis of a protein in the wheat germ system.[198]

The satellite RNA of SLRV (0.4×10^6 M_r) induces the synthesis of a 38,000 M_r protein in rabbit reticulocyte lysates.[193] The size of this product corresponds closely to the maximum coding capacity of the satellite RNA.

3. In Vivo Observations

The genome expression of TBRV has been studied in tobacco protoplasts.[192,197,199] TBRV RNA 1 replicates and produces genome-linked protein independent of RNA 2 in tobacco mesophyll protoplasts.[192] This is evidence that RNA 1 contains all the information necessary for its own replication and for the synthesis of the genome-linked protein. RNA 2 is not capable of replicating by itself under similar conditions. Heterologous combinations of RNAs from two strains of TBRV can induce the synthesis of virus-specific products in inoculated protoplasts.[199] A mixture of RNA 1 from TBRV strain A and RNA 2 from TBRV strain G induces the synthesis of TBRV (G) capsid protein which assembles into empty protein shells. However, the multiplication of potentially infective RNA-2 (of strain G) is not detected. The observations that RNA 1 (of strain A) is not assembled into virions with the strain G coat protein and that RNA 2 (of strain G) directs the synthesis of significant amounts of

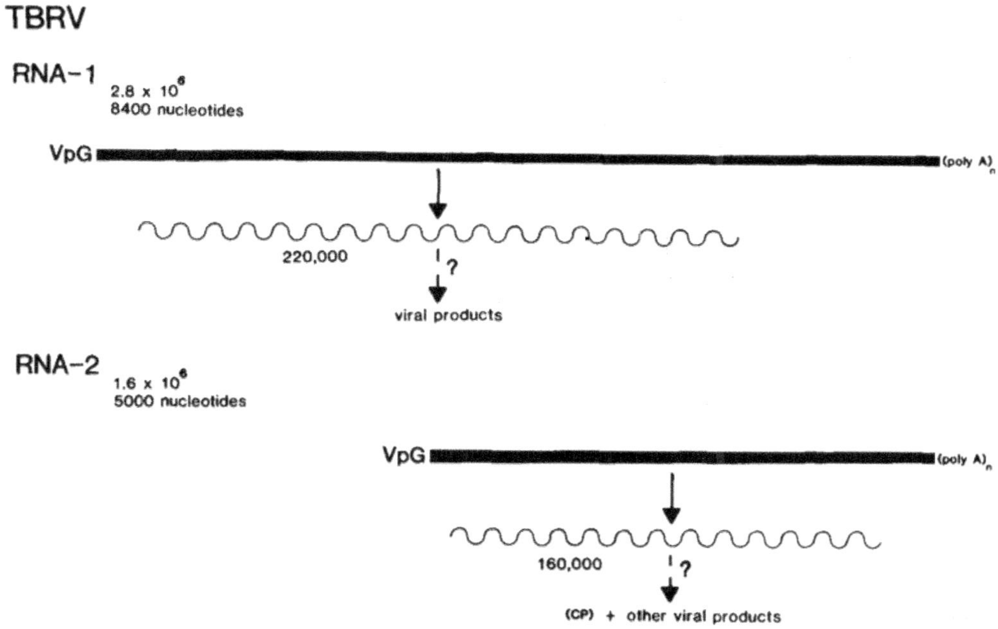

FIGURE 8. A scheme of the genomic organization and expression strategy of the nepovirus tomato black ring virus (TBRV) RNAs. See Figure 1 for an explanation of the figure.

capsid protein without the concurrent multiplication of infectious strain G RNA 2 in the pseudorecombinant TBRV raise some interesting questions. It has been proposed[199] that (1) RNA 2 (of strain G) is synthesized but lacks the genome-linked protein coded for by the heterologous RNA 1, thus the RNA is not infectious and not assembled into virions and (2) the capsid protein from strain G does not recognize the genome-linked protein of strain A RNA 1 and therefore RNA 1 is not assembled into virions.[199] These proposals need further investigation. This is, however, an example of the usefulness of the analyses of pseudorecombinants in mapping gene function.

In summary, both RNAs of the nepovirus genome are translated into large polypeptides (Figure 8). These polyproteins may then be cleaved into functional proteins although processing of nepovirus polypeptides has not been demonstrated to date.

The nepoviruses share a number of properties with the comovirus group; the genome is bipartite; both are translated into two large polypeptides representing nearly all the coding capacity, and both genomes have similar structures at the termini. They differ in that the capsid protein is comprised of one polypeptide 55,000 to 60,000 in molecular weight for nepoviruses and two polypeptides with molecular weights of 24,000 and 40,000 for comoviruses; the genome-linked protein is essential for nepovirus infectivity but not for comoviruses; and the association of satellite RNAs with many nepoviruses has not been demonstrated for any comovirus. SLRV, a tentative member of the nepovirus group, has two capsid protein subunits and therefore may represent a transition virus in the evolution of these two virus groups.

B. Pea Enation Mosaic Virus Group

Pea enation mosaic virus (PEMV) is a monotypic virus whose genomic RNAs are encapsidated separately into polyhedral particles with a diameter of 28 nm.[4] The capsid consists of protein monomers with a molecular weight of 22,000.[200]

1. Viral Genome Structure

The PEMV genome consists of two, messenger-sense, single-stranded RNAs with molecular weights of 1.8×10^6 (RNA 1) and 1.4×10^6 (RNA 2).[201] PEMV RNAs have a covalently linked protein (18,000 M_r) attached to the 5′ termini.[202] The 3′ termini do not contain polyadenylate sequences and cannot be aminoacylated.[203]

2. Cell-Free Translation Studies

The translation of PEMV RNAs have been studied in a rabbit reticulocyte lysate and wheat germ extract.[201] Translation of PEMV RNAs has been optimized in the rabbit reticulocyte lysate system using 0.8 mM magnesium acetate, 130 mM potassium acetate, and 100 μg RNA per milliliter, RNA 1 codes for two major products with molecular weights of 88,000 and 36,000 and a minor product with a molecular weight of 147,000. RNA 2 codes for a single major product of 45,000 M_r. Identical products are produced in the wheat germ extract.

No product comigrating with capsid protein has been detected. However, antiserum to PEMV capsid protein does immunoprecipitate the 88,000 M_r product of RNA 1. The 88,000 M_r product is an unstable protein and readily degrades within 8 hr of incubation.[201]

The kinetics of RNA 1 product formation, in time-course and pulse-chase experiments, show that the 36,000 M_r product appears first. No overlapping amino acid sequences are detected in tryptic digests of the 36,000 and 88,000 M_r products.[201] It is not known whether the two major products of RNA 1 are derived from (1) two independent initiation events, (2) from subgenomic RNAs contaminating RNA 1 preparations, or (3) from proteolytic processing of the 147,000 M_r product.

3. In Vivo Observations

PEMV has been cultured in isolated tobacco protoplasts[204] and in suspension cells from embryonic ovary tissue of the aphid *Myzus persicae*.[205] Viral antigen is the only virus-specific protein monitored in these studies. Two double-stranded replicative forms of PEMV RNA with molecular weights of 3.3×10^6 and 2.2×10^6 also have been resolved by gel electrophoresis in extracts from infected plant tissue.[206]

The translation strategy for the bipartite PEMV genome may involve processing of polyproteins and/or subgenomic messengers (Figure 9). The 147,000 M_r product of RNA 1, representing about 80% of the coding information of RNA 1, may be a polyprotein. The 88,000 M_r product of RNA 1 reacts with virion antiserum but is much larger than the capsid protein subunit. Thus the 88,000 product may be a polyprotein which requires processing. The 45,000 M_r product of RNA 2 represents only 32% of potential coding capacity of RNA 2. Since the remaining 68% of the RNA 2 coding capacity does not appear to be open to translation in vitro, the information may be made available for translation in vivo via subgenomic mRNAs.

The location of the capsid protein gene on RNA 1 is based on the translation data of Gabriel and de Zoeten[201] and is in agreement with the report of Hull and Lane,[207] who found that electrophoretic mobility of the PEMV virions is correlated with the parent RNA 1 component in pseudorecombinant experiments with PEMV strains.

C. Dianthovirus Group

Carnation ringspot virus is the type member of the dianthovirus group which contains three other members, red clover necrotic mosaic virus (RCNMV), clover primary leaf necrosis virus, and sweet clover necrotic mosaic virus.[4,,207a] The two dianthovirus RNAs are encapsidated into polyhedral particles, 31 to 34 nm in diameter, in such ways that no particle heterogeneity is evident in electron microscopic or sedimentation analyses.[208] The capsid protein subunit has a molecular weight ~ 40,000.

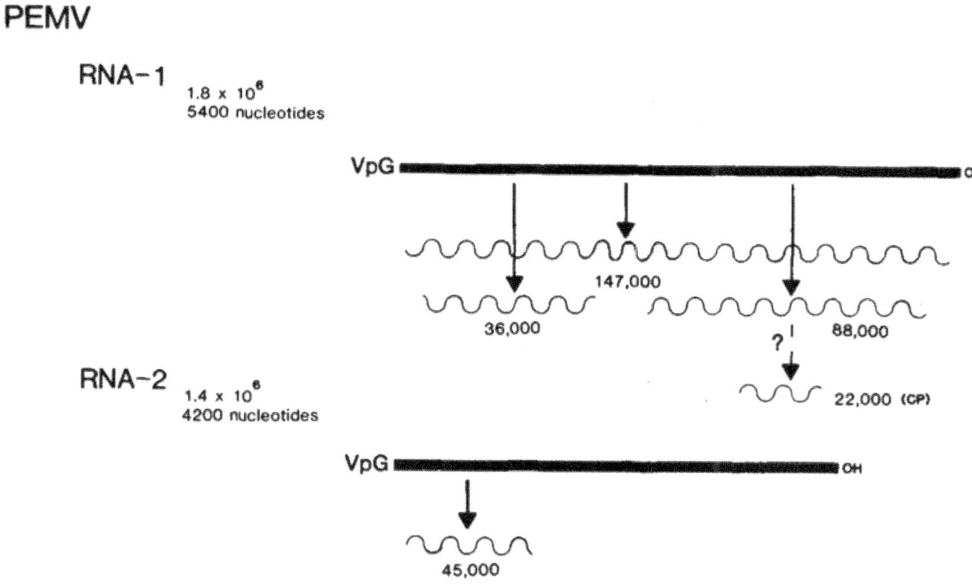

FIGURE 9. A scheme of the genomic organization and expression strategy of pea enation mosaic virus (PEMV) RNAs. See Figure 1 for an explanation of the figure.

1. Viral Genome Structure

The dianthovirus genome consists of two messenger-sense, single-stranded RNAs of molecular weight 1.5×10^6 (RNA 1) and 0.5×10^6 (RNA 2).[70,208] Both RNAs 1 and 2 are required for infectivity.[70,208] The RNAs of these viruses can be readily exchanged in vitro to form infectious pseudorecombinants.[209] Such studies have shown that the genetic information for capsid protein is carried on RNA 1. Hybridization analysis using complementary DNA prepared to RCNMV RNAs indicate little or no sequence homology between RNAs 1 and 2.[210]

2. Cell-Free Translation Studies

RCNMV RNAs 1 and 2 have been translated in a rabbit reticulocyte lysate system.[211] Translation of unfractionated RNAs produces three major products with molecular weights of 39,000, 36,000, and 34,000. The 39,000 M_r product has been identified as capsid protein by peptide mapping and by immunoprecipitation. The RNAs encoding these products have been identified by hybrid-arrested translation. Synthesis of the virus capsid protein and the 36,000 M_r protein is arrested by hybridization of the unfractionated RNA with complementary DNA prepared to RNA 1 while the synthesis of the 34,000 M_r protein is not affected. The results of the translation of RCNMV RNA fractions obtained by rate-zonal centrifugation of heat-denatured RNA using formamide-sucrose gradients present a less clear picture of the RNA-product assignments. The synthesis of the 36,000 M_r product is associated only with the translation of the gradient fractions presumed to be RNA 1. However, the capsid protein (39,000 M_r) and 34,000 M_r protein are synthesized in the translation of sucrose gradient fractions containing either RNA 1 or 2. This suggests that the mRNA for 34,000 M_r protein may be distributed throughout the gradient. A similar conclusion can be made for the capsid protein mRNA. It would be interesting to test the 56,000 and 130,000 M_r products, associated with the translation of presumed RNA 1 fractions, to see if they contain capsid protein sequences. The 130,000 M_r product may represent the total coding capacity of RNA 1. A 30,000 M_r product associated with translation of RCNMV RNA-2 also needs to be analyzed.

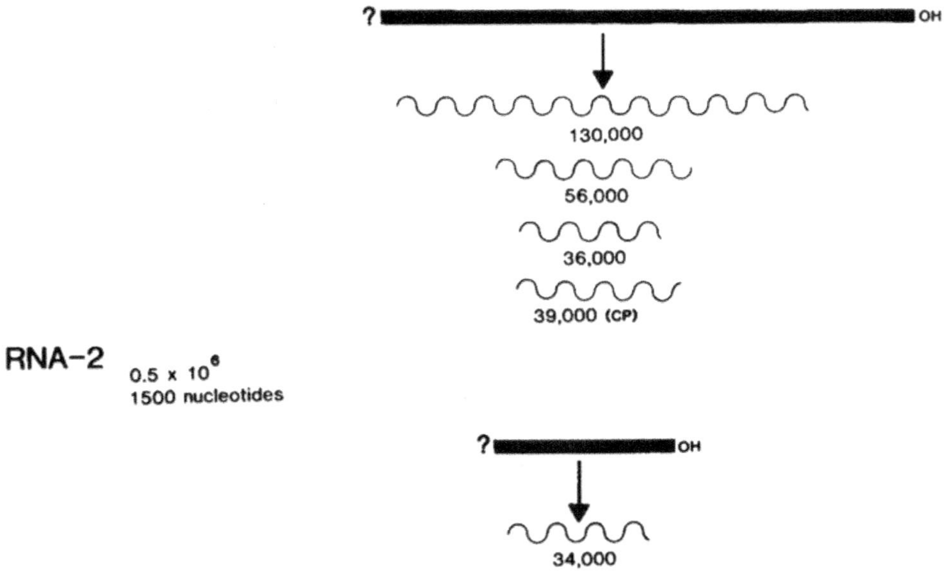

FIGURE 10. A scheme of the genomic organization and expression strategy of the dianthovirus red clover necrotic mosaic virus (RCNMV) RNAs. See Figure 1 for an explanation of the figure. The initiation site and interrelationships of the products of RNA 1 is unknown.

The genome structure and the relationship of the viral cell-free translation of RCNMV is presented in Figure 10.

3. In Vivo Observations

The in vivo genome expression of dianthoviruses has been limited to genetic studies involving heterogeneous combinations of RNAs 1 and 2.[209] Such studies have located the gene for the capsid protein on RNA 1. These viruses appear to be pliable models for such studies because the heterologous pseudorecombinants are highly infectious and stable. No other information is available on the expression of these viruses during infection of host tissue.

Information about the molecular biology of the dianthovirus group is insufficient to relate its strategy of genomic expression to other plant virus groups with any degree of authority. The location of the capsid protein cistron on the larger RNA of a bipartite genome is similar to the genomic organization of the pea enation mosaic virus group.[199,205]

D. Comovirus Group

The comoviruses are spherical plant viruses with a bipartite genome consisting of two single-stranded RNA molecules. The expression of this genome is by translation of each of these RNAs into large polyproteins which are posttranslationally processed by proteases into functional viral proteins. A comprehensive treatise of this subject by Drs. R. Goldbach and A. van Kammen is found in Chapter 3 of this volume.

E. Tobravirus Group

Tobacco rattle virus (TRV) is the type member of the tobravirus group which contains two other members, pea early-browning virus and peanut clump virus.[4,212] Two genomic RNAs are encapsidated into rod-shaped particles of 20-nm diameter and lengths varying according to the molecular weight of the RNA species.[213] The relationship between length and RNA size has been determined to be 40 nucleotides per nanometer of nucleoprotein particle length.[213] The capsid protein monomers have a molecular weight of 29,000 as estimated by SDS-PAGE. The actual molecular weight of the capsid protein subunit has been calculated to be 22,000.[214]

1. Viral Genome Structure

The tobravirus genome consists of two strands of messenger-sense, single-stranded RNAs with molecular weights of 2.4 × 10^6 (RNA 1) and 0.6 to 1.4 × 10^6 (RNA 2).[212,215] The capsid protein gene has been assigned to RNA 2 by experiments with pseudorecombinants.[216] Additionally, a gene which controls the phenotypic expression of systemic symptoms in solanaceous species has been located on RNA 2.[216] In one TRV strain, subgenomic messenger RNAs for capsid protein (RNA 3, 0.6 × 10^6 M_r) and for an unrelated 29,000 M_r protein (RNA 4, 0.54 × 10^6 M_r) are detected in isolated nucleoproteins.[217]

Tobravirus RNAs have a m^7Gppp cap structure at the 5′ end and lack a polyadenylate at the 3′ end.[217] The nucleotide sequences of the genomic RNAs of three strains of TRV have been compared by hybridization using complementary DNA.[213] No nucleotide sequence homology between the RNA 2 species of the three strains has been detected. RNA 1 molecules of two strains were almost identical but they had no detectable sequence in common with RNA 1 of a third strain. The size of RNA 1 does not vary much between the different strains of TRV. However, there is considerable size variation in RNA 2, ranging from 0.6 to 1.4 × 10^6.[215,217,218] Several subgenomic mRNA and less-than-full-length RNA species have been detected in TRV preparations.[217,218]

2. Cell-Free Translation Studies

The TRV genome has been translated using extracts from mouse L cells,[219] wheat germ,[218,220,222] and rabbit reticulocytes.[215,217,221] Translation of the RNA 1 of TRV strain PRN produces two products with molecular weights of 170,000 and 140,000 in both the wheat germ and rabbit reticulocyte lysate cell-free translation systems (Figure 11).[221] The molecular size of these two products together exceed the coding capacity of the RNA 1. Therefore, they may share a common amino acid sequence and the synthesis of the larger products may be due to a read-through of the termination site for the smaller product. Evidence has been presented for a subgenomic messenger (RNA 4) which is derived from RNA 1 and codes for a 31,000 M_r protein.[215,217] The 170,000 and 31,000 M_r proteins would account for 80% of potential coding capacity of RNA 1.

Two products with molecular weights of 29,000 (similar to capsid protein) and 31,000 are detected when RNA 2 of the PRN strain of TRV is translated.[221] The 29,000 M_r product has been identified as capsid protein by serology, peptide mapping, and specific aggregation with authentic capsid protein to form 36S disk aggregates.[221] RNA 2 preparations may be contaminated with a subgenomic RNA (RNA 3, 0.55 × 10^6 M_r) which serves as a mRNA for a 31,000 M_r protein. Peptide mapping and serological tests have demonstrated that this protein is not related to the coat protein.[217,221] RNA 2 of the CAM strain of TRV only directs the synthesis of capsid protein.[220] A virus-specific RNA (0.5 × 10^6 M_r) has been isolated from tissue infected with TRV CAM strain.[218] This subgenomic messenger RNA is derived from RNA 2 and codes for a 30,000 M_r protein. No messenger activity has been detected in cell-free translation experiments with RNA 2 isolated from TRV strain SYM.[215] However, two encapsidated subgenomic messengers RNAs have been identified. One is derived from RNA 2 with a molecular weight of 0.6 × 10^6, and is a subgenomic messenger RNA for capsid protein. The other, derived from RNA 1 with a molecular weight of 0.54 × 10^6, is a subgenomic mRNA for a 29,000 M_r protein which is distinct from the capsid protein. The conflicting translation results described for RNA 2 may be due to variations found in the different TRV strains and/or to problems associated with isolating RNA 2 free from the subgenomic RNAs and fragmented RNA 1. For the CAM strain of TRV, the two proteins assigned to RNA 2 account for about 75% of its potential coding capacity.

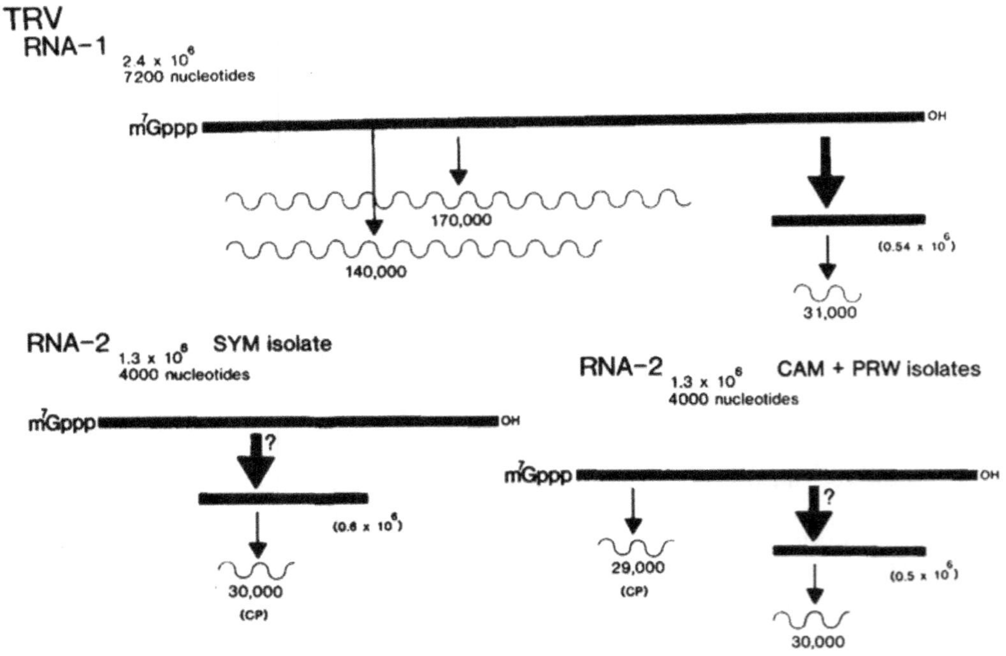

FIGURE 11. A scheme of the genomic organization and expression strategy of the RNAs of three strains of the tobravirus tobacco rattle virus (TRV). The expression of RNA-1 of the SYM, CAM, and PRW appears similar; however expression results of RNA-2 of the strains are different. See Figure 1 for an explanation of the figure.

3. In Vivo Observations

The genomic expression of TRV RNA has been studied in tobacco protoplasts.[222,223] RNA 1 can replicate independently of RNA 2 in protoplasts and this type of infection also occurs in plants in nature.[223] Three polypeptides have been associated with TRV CAM strain infection in protoplasts.[223] Two proteins with molecular weights 187,000 and 142,000 are specific to RNA 1 infections and comigrate with the cell-free translation products of RNA 1. Thus, the read-through mechanism also may operate in vivo and both products may have some functional significance. The third polypeptide detected in infected protoplasts has been identified as capsid protein (22,000 M_r) and it is synthesized if both RNAs 1 and 2 are in the inoculum.

TRV specific RNAs have been identified in TRV-infected tissue.[223a] The two genomic RNAs and five other TRV specific-RNAs have been identified. One of the less-than-full-length RNAs has a molecular weight of 1.1×10^6 and hybridizes with complimentary DNA to TRV RNA 1. The other RNAs (0.16×10^6; 0.22×10^6; 0.36×10^6; 0.50×10^6 M_r) all hybridize with complimentary DNA to TRV RNA-2. Only the 0.50×10^6 M_r RNA has been associated with mRNA activity (30,000 M_r protein) and further analysis is required to determined if these other TRV RNA species are authentic subgenomic mRNAs, fragmented TRV RNAs, or hybridization artifacts.

The translation strategy of TRV RNA 1 (Figure 11) bears some resemblance to the translation of the TMV (tobamovirus) genome. The translation products are similar in size and the synthesis of one of the products of both viruses may involve a mechanism of reading through a termination codon. The detection of subgenomic messenger for capsid protein of TRV SYM strain also resembles the TMV strategy for capsid protein gene expression. The genomes of both virus groups also code for a 30,000 M_r protein of unknown function.

Tobraviruses may represent a transitional variation of the monopartite genome as seen in the tobamovirus group, to a more versatile bipartite genome strategy.

IV. VIRUSES WITH TRIPARTITE GENOMES

The genetic complement of the six virus groups in this section is found on three RNAs. For infectivity, the bromoviruses, cucumoviruses, and hordeivirus require RNAs 1 to 3. The ilarvirus and alfalfa mosaic virus and possibly tomato spotted wilt virus require activation of their tripartite genome by capsid protein. This activation may be accomplished either by the addition of the proper capsid protein or by including the monocistronic mRNA for the capsid protein (RNA 4) in the inoculum.

A. Alfalfa Mosaic Virus Group

Alfalfa mosaic virus (AlMV) is the sole member of the alfalfa mosaic virus group.[227] Four RNAs can be isolated from purified virions. The genomic RNAs 1 to 3 are encapsidated separately into three distinct nucleoproteins with a bacilliform morphology. Two copies of RNA 4 are encapsidated in an ellipsoidal particle. All capsids are built from a single type of protein subunit with a molecular weight of 24,300.[225] A detailed account is given by E. M. J. Jaspers in Volume 1 of this series.

Although the three genomic RNAs contain all the AlMV genetic information, the three together are only infectious with capsid protein molecules.[226] The capsid protein subunits are not involved in the penetration of genomic RNAs into the cell or protection during infection. Rather, the binding of capsid protein near the 3′ termini of the genomic RNAs appears to be required for proper viral replicase recognition of the genomic RNAs.[226,227] RNA 4, the subgenomic mRNA for the capsid protein, may be substituted for AlMV capsid protein in infectivity studies with genomic RNAs.

1. Viral Genome Structure

The three genomic RNAs are messenger-sense, single-stranded RNAs with molecular weights of 1.1×10^6 (RNA 1), 0.8×10^6 (RNA 2), and 0.7×10^6 (RNA 3). The subgenomic RNA (RNA 4) has a molecular weight of 0.3×10^6 and is the messenger RNA for capsid protein. The sequence of RNA 4 is also present on the 3′ terminal half of RNA 3. There is a cap structure (m^7Gppp) at the 5′ terminus of each RNA. Hybridization studies have revealed that RNAs 1 to 3 have limited sequence homology, but the sequence of RNA 4 also is present in RNA 3.[228,229] Sequence analyses indicate that all the RNAs have a similar sequence of about 150 nucleotides at the 3′ terminus.[230-232] This homologous sequence on the RNAs has been implicated as the capsid protein subunit binding site involved in the activation of the genomic RNAs and as the replicase recognition site.[226,227] The complete nucleotide sequences of RNAs 1 to 4 have been determined.[233-238] Some of the sequence features are illustrated in Figure 12. The leader sequences of RNAs 1 to 4 have been compared for several AlMV isolates.[239,246]

2. Cell-Free Translation Studies

All four AlMV RNAs have been translated in various cell-free systems. In the rabbit reticulocyte cell-free system, AlMV RNA 1 directs the synthesis of a 115,000 M_r protein (at low mRNA concentration) or a mixture of 58,000 and 62,000 M_r proteins (at high mRNA concentration).[241,242] All three products have the same N terminus.[241] Translation in the presence of low RNA concentrations and/or high concentrations of glutamine results in the suppression of termination and the predominant synthesis of the 115,000 M_r product.[241] Whether 58,000 or 62,000 M_r proteins are functional viral proteins remains to be determined. It may be that these two proteins are synthesized at the expense of 115,000 M_r protein

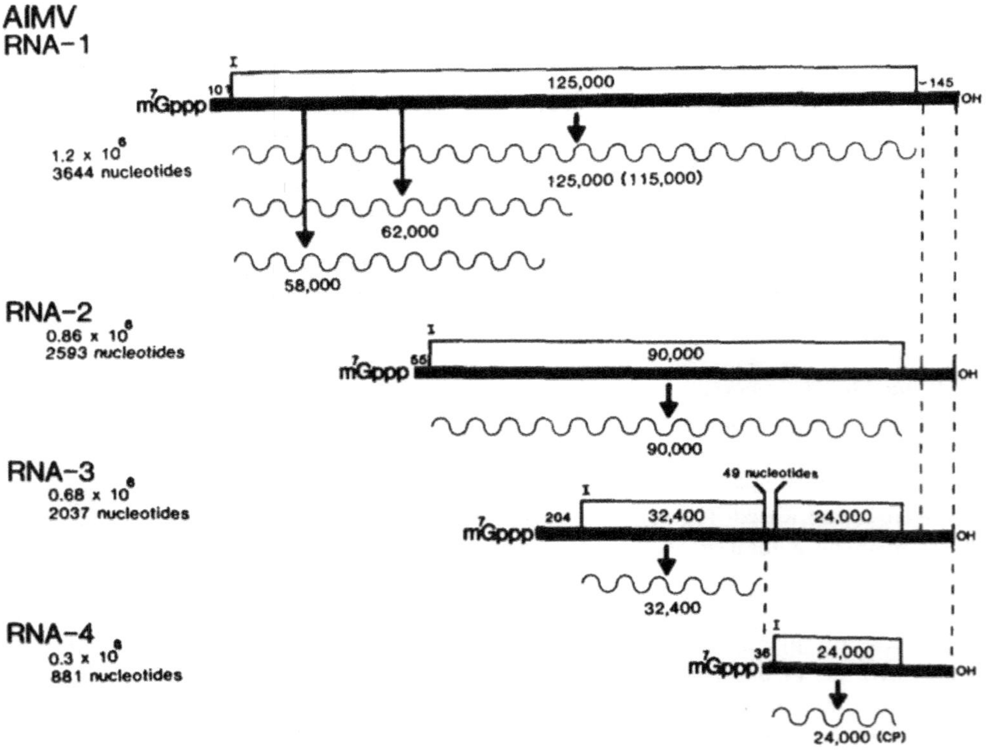

FIGURE 12. A scheme of the genomic organization and expression strategy of alfalfa mosaic virus (AlMV) RNAs. See Figure 1 for an explanation of the figure.

synthesis because some component (tRNA, ribosomes, initiation, or elongated factors) of the cell-free translation system is limiting.[241] The 115,000 M_r protein is not synthesized by a read-through mechanism of the termination signals for the two smaller proteins as UGA termination codons for the smaller proteins are not found in the nucleotide sequence of RNA 1. In the wheat germ or the Krebs-II ascites cell-free system, the 115,000 M_r product is formed as well as a number of smaller products.[243,244] The open-reading frame found in the nucleotide sequence of RNA 1 predicts a protein with a calculated molecular weight of 125,685 (1126 amino acids).[233] This protein probably corresponds to the 115,000 M_r product as estimated by polyacrylamide gel electrophoresis (PAGE).

AlMV RNA 2 is a monocistronic messenger for a protein with an estimated molecular weight of 100,000. The open reading frame found in the nucleotide sequence of RNA 2 predicts a 89,753 M_r product. Translation studies using the rabbit reticulocyte,[241,244] the Krebs II ascites,[244] and wheat germ cell-free systems,[243,244] reveals similar virus-specific proteins except that differing amounts of low-molecular-weight products are detected.

The products of AlMV RNA 3 translation have been analyzed using wheat germ,[243,245] rabbit reticulocyte,[241] and Krebs II[244,247] and ascites cell-free systems. The primary translation product produced in all systems has a molecular weight of 35,000. The nucleotide sequence for RNA 3 contains an open reading frame, proximal to the 5′ terminus, for a 32,400 M_r protein. Evidence has been presented for the synthesis, in the wheat germ system, of minor amounts of a product comigrating with capsid protein and serologically related to capsid protein.[243] In addition, trace amounts of a 65,000 M_r product, reactive with capsid protein antiserum, have been detected. This may be due to a read-through of the termination site for the 35,000 M_r protein and thus may represent the total coding capacity of RNA 3.[243] However, the tryptic peptide mapping of these products did not present an unambiguous picture of the product origin. It is clear that, even though RNA 3 is dicistronic on the basis of its genetic content, it is a monocistronic messenger in the cell-free translation systems.

The cell-free translation product of AlMV RNA 3 (35,000 M_r protein) has been used as an antigen to prepare an antiserum against the AlMV nonstructural protein.[246] AlMV RNA 3 was translated in a large-scale mRNA-dependent, rabbit reticulocyte translation mixture. About 75 μg of the 35,000 M_r protein was partially purified and used to immunize a rabbit. The resulting antiserum reacts with the 35,000 M_r translational product of AlMV RNA 3 but not with the corresponding products directed by RNA 3 from three other viruses with a tripartite genome (tobacco streak, cucumber mosaic, and brome mosaic). This approach may be useful for the development of probes for the serological detection of nonstructural proteins in vivo using the "Western" blotting technique.

AlMV RNA 4 is an efficient monocistronic messenger for capsid protein in the cell-free protein synthesizing systems such as wheat germ,[243,245,247,248] rabbit reticulocyte,[241,244] Krebs II ascites,[244] and *E. coli*.[249,250] The 24,300 M_r product represents the total coding capacity of AlMV RNA 4. Translation in the *E. coli* system may be initiated at two sites near the 5′ terminus of RNA 4 and proceeds in the same reading frame along the monocistronic mRNA.[249] The product initiated by an acetylphenylalanine-accepting tRNA is 11 amino acids longer than that initiated with the formylmethionine accepting tRNA in the *E. coli* system. In the other cell-free systems, the product of RNA 4 translation appears identical to the AlMV capsid protein. However, experiments demonstrating that the cell-free synthesized capsid protein is competent in genome activation of RNAs 1 to 3 through infectivity experiments have not been reported.

The nucleotide sequence of the 5′-terminal untranslated region of AlMV RNAs has been compared in an attempt to analyze translational efficiency.[239] The structure of 5′-terminal regions of AlMV RNAs 1 and 2 are highly conserved in three strains of AlMV. On the other hand, significant differences were found near the 5′ terminus in RNA 3 for one isolate and long leader sequence lengths are resolved in RNA 3 for all three isolates. The 5′-noncoding regions of RNA 4 of seven strains of AlMV are highly conserved except for nucleotide position 26.[240] In the wheat germ translation system, competitive initiation rank of the AlMV RNAs is RNA 4 > RNA 2 > RNA 1 > RNA 3.[239] The free energy of base-pairing increases in the order of the competitive hierarchy of these RNAs and the secondary structure of the untranslated region has been proposed to be the most likely factor in modulating translational efficiency. However, the length of the leader sequences also may influence the efficiency with which the mRNAs are translated.[251]

3. In Vivo Observations

The expression of AlMV RNAs has been examined in *Xenopus* oocytes.[243] Similar virus-specific proteins are synthesized using oocytes, as with the various cell-free systems, with the exception of AlMV RNA 2. No virus-specific products are detected when AlMV RNA 2 is injected into *Xenopus* oocytes.[243]

In vivo genomic expression of AlMV RNAs also has been studied using the cowpea protoplast system.[252,253] RNAs 1 and 2 can replicate independently of RNA 3 in protoplasts.[253] In this incomplete infection both negative-strand and positive-strand of RNAs 1 and 2 are produced but the ratio of positive-strand to negative-strand is shifted to 1000:1 in infections involving all three RNAs.[254] From these results it appears that genomic RNAs 1 and 2 products are involved in replicase activity and that an RNA 3-encoded product is responsible for a switch to the production of positive-strand RNA. The addition of RNA 4 to an inoculum with RNAs 1 and 2 does not result in the synthesis of positive-strand RNA 4 or capsid protein.[253] Apparently, RNA 4 is not replicated but is transcribed from the negative-strand of RNA 3.[254] RNAs 1 to 3 are readily detected in infected protoplasts but their corresponding products have not been identified.[252] Capsid protein synthesis is detected as early as 6 hr after infection in the protoplast system.

Complementation studies with temperature-sensitive mutants have shown each genome segment of AlMV encodes at least two independently mutable functions.[255-257] A large number of temperature-sensitive mutants of RNA 2 have been studied.[257] Based on the genetic interaction between these mutants RNA 2 encodes a protein which is active in a multimeric form with two functional domains. These mutation studies and the in vivo translation analyses suggest that the AlMV genome has at least five gene complementation groups. The genomic organization of the AlMV genome and its relationship to virus specific proteins is presented in Figure 12.

B. Ilarvirus Group

Tobacco streak virus (TSV) is the type member of the ilarviruses (11 members).[4,259] Four RNAs are isolated from purified virions. Viral RNAs are encapsidated into particles of quasi-isometric or occasionally bacilliform morphology. Particles with different RNA components differ in size, ranging from 26 to 35 nm in diameter. Infectivity requires all four RNAs or the three largest RNAs and capsid protein[259-261] analogous to AlMV (see below).

1. Viral Genome Structure

The TSV genome consists of three messenger-sense, single-stranded RNAs with molecular weights of 1.1×10^6 (RNA 1), 0.9×10^6 (RNA 2), and 0.7×10^6 (RNA 3). A fourth RNA ($0.3 \times 10^6 M_r$) is a subgenomic messenger for capsid protein (25,000 M_r). The genome structure of ilarviruses is similar to that described for AlMV.[260,238] The capsid proteins of the ilarviruses and AlMV can be interchanged in infectivity studies and are equally capable of "activating" their own as well as each other's genome.[262]

2. Cell-Free Translation Studies

The RNAs of TSV have been translated in the wheat germ and rabbit reticulocyte cell-free systems.[6,246] Translation of a mixture of four virion RNAs produces proteins with molecular weights of 120,000, 100,000, 34,000, and 25,000 (capsid protein). The translation product pattern is similar to that observed with AlMV RNA translations. The cell-free translation product (35,000 M_r) of TSV RNA 3 serologically is unrelated to the 35,000 M_r cell-free translation product of AlMV RNA 3 using antibodies prepared to the AlMV RNA 3 translation product.[246]

3. In Vivo Observations

Information about the in vivo genome expression of ilarviruses is not available.

Although not studied extensively, the ilarviruses appear to share many properties with AlMV and a number with the bromo- and cucumovirus groups.[258] The 3′ termini of the RNAs of both the ilarviruses and AlMV lack a polyadenylate region and are not aminoacylated. The capsid protein of AlMV and the ilarviruses are interchangeable in the activation of each other's genomes as well as their own.[258] High affinity, capsid protein binding nucleic acid sequence are present in the RNAs near the 3′ termini of both groups.[238] The RNAs for the AlMV, ilarviruses, bromoviruses, and cucumoviruses are similar in size. The translation strategy of these viral RNAs appears to be similar for the four groups.

C. Bromovirus Group

Brome mosaic virus (BMV) is the type member of the bromovirus group which contains two other members, broad bean mottle virus (BBMV) and cowpea chlorotic mottle virus (CCMV).[4] RNAs 1 and 2 are encapsidated separately, and RNAs 3 and 4 together in polyhedral particles about 26 nm in diameter.[264]

1. Viral Genome Structure

The BMV genome consists of three messenger-sense, single-stranded RNAs with estimated molecular weights of 1.1×10^6 (RNA 1), 1.0×10^6 (RNA 2), and 0.7×10^6 (RNA 3).[265] A fourth RNA (RNA 4), 0.28×10^6 M_r is a subgenomic messenger RNA for the capsid protein subunit (20,000 M_r). The 5′ terminus of each RNA has a cap structure (m^7Gppp).[266] The 3′ terminus can be aminoacylated with tyrosine both in vitro and in vivo.[267] The 3′-terminal sequences of all four bromovirus RNAs are highly conserved for 170 to 190 nucleotides.[268-271] These conserved sequences theoretically can be folded into extensively base-paired secondary structures with similar configurations, which may include binding sites for the viral RNA replicase and for the capsid protein in assembly.[268,269] The nucleotide sequence of BMV RNA 3 has been determined.[272] The leader sequence of BMV 3 is 91 nucleotides long, followed by an open reading frame of 909 nucleotides (coding for 33,500 M_r nonstructural protein). A 250-nucleotide intercistronic noncoding region is present between this cistron and the capsid protein cistron. An internal polyadenylate sequence of 16 to 22 bases occurs in the intercistronic region approximately 20 nucleotides upstream of the start of the capsid protein gene sequence.[272,273] The 879 nucleotides sequence of the subgenomic messenger RNA for capsid protein (RNA 4) overlaps the intercistronic region by 9 nucleotides[272,274] and possesses an untranslated 3′ sequence 297 nucleotides long.[274] RNA 4 appears in progeny virus when it is omitted from a viral RNA inoculum; thus RNA 4 appears to be transcribed from RNA 3 in vivo.

2. Cell-Free Translation Studies

Bromovirus RNAs are efficient messenger RNAs in several cell-free translation systems.[217,248,275-282] For BMV RNAs, the wheat germ[248,275,276] cell-free protein synthesizing system is a homologous system. BMV RNAs 1 and 2 each direct the synthesis of a single protein with molecular weights of 110,000 and 105,000, respectively.[6,277] These products represent over 90% of the genetic potential of these RNAs assuming no overlapping genes. BMV RNA 3 directs the synthesis of 35,000 M_r protein which has no tryptic peptides in common with the capsid protein.[275] The cistron located proximal to the 3′ terminus of BMV RNA 3 and coding for capsid protein is not active in cell-free translation systems. BMV RNA 4 directs the synthesis of the 20,000 M_r capsid protein.[275] This subgenomic messenger RNA is preferentially translated in the presence of the other BMV RNAs.[278,279]

BMV RNAs also have been translated in other cell-free systems such as *E. coli* extracts,[280] mouse L cell extracts,[217] *Drosophila melanogaster* extracts,[281] and rabbit reticulocyte lysates[282] with varying degrees of efficiency. There are no reports of proteins differing significantly from those obtained in the homologous system described above.

CCMV, which is serologically related to both BMV and BBMV, has been analyzed in wheat germ and the rabbit reticulocyte lysate cell-free systems.[284-286] Both CCMV RNAs 1 and 2 code for polypeptides with an estimated molecular weight of 105,000. Since RNA 1 is larger than RNA 2 it presumably has larger noncoding regions. RNA 3 directs the synthesis of a 32,000 M_r product while RNA 4 is an efficient subgenomic messenger RNA for capsid protein (19,000 M_r).

The in vitro translation information of BBMV, a virus that is distantly related to BMV and CCMV,[287] is limited. The translations of BBMV RNAs 1 and 2 together, RNAs 3 and 4 alone, in the wheat germ system have resulted in similar products as those reported for the BMV RNAs.[248] However, BBMV RNAs are less efficient in translation assays and the products are more heterogeneous in comparison with those for CCMV and BMV RNAs.

3. In Vivo Observations

In vivo genomic expression of BMV has been studied in tobacco and barley protoplasts and in tobacco protoplasts for CCMV.[288-292] Three products with molecular weights of

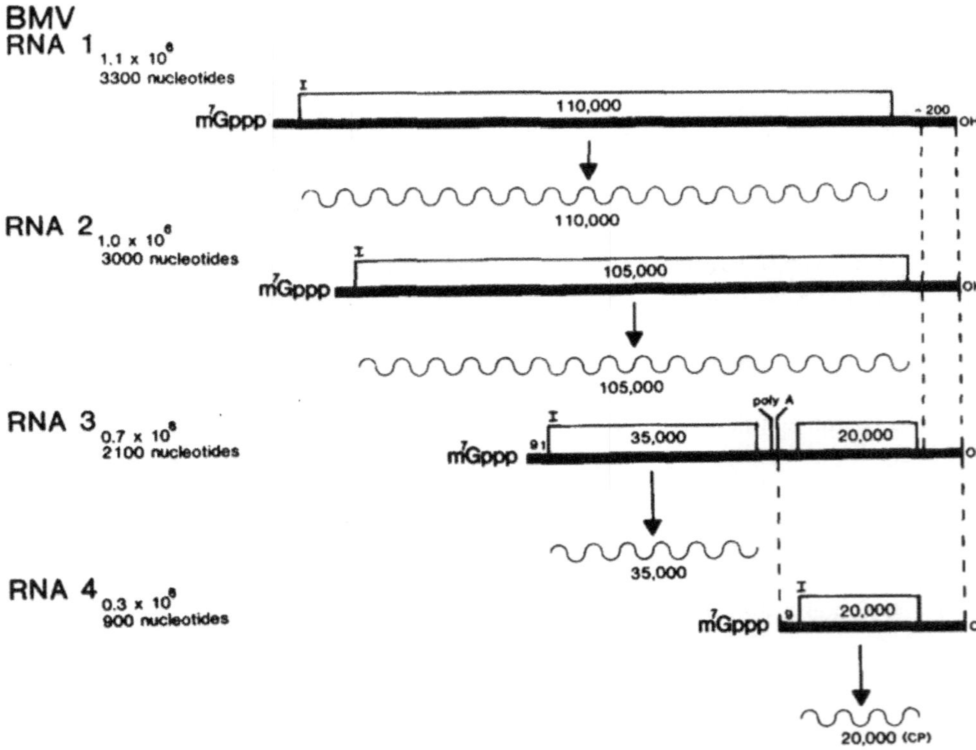

FIGURE 13. A scheme of the genomic organization and expression strategy of the bromovirus RNAs of brome mosaic virus (BMV). See Figure 1 for an explanation of the figure.

19,000, 34,000, and 100,000 are detected in CCMV-infected tobacco protoplasts.[290] These products are similar in size to those detected in the cell-free translations studies and thus may represent the expression of the entire genome. The products detected in BMV-infected tobacco protoplasts are in agreement with the four products obtained in the cell-free translations analysis of BMV RNAs.[288] BMV predominates in mixed infections of CCMV and BMV in tobacco protoplasts and prevents the synthesis of infectious CCMV.[289] Inoculation of barley protoplasts with BMV RNAs 1 and 2 (without RNA 3), results in the transcription of these RNAs and in the synthesis of 110,000 and 100,000 M_r proteins.[291] Virus-specific proteins are not detected when either RNA 3 alone or RNA 4 alone are tested in barley protoplasts.[291] These findings implicate BMV RNAs 1 and/or 2 in early infection events such as replication and in the coding for viral proteins involved in these early events. Preparations having template-dependent and template-specific polymerase activity have been isolated from BMV-infected barley.[293,294] PAGE analysis of this preparation reveals a 110,000 M_r protein similar in size to the translation product of RNA 1. The tryptic peptide profile of the 110,000 M_r protein isolated from BMV-infected barley is similar to that from the product of cell-free translation of RNA 1.[293] However, the presence of a virus-specific protein in RNA-polymerase-containing extract from infected tissues does not prove conclusively a functional role.

The structural organization of the BMV genome and its relationship to BMV-specific proteins is presented below (Figure 13). Recently, a model has been proposed for BMV for the regulation of transcription, translation, and encapsidation.[294a] Once BMV RNAs infect the host cell, the 3′-terminal nucleotide becomes aminoacylated with tyrosine and binds elongation factor-1 (EF-1). This blockage of the 3′ terminus results in the primary involvement of BMV RNAs in the synthesis of BMV specific proteins. As the viral replicase attains a certain intracellular level, it is able to compete with the EF-1, displace it from the 3′ termini of the BMV RNAs, and initiate viral RNA replication. The synthesis of BMV RNA-4 ultimately results in the production of high intracellular levels of capsid protein which then outcompetes EF-1 and replicase for the 3′ terminal region of BMV RNA and thus results in encapsidation. This model may explain the involvement of the highly conserved untranslated 3′ terminus in the regulation of viral gene expression and replication.

The bromoviruses share many properties with the cucumovirus group. The genomes and the translational strategies of these two virus groups are remarkably similar. The 3′ RNA secondary structures of these two virus groups are also very similar. They both have tRNA-like 3′ termini which accept tyrosine. However, viruses in these two groups differ in a number of biological properties such as insect transmission and host range. In addition, the properties of the bromoviruses show similarities to some of the properties of ilarvirus and AlMV groups. The similarities and differences in these four virus groups have been summarized by van Vloten-Doting et al.[258]

D. Cucumovirus Group

Cucumber mosaic virus (CMV) is the type member of the cucumovirus group which contains two other members, peanut stunt virus (PSV) and tomato aspermy virus (TAV).[4] Four RNAs can be isolated from purified virions; however, only the three largest species are required for infection. The two largest RNAs (RNAs 1 and 2) are encapsidated separately, while the two smaller RNAs (RNAs 3 and 4) are packaged together in polyhedral particles, about 29 nm in diameter.[295] A single-stranded satellite RNA (0.1×10^6 M_r) has been associated with some isolates of CMV and PSV.[183]

1. Viral Genome Structure

The CMV genome consists of three messenger-sense single-stranded RNAs with molecular weights of 1.27×10^6 (RNA 1), 1.13×10^6 (RNA 2), and 0.75×10^6 (RNA 3). A fourth RNA (RNA 4), molecular weight of 0.35×10^6, is a subgenomic messenger RNA for the capsid protein (24,000 M_r).[296] All four CMV RNAs have a "capping" structure (m^7Gppp) at the 5′ terminus.[297] The 3′ termini of CMV RNAs have a tRNA-like structure which can be aminoacylated with tyrosine.[298] The nucleotide sequence which makes up the 3′ half of RNA 3 of CMV (Q strain) is identical to RNA 4. Additionally, the 3′ terminal 138 nucleotide residues of all four RNAs are identical except for a few nucleotide changes.[299] The 172 nucleotides located adjacent to the 3′ terminal nucleotide of CMV (Q strain) RNA 4 can be aligned to show 73% sequence homology with the 3′ terminal nucleotide of the RNAs of TAV, another cucumovirus.[300] The complete nucleotide sequences of CMV RNAs 3 and 4 have been determined.[301] The leader sequence of CMV RNA 3 is 91 nucleotides long. This is followed by an open reading frame of 1002 nucleotides which codes for a protein with 333 amino acids (36,700 M_r). An intercistronic region of 123 nucleotides separates this coding region from the second coding region which is identical to RNA 4. Both coding regions in RNA 3 are in the same reading frame. CMV RNA 4 has a 53 nucleotide leader sequence, a 711 nucleotide open reading frame (coding for a 26,200 M_r protein), and a 263 nucleotide untranslated region at the 3′ end.[300] A 0.1×10^6 M_r RNA (referred to as RNA 5 or CARNA 5) which possesses a cap structure at the 5′ terminus, has been associated with certain strains of cucumoviruses.[296] CARNA 5 is encapsidated together with genomic RNAs and is dependent on CMV for replication, encapsidation, and transmission. This satellite RNA is of interest because it is capable of modifying disease expression in the presence of the helper virus. The nucleotide sequences of three different satellite RNAs have been analyzed and share 88 to 93% sequence homology with each other. The small differences in their nucleotide sequences are associated with significant biological modifications in CMV

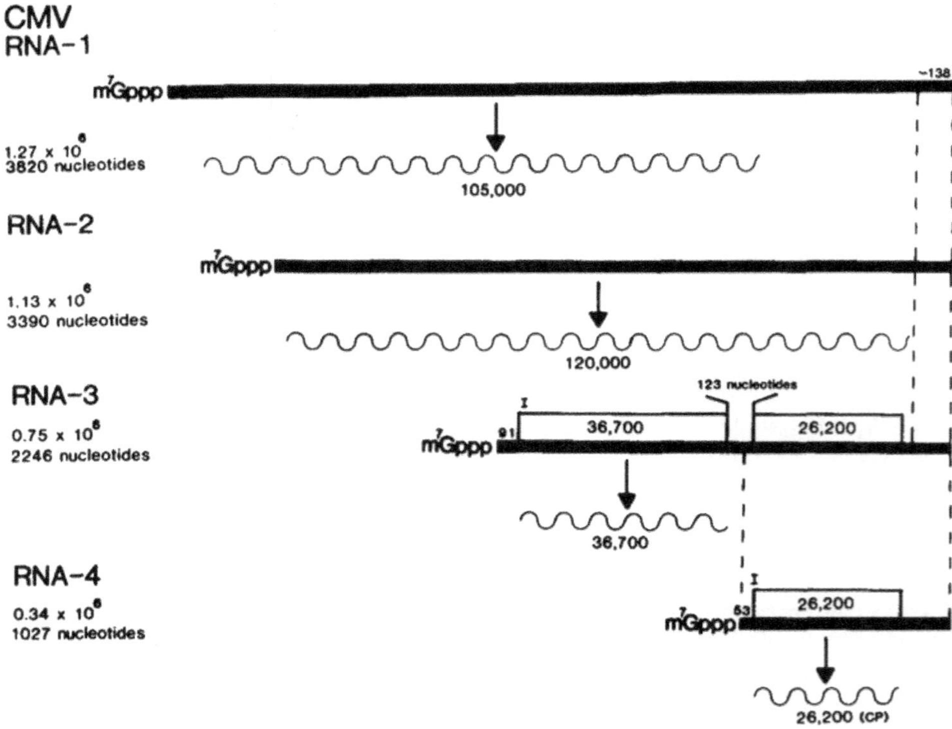

FIGURE 14. A scheme of the genomic organization and expression strategy of the cucumovirus RNAs of cucumber mosaic virus (CMV). See Figure 1 for an explanation of the figure.

infections in certain hosts.[302-305] The proposed 3′-terminal structure of the satellite RNA associated with CMV strain Q shows sequence and structure homology with parts of the 3′ terminal tRNA-like structure of the CMV genomic RNAs.[304]

2. Cell-Free Translation Studies

The cell-free translation of CMV RNAs has been studied in wheat embryo[305,306] and rabbit reticulocyte lysate[306] cell-free protein synthesizing systems. CMV RNAs 1 and 2 direct the synthesis of 105,000 and 120,000 M_r products, respectively. RNA 3 directs the synthesis of a 34,000 M_r product in the rabbit reticulocyte lysate system, and three products, 34,000 to 39,000 M_r in size, in the wheat cell-free system. The capsid protein gene on the 3′ end of RNA 3 is not available for translation in the cell-free analysis. RNA 4 directs the synthesis of the 24,500 M_r capsid protein in the wheat and reticulocyte cell-free systems. An additional product, 39,000 M_r in size, has been detected in the reticulocyte cell-free system with RNA 4. To account for these conflicting results for the translation of RNAs 3 and 4 in the different translation systems possible translation errors of the cell-free translation systems such as read-throughs of the termination codons or incorrect initiations have been proposed. The structural organization of CMV RNAs and their relationship to CMV specific proteins is presented below (Figure 14).

The satellite RNA (0.1×10^6 M_r) associated with CMV strain S (S-CARNA 5) has been tested for messenger activity in the wheat germ cell-free system.[307] S-CARNA 5 is an active messenger in wheat germ extracts but the translation of S-CARNA 5 is suppressed in the presence of CMV RNAs 1-4. Two products with estimated molecular weights of 3800 and 5200 are detected. It is not known whether these two products represent unique amino acid sequences or share a common terminus. The two products represent less than the theoretical coding potential of S-CARNA 5. However, it would be unusual for the initiation of protein synthesis to occur at two distinct sites on the same template in an eukaryotic system. The complete nucleotide sequence of a closely related satellite RNA has been presented and two nonoverlapping, open reading frames coding for proteins of molecular weight 2800 and 4800 can be resolved.[303] The nucleotide sequence of a satellite RNA associated with CMV strain Q has also been presented.[304] This satellite RNA (Q-CARNA 5) has 88% sequence homology with the S-CARNA 5 but they differ significantly in their ability to modify CMV symptoms in certain host plants.[296] Q-CARNA 5 lacks detectable messenger activity in rabbit reticulocyte lysates but stimulates the synthesis of 3000 M_r protein in wheat germ extracts.[304] This product size is consistent with the one detectable open reading frame (coding for 17 amino acids) resolved in the Q-CARNA 5 nucleotide sequence.

The cell-free translation of TAV RNAs has received limited attention. The major translation product which has been synthesized in the wheat germ system by total TAV RNA preparations or TAV RNA 4 is similar to the TAV-capsid protein.[5] An additional product around 40,000 M_r, similar in size to that observed in CMV RNA 4 translation in the rabbit reticulocytes, also has been detected. There are no reports regarding the cell-free translation of PSV RNAs.

3. In Vivo Observations

CMV RNAs have been translated in *Bufo marinus* (toad) oocytes.[306] CMV RNA 1 and 2 are not translated in toad oocytes. CMV RNA 3 and 4 are translated and direct the synthesis of a 34,000 and 24,500 M_r proteins, respectively.

In vivo genome expression has been studied in CMV-infected tobacco and cowpea protoplasts,[308] in virus-infected tobacco leaves, and in pre-infected tobacco cells.[309] Capsid protein synthesis is detected readily in all systems but products of RNAs 1 to 3 are not resolved in any of the systems.

An extensively purified RNA-dependent-RNA-polymerase activity, induced in CMV-infected cucumber seedlings, contains a major polypeptide with a molecular weight of 100,000 and two other proteins, 110,000 and 35,000, in smaller and varying amounts.[310] These polypeptides have electrophoretic mobilities similar to the major translation products of CMV RNAs 1 to 3. The polypeptides from the polymerase preparations have been compared with the comigrating, cell-free translation products of CMV RNAs 1 to 3 by peptide mapping studies. Results of these studies indicated that the three components of the purified RNA polymerase preparations are not related to the translation products of CMV RNAs.[310]

The cucumoviruses share many properties with the bromoviruses and a number with the ilarvirus and AlMV groups. However, cucumoviruses differ from these groups in some important biological properties. Cucumoviruses, unlike bromoviruses, are readily transmitted by aphids in a nonpersistent manner. Another difference is the presence of satellite RNAs in certain cucumovirus strains. These satellite RNAs are capable of modifying disease expression and altering virus accumulation in infected tissues.[296,311]

E. Hordeivirus Group

Barley stripe mosaic virus (BSMV) is the type member of the hordeivirus group, which has two other members, *Lychnis* ringspot and *Poa* semilatent.[4] Three viral RNAs are encapsidated into rod-shaped particles with a 20-nm diameter and lengths ranging from 100 to 150 nm, depending upon the size of encapsidated RNA. The capsid protein subunit has a molecular weight of 25,000.[312]

1. Viral Genome Structure

The BSMV genome consists of three messenger-sense, single-stranded RNAs with molecular weights of 1.5×10^6 (RNA 1), 1.35×10^6 (RNA 2) and 1.05 to 1.35×10^6 (RNA 3).[313] A subgenomic RNA (0.28×10^6 M_r), coding for a 20,000 M_r protein, is a minor component of BSMV RNAs.[314] The 5′ termini of BSMV RNAs are blocked with a cap structure (m^7Gppp) and the 3′ termini have a tRNA-like structure which can bind tyrosine in vitro and in vivo.[267,315,316] BSMV RNAs contain an internal poly(A) sequence of 8 to 30 residues approximately 210 nucleotides from the 3′ terminus.[316] The nucleotide sequence relationships of the RNAs of three BSMV strains have been studied by hybridization analysis of the viral RNA species using recombinant complementary DNA molecules to the corresponding RNA species.[313] The genomes of two strains (North Dakota-18 and Norwich) consist of three distinct RNAs with a high proportion of unique sequence. The other strain (type) also is composed of three distinct RNAs but two of the RNAs have similar electrophoretic mobility in gels. This observation resolves the confusion in previous reports of the existence of two- and three-component strains of BSMV.[312] The Argentina mild strain of BSMV has four RNA species but the fourth RNA (1.0×10^6 M_r), appears to be a partial transcript of the genomic RNA 3 (minus 300 nucleotides), and is not necessary for infectivity.[317]

2. Cell-Free Translation Studies

The BSMV RNAs have been translated in wheat germ[314,318,319] (homologous system) and in rabbit reticulocyte lysates.[318,319] RNA 1 of all BSMV strains directs the synthesis of a 120,000 M_r protein. This accounts for more than 70% of RNA 1 coding capacity. The translation of RNA 2 represents a more complex picture. The primary product of RNA 2 translation has a molecular weight of 23,000 to 25,000 and comigrates with the capsid protein in gel electrophoresis experiments. This protein accounts for only 15% of the potential genetic information in RNA 2. RNA 2 of the Norwich strain directs the synthesis of an additional product (25,000 M_r) serologically unrelated to capsid protein.[319] RNA 3 of the type strain (comigrates with RNA 2) directs the synthesis of a 82,000 to 85,000 M_r protein.[319,320] RNA 3 of the other BSMV strains (smaller than the type strain RNA 3) directs the synthesis of a 71,000 to 75,000 M_r protein.[319,320] The Argentina mild strain has an additional RNA (RNA 4) and this RNA directs the synthesis of a 55,000 M_r protein. Tryptic fingerprint analysis shows that the 82,000 to 85,000, 71,000 to 75,000, and 55,000 mol wt products share amino acid sequences.[319] The subgenomic RNA (0.28×10^6 M_r) directs the synthesis of two proteins with molecular weights of 20,000 to 21,000.[314] These two proteins may have overlapping amino acid sequences or they may be directed by two distinct subgenomic RNAs of similar size. Complementary DNA of the subgenomic RNA hybridizes with RNA 3. This subgenomic RNA may be derived from 3′ end of RNA 3.[321] The 71,000 to 75,000 or 82,000 to 85,000 M_r protein and the 20,000 to 21,000 M_r protein would account for 70 to 90% of potential coding capacity of RNA 3.

3. In Vivo Observations

The BSMV genome expression has been studied by comparing proteins extracted from BSMV-infected barley with proteins extracted from noninoculated barley.[320,322] Three BSMV-specific proteins, 120,000, 67,000, and 25,000 (capsid protein) M_r have been detected.[320] The 67,000 M_r product has been proposed to be the result of in vivo processing of the 71,000 to 75,000 or 82,000 to 85,000 M_r proteins.[320] Increased amounts of proteins, with molecular weight of 70,000 to 83,000 as well as capsid protein, have been detected in extracts of BSMV-infected plants.[322]

The genomic structure of the type strain of BSMV is presented below (Figure 15) with a possible expression strategy. The hordeivirus genome presents a distinct variation in the expression of capsid protein gene among the tripartite genome viruses. The hordeivirus coat protein appears to be synthesized from a genomic RNA while the capsid proteins from other tripartite viruses are synthesized via subgenomic messenger RNAs. The heterogeneity in

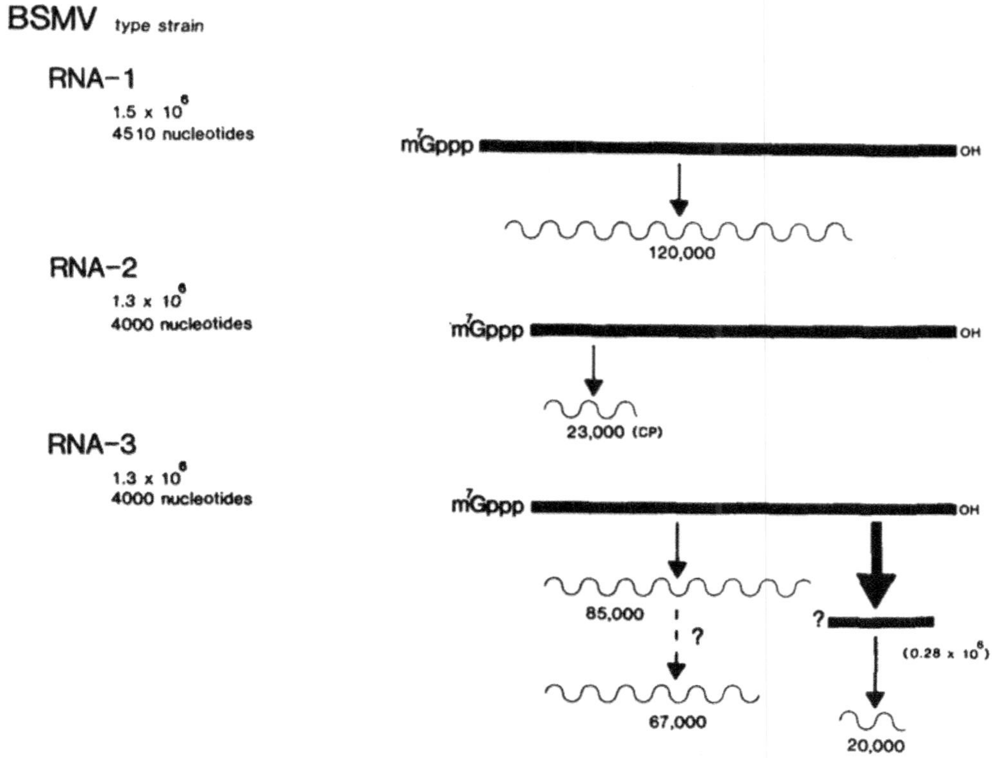

FIGURE 15. A scheme of the genomic organization and expression strategy of the hordeivirus RNAs of barley stripe mosaic virus (BSMV). See Figure 1 for an explanation of the figure.

RNA components isolated from various BSMV strains is also unusual. Two independent deletion events in RNA 3 have ben proposed as a mechanism for generating the heterogeneity observed in the RNA components of certain BSMV strains.

F. Tomato Spotted Wilt Virus Group

Tomato spotted wilt virus (TSWV) is the sole member of the tomato spotted wilt virus group.[4] TSWV has isometric morphology and is membrane bound.[323-325] Ribunucleoproteins are contained within the viral membrane envelope and are comprised of three unique, single-stranded RNA molecules and a capsid protein with a subunit molecular weight of 29,000. The ribonucleoproteins are surrounded by a membrane-like structure which contains three virus-specific glycoproteins with molecular weights of 52,000, 58,000, and 78,000.[326,327] The membrane-bound and unstable nucleocapsid is approximately 85 nm in diameter and has a sedimentation coefficient of 520 to 530S. The virus can be transmitted mechanically (ribonucleoproteins and the virion are infectious) or by thrips in a persistant manner.[323-325]

1. Viral Genome Structure

The TSWV genome consists of three single-stranded RNA molecules with molecular weights of 2.7 $\times$ 10^6 (RNA 1), 1.7 $\times$ 10^6 (RNA 2), and 1.1 $\times$ 10^6 (RNA 3). A fourth RNA (3.6 $\times$ 10^6 M_r) has been detected in one isolate.[325,328,329] Purified RNA is not infectious; however, ribonucleoprotein containing the three RNA species and the capsid protein is infectious.[325,329,330] All three RNA species are believed to lack a polyadenylate region as they do not bind to oligodeoxythymidylate cellulose under high salt condition.[325,329]

2. Cell-Free Translation Studies

TSWV RNA has been translated in a rabbit reticulocyte lysate and a wheat germ extract.[325,329] Translation in wheat extracts is dependent on the addition of spermidine. Analysis of TSWV cell-free translation products has revealed six major and a number of minor products. One of these products has a molecular weight of 29,000, comigrates with TSWV capsid protein, and is immunoprecipitated by antisera to TSWV. A major cell-free translation product with a molecular weight of 60,000 is unreactive with antiserum to TSWV. Three other major proteins, with molecular weights of approximately 40,000, 35,000, and 33,000, also are detected and are unreactive with the antiserum to TSWV. A minor cell-free translation product, comigrating with TSWV 78,000 M_r envelope protein in electrophoretic analyses, and a 110,000 M_r product, also are unreactive with antiserum to TSWV. The lack of a positive serological test may be due to the use of low titer antiserum or due to unglycosylated cell-free translation products not being recognized by the antiserum. Translational analysis of TSWV RNA in the rabbit reticulocyte lysate reveals similar products to those observed in wheat germ extracts, except for the two minor high-molecular-weight products.

An attempt has been made to separate the three TSWV RNA species and determine the coding assignments for each RNA. This has been only partially successful because of RNA 3 contamination. RNA 3 apparently directs the synthesis of both the 60,000 and 29,000 M_r proteins. It has been hypothesized that the two minor large-molecular-weight proteins of the wheat germ system are gene products of RNA 1 and 2.

3. In Vivo Observations

The positive-stranded nature of the TSWV genome, exhibited by the messenger activity of TSWV RNA in the cell-free translation systems, is supported by the lack of hybridization between RNA from TSWV-infected, plant polysome fractions, when labeled TSWV RNA is used as a hybridization probe.[325,329] Additionally, there does not appear to be any transcriptase activity in purified TSWV preparations.[325] This evidence would support TSWV RNA being positive-stranded and not negative-stranded as infection assays with purified TSWV RNA would indicate.[328] The negative-stranded concept for TSWV RNA was based on the failure to detect infectivity with purified TSWV RNA. To rationalize this, it has been proposed that purified TSWV RNA was not infectious because of instability or TSWV RNA may be similar to AlMV RNA in the requirement of capsid protein binding for infection.[325]

There have been attempts to identify TSWV-specific proteins during infection.[325] Only the nucleocapsid protein (29,000 M_r) has been detected in the analysis of infected plant material. However, membrane fractions have not been examined closely for the presence of membrane-bound glycoproteins. Polyribosomes have been isolated from infected and noninfected plant tissue and analyzed for translational activity in the wheat germ cell-free system. Two translational proteins are unique to polysomes from infected tissue. They have molecular weights of 27,000 and 14,000, but neither of the proteins is immunoprecipitated with antiserum to TSWV.[325]

Defective forms of the virion have been used successfully to analyze the genome of TSWV.[325,331] A number of defective forms which do not form mature enveloped virions have been identified by electron microscopy studies. Instead, amorphous masses of TSWV RNA and nucleocapsid protein are observed. Analysis of infected plant tissue by enzyme-linked-immunoabsorbent assays and gel electrophoresis reveal that the 78,000 M_r membrane glycoprotein is not present. Gel electrophoretic analysis of the genomic RNAs of the defective forms reveals the RNA 2 migrates faster than RNA 2 from wild-type TSWV and may represent a sequence deletion in RNA 2. Other defective forms, unable to assemble into mature, fully formed virions have been examined and in all cases, RNA 1 and RNA 3 comigrate with the corresponding RNA from wild-type TSWV while RNA 2 migrates faster, indicating a segment of RNA 2 has been lost in these defective forms.

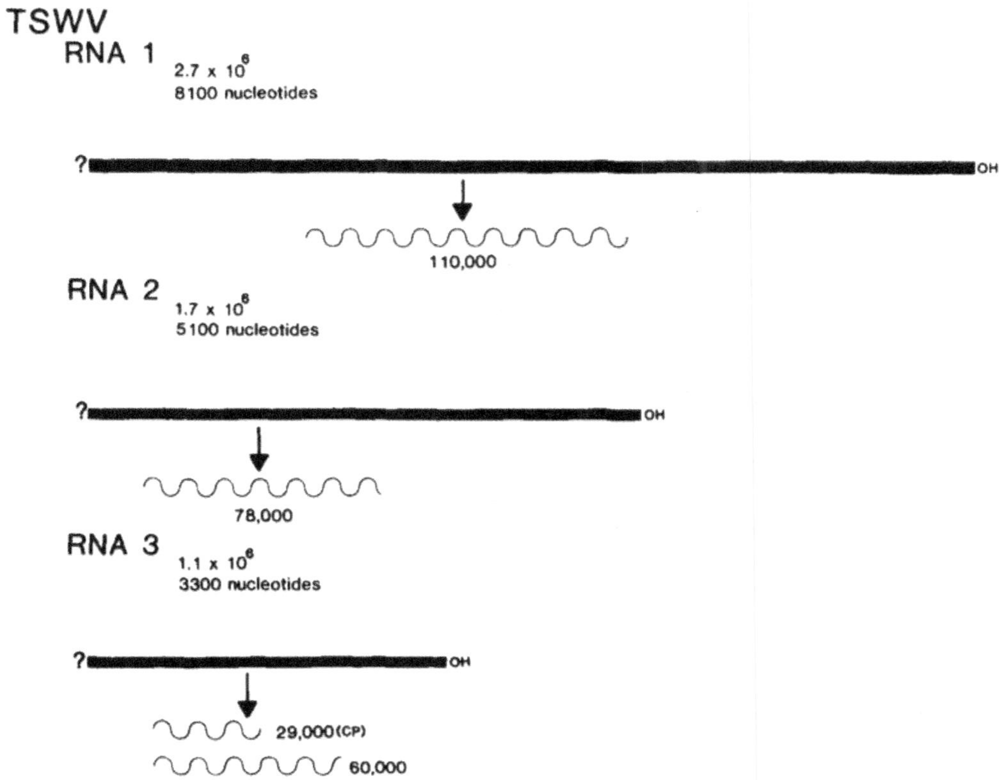

FIGURE 16. A scheme of the genomic organization and expression strategy of tomato spotted wilt virus (TSWV) RNAs. See Figure 1 for an explanation of the figure.

The genomic structure and the expression of viral RNAs based in cell-free translation and genetic studies is presented below (Figure 16).

TSWV is structurally unique when compared to other plant viruses. It does bear some resemblance to the animal bunyaviruses in that both viruses are tripartite and enveloped.[325] However, they differ in that the bunyaviruses have an ordered nucleocapsid structure and a negative-stranded genome while the nucleocapsid of TSWV appears to lack organization and contain a positive-stranded genome.

V. VIRUSES WITH DOUBLE-STRANDED RNA GENOME

Plant viruses which have double-stranded RNA molecules for a genome have been found infrequently in nature and have not been studied extensively.

A. Phytoreovirus Group

Reoviruses which infect plants have been classified into two subgroups. Subgroup 1 (phytoreoviruses) members have a double-stranded RNA genome that is divided into 12 segments, and are transmitted by leafhoppers. Subgroup 2 (fijiviruses) members possess a double-stranded RNA genome which has 10 segments, and are transmitted by planthoppers. Only wound tumor virus (WTV), a member of subgroup 1, will be discussed in this section.[4,322]

WTV is an isometric unicomponent virus possessing a diameter of 60 to 80 nm.[332,333] The structural organization of the virus is complex. In addition to encapsidating a genome of 12 double-stranded RNA segments,[334] numerous enzymatic activities also are associated

Table 1
INTERRELATIONSHIP OF WTV GENOME SEGMENTS AND GENE PRODUCTS

Segment	Mol wt of genome segment ($\times 10^6$)	Mol wt of gene products		Location	Nomenclature
		In vivo	In vitro		
1	2.90	155,000	155,000	Nucleoprotein core	P1
2	2.40—2.48	130,000	130,000	Outer protein coat	P2
3	2.20—2.25	108,000	108,000	Nucleoprotein core	P3
4	1.78—1.80	72,000	74,000		Pns4
5		76,000	76,000	Outer protein coat	P5
6	1.10—1.15	57,000	57,000	Nucleoprotein core	P6
7	1.05—1.15	52,000	52,000		Pns7
8	0.83—0.90	42,000	42,000	Capsid	P8
9	0.57—0.63	41,500	41,500	Capsid	P9
10	0.55—0.61	39,000	39,000		Pns10
11	0.54—0.60	35,000	35,000		Pns11
12	0.32—0.35	19,000	19,000		Pns12

with the virion structure. The RNAs are contained in a nucleoprotein core structure. The core is encapsidated by two protein layers. A capsid-type structure, comprised of two protein subunits with molecular weights of 42,000 and 41,000, surrounds the core and this capsid structure in turn is surrounded by an amorphous protein layer composed of two protein subunits with molecular weights of 131,000 and 96,000.[332,333,335] WTV is able to replicate in both animal and plant cells. However, it is only capable of infecting dicotyledonous plants and induces vein swelling and tumor formation.[333]

1. Viral Genome Structure

The genome of WTV consists of 12 double-stranded RNA segments ranging in molecular weight from 2.9×10^6 to 0.32×10^6 with a combined molecular weight of 16×10^6 (Table 1).[333] Detailed structural analysis of the WTV genome has not been carried out.

2. Cell-Free Translation Studies

WTV mRNA transcripts have been synthesized in vitro and have been translated in wheat germ extracts and rabbit reticulocyte lysates.[330] The 12 mRNAs, which are capped at the 5′ terminus and possess a penultimate methylated nucleotide,[337] serve as monocistronic mRNAs in the cell-free system. Seven proteins, also found associated with the virion structure, and five other virus-specific proteins are synthesized.[336] Table 1 shows the products and the genome segment interrelationship. These products are identical to the proteins observed during infection with one exception. The nonstructural protein translated from a transcript of segment 4 has a molecular weight of 74,000, while it is a 72,000 M_r protein when identified in infected plant tissue.

A coupled transcription, translation, cell-free system, using wheat extracts, has been developed for WTV.[336] The system generates viral proteins identical to proteins synthesized by translation of viral transcripts.

WTV mRNA transcripts have been synthesized in cell-free systems. Full-length transcripts of each of the RNA genomic segments are synthesized by a transcriptase which is contained in the virion. The transcripts have a 5′-terminal blocking structure consisting of a 7-methyl guanosine linked by a pyrophosphate bridge to an adenosine. Unlike most plant viral RNAs this adenosine is methylated.[333,337] A methylase and the capping enzymes are virus coded

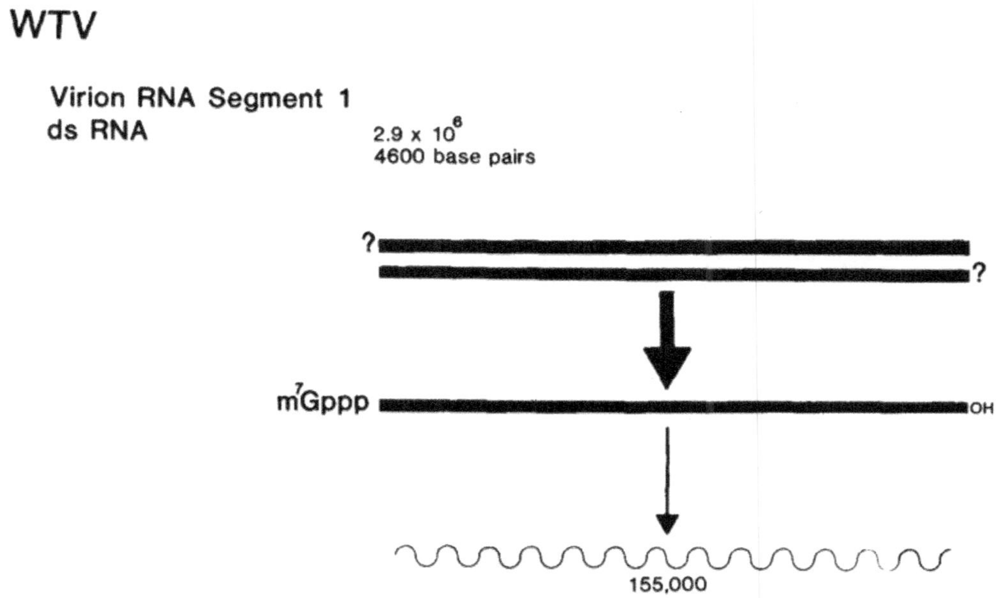

FIGURE 17. A scheme of the genomic organization and expression strategy of one of the ten double-stranded RNA segments of the wound tumor virus (WTV) genome. WTV genome segment 1 and its translation product is presented. See Figure 1 for an explanation of the figure.

and are contained within the virion. The transcripts have been analyzed by gel hybridization studies and have been assigned to particular RNA genome segments. Segments 4 and 5 of the WTV genome are difficult to separate by electrophoresis.[338-340] Likewise, separation of the mRNA transcripts is also difficult. A WTV isolate (−S 5), which is no longer transmitted by leafhoppers, lacks segment 5 or a major portion of it. The use of this isolate has identified the transcript and coded protein derived from segment 5.[338,339]

3. In Vivo Observations

WTV replicates in both insect and plant cell cultures. Analysis of protein synthesized during the replicative cycle has been carried out in an insect cell line synchronously infected with WTV.[336] Samples taken at 8, 12, 16, or 22 hr after infection contained up to 11 new proteins. These proteins are similar in molecular weight to the proteins detected when WTV transcripts are translated in the cell-free systems. Only the 41,000 M_r capsid protein has not been detected in these studies. This may have been due to the fact that this protein frequently co-electrophoreses with the 42,000 M_r viral protein.

The genomic structure and expression strategy of segment 1 of WTV genome is presented in Figure 17.

VI. DISCUSSION

In this review chapter, dealing with RNA plant virus gene expression, we have attempted to describe the results of various in vitro analyses of plant viral RNA and to correlate these results with in vivo studies. It is apparent that there are a number of viral systems where little information is available or seemingly contradictory results have been obtained.

Some of the problems in interpreting the cell-free translation literature are the varied and sometimes contradictory results in the translation analyses of certain plant viral RNAs. We have experienced, in our cell-free analyses, variation in results that can be directly attributed to the cell-free system or the viral RNA. Over a 50-fold difference in incorporation of labeled amino acid into protein may be encountered between different lysates or between different wheat germ extracts. Often accompanying this difference is a change in the product gel profile. Another source of variation in cell-free analysis is the RNA used to program the system. This is especially a problem with unstable virions or with viruses from which it is difficult to isolate RNA. The translation of the two full-length genomic CPMV RNAs results in the synthesis of a two high-molecular-weight polyproteins. Fragmentation of CPMV RNAs results in RNA species which apparently are capable of directing the synthesis of numerous unique viral polypeptides in the wheat germ system.[183] Genomic length TMV RNA also directs the synthesis of a large-molecular-weight product in cell-free translation assays. However, fragmentation of TMV RNA does not make previously "closed cistrons" available for translation.[183] Presently, this observation is not understood however, the cell-free expression of other plant virus RNAs which have been fragmented may be analogous to the expression of fragmented CPMV RNA. Because of the presence of apparently translationally active RNA fragments and the difficulty in separating these RNAs from the authentic viral RNA species, it maybe difficult to ascertain if viral gene expression is from a genomic RNA by an internal initiation events or from discrete subgenomic mRNAs or from fragmented viral RNAs. Until this is resolved, caution should be exercised in the interpretation and extrapolation of cell-free translation results to the possible authentic expression strategy used by viral RNAs during the replicative cycle.

The relationship of replicative forms to the strategy of viral gene expression did not provide an insight into gene expression. In virus systems (TNV) where one replicative form is anticipated, three replicative forms have been isolated. In virus systems where two replicative forms are expected (TYMV) only one replicative form is detected. In other systems (TMV) where 3 replicative forms are expected, 4 to 12 are detected. These somewhat contradictory results raise the question, "what is the function of a replicative form?" Are replicative forms extraction artifacts of complexes active in replication or may they represent a means by which mRNA activity or replication is controlled?

As with the lack of correlation between replicative forms and strategy of gene expression, an analysis of genomic terminal structures (5′ and 3′) did not reveal an obvious relationship to a particular type of genome expression.

However, from this overview of plant viral gene expression, we have developed four general strategies of gene expression. These are presented below.

A. Strategy One: Viral Genomic RNA Serves as Polycistronic mRNA

This model of viral gene expression requires the viral RNA to serve as a functional and translationally active polycistronic messenger RNA. This type of gene expression is common in prokaryotic gene expression and has been demonstrated for RNA bacteriophages. Ribosomes bind to the viral RNA and initiate translation at internal ribosome binding sites in addition to the 5′ terminal ribosome binding site, resulting in the synthesis of a number of discrete viral polypeptides from a single RNA species. Two spherical plant viruses, TNV and CarMV, and an elongated plant virus, PMV, have been proposed to express their genomes in this manner. Four TNV gene products, three CarMV gene products, and three PMV gene products are translated in each case from a single genome-length RNA species. The definitive experiments to confirm this type of expression have yet to be carried out. In view of the possibility of fragmented RNA or subgenomic mRNAs, present in minor amounts, initiating the synthesis of authentic viral proteins, these systems should be reexamined.

B. Strategy Two: Viral Genomic RNA Is a Monocistronic mRNA for a Polyprotein

This model of gene expression requires ribosomes to bind to the viral genomic RNA at the 5′ terminus and to express the total genetic capacity of the viral RNA by synthesizing a single polyprotein. This polyprotein is processed proteolytically into functional viral gene products by proteases (viral and/or host) or by an autocatalytic event. This mode of gene expression has been demonstrated for members of the picornaviridae and for members of the comovirus group. Although speculative at this time, it appears that members of the nepovirus group may express their genome using a similar strategy.

C. Strategy Three: Expression of Viral Genomic RNA Is Limited to the 5′ Proximal Gene and Involved Synthesis of Subgenomic mRNAs to Express other Viral Products

In this model of expression, ribosomes bind to the 5′ terminus of the genomic RNA and translate the gene proximal to the 5′ terminus. Other reading frames on the genome are first transcribed into a subgenomic mRNAs with the gene to be expressed located proximal to an open 5′ terminus. This subgenomic RNA species then serves as an efficient messenger for a single protein. A number of plant viral genomes are apparently expressed in this manner. The multicomponent viruses, such as the bromoviruses, cucumovirues, alfalfa mosaic virus, and ilarviruses have partitioned their genomes into multiple RNAs which also serve as monocistronic mRNAs. Only RNA 3 is dicistronic and the capsid protein gene, which is proximal to the 3′ terminus of RNA 3, is expressed by transcribing this part of RNA 3 into the monocistronic RNA 4 coding for capsid protein.

This type of gene expression also has been proposed for TMV and possibly TRV. With these viruses, a nested set of mRNA species maybe transcribed which are 3′ coterminal and colinear and only differ in length and the presence of an additional coding sequence at the 5′ terminus. Specifically, this has been demonstrated for the TMV 30,000 molecular weight protein gene and, the TMV capsid protein gene. The synthesis of the discrete subgenomic mRNAs is presumed to be via a specific transcription event; however, nucleolytic processing has not been ruled out.

D. Strategy Four: Expression of the Viral Genomic RNA by Polyprotein Synthesis and Proteolytic Processing and Synthesis and Translation of Subgenomic RNAs

This mode of expression makes use of the strategies described in two and three above. This mode of genome expression has been proposed for TYMV and TRosV RNA. Ribosomes bind to the 5′-terminal region of the genomic RNA and synthesize a high-molecular-weight polyprotein which is cleaved posttranslationally into functional viral gene products. The capsid protein gene, which is located proximal to the 3′ terminus and is required in large amounts, is not translated from the genomic RNA. Instead, this segment of the genomic RNA is transcribed into a discrete monocistronic mRNA which is subsequently translated. Presently, it is unclear how this transcription event occurs or how it is regulated.

The genome expression of only a limited number of plant viruses has been elucidated. For most plant virus groups, insufficient information is available to make a definitive statement about the strategy of gene expression. The challenge over the next decade will be not only to unfold the strategies that are involved in plant virus gene expression but also to develop an understanding of how the virus replicative cycle is regulated. This information is necessary to address the far more complex questions, namely, what are the biological functions of the viral gene products and how do these products and the viral genome interact with a host plant in eliciting a pathogenic or resistance response?

ABBREVIATIONS

AlMV	Alfalfa mosaic virus
BBMV	Broad bean mottle virus
BMV	Brome mosaic virus
BSMV	Barley stripe mosaic

BYDV	Barley yellow dwarf virus
CarMV	Carnation mottle virus
CCMV	Cowpea chlorotic mottle virus
CGMMV	Cucumber green mottle mosaic virus
CLMV	Clover yellow mosaic virus
CMV	Cucumber mosaic virus
CPMV	Cowpea mosaic virus
CTV	Citrus tristaza virus
EMV	Eggplant mosaic virus
NMV	Narcissus mosaic virus
PeMV	Pepper mottle virus
PEMV	Pea enation mosaic virus
PLRV	Potato leaf roll virus
PMV	Papaya mosaic virus
PSV	Peanut stunt virus
PVX	Potato virus X
RCNMV	Red clover necrotic mosaic virus
SBMV-B	Southern bean mosaic virus — bean strain
SBMV-C	Southern bean mosaic virus — cowpea strain
SLRV	Strawberry latent ringspot virus
STNV	Satellite tobacco necrosis virus
TAV	Tomato aspermy virus
TBSV	Tomato bushy stunt virus
TBRV	Tomato black ring virus
TCV	Turnip crinkle virus
TEV	Tobacco etch virus
TMV	Tobacco mosaic virus
TNDV	Tobacco necrotic dwarf virus
TNV	Tobacco necrosis virus
TRosV	Turnip rosette virus
TRSV	Tobacco ringspot virus
TRV	Tobacco rattle virus
TSWV	Tomato spotted wilt virus
TSV	Tobacco streak virus
TVMV	Tobacco vein mottling virus
TYMV	Turnip yellow mosaic virus
WMV-I	Watermelon mosaic virus-I
WTV	Wound tumor virus

ACKNOWLEDGMENTS

We would like to thank Drs. Andrew Jackson, L. van Vloten-Doting, G. A. de Zoeten, M. A. Mayo, E. Hiruki, L. Pinck, S. Joshi, M. AbouHaidar, J. W. Davies, and M. Morch for providing unpublished manuscripts and data. We are especially grateful to Don Nuss who provided much of the WTV information and to Dan Purcifull for his comments. We would also like to thank Tina Nowaczyk for her excellent typing. Preparation of this chapter and the work in our laboratories has been supported in part by the National Science Foundation and the U.S. Department of Agriculture Competitive Research Grants Office.

REFERENCES

1. **Davies, J. W. and Kaesberg, P.,** Translation of virus mRNA:synthesis of bacteriophage Qβ proteins in a cell-free extract from wheat embryo, *J. Virol.*, 12, 1434, 1973.
2. **Efron, D. and Marcus, A.,** Translation of TMV-RNA in a cell-free wheat embryo system, *Virology*, 53, 343, 1973.
3. **Roberts, B. E. and Paterson, B. M.,** Efficient translation of tobacco mosaic virus RNA and rabbit globin 9s RNA in a cell-free system from commercial wheat germ, *Proc. Natl. Acad. Sci. U.S.A.*, 70, 2330, 1973.
4. **Matthews, R. E. F.,** Classification and nomenclature of viruses. Third report of the international committee on taxonomy of viruses, *Intervirology*, 12, 129, 1979.
5. **Atabekov, J. G. and Morozov, S. Yv.,** Translation of plant virus messenger RNAs, in *Advances in Virus Research*, Vol. 25, Smith, K. M. and Lauffer, M. A., Eds., Academic Press, New York, 1979, 1.
6. **Davies, J. W.,** Translation of plant virus ribonucleic acids in extracts from eukaryotic cells, in, *Nucleic Acids in Plants*, Vol. 2, Hall, T. C. and Davies, J. W., Eds., CRC Press, Boca Raton, 1979, 111.
7. **Van Vloten-Doting, L. and Neeleman, L.,** Translation of plant virus RNAs, in *Genome Organization and Expression in Plants*, Vol. A29, Leaver, C. J., Ed., NATO Advanced Study Institute Series, Plenum Press, New York, 1980, 511.
8. **Davies, J. W., and Hull, R.,** Genome expression of plant positive-strand RNA viruses, *J. Gen. Virol.*, 61, 1, 1982.
9. **Kozak, M.,** Comparison of initiation of protein synthesis of prokaryotes, eukaryotes, and organelles, *Microb. Rev.*, 47, 1, 1983.
10. **Lane, L. C.,** The nucleic acids of multipartite, defective, and satellite plant viruses, in *Nucleic Acids in Plants*, Vol. 2, Hall, T. C. and Davies, J. W., Eds., CRC Press, Boca Raton, 1979, 65.
11. **Finch, J. T. and Klug, A.,** Arrangement of protein subunits and the distribution of nucleic acid in turnip yellow mosaic virus. II. Electron microscopic studies, *J. Mol. Biol.*, 15, 344, 1966.
12. **Peter, R., Stehelin, D., Reinholt, J., Collot, D., and Duranton, H.,** Primary structure of turnip yellow mosaic virus coat protein, *Virology*, 49, 615, 1972.
13. **Matthews, R. E. F.,** Properties of nucleoprotein fractions isolated from turnip yellow mosaic virus preparations, *Virology*, 12, 521, 1960.
14. **Derosier, D. and Haselkorn, R.,** Minor components associated with turnip yellow mosaic virus, *Virology*, 30, 705, 1966.
15. **Higgins, T. J. V., Whitfield, P. R., and Matthews, R. E. F.,** Size distribution and *in vitro* translation of the RNAs isolated from turnip yellow mosaic virus nucleoproteins, *Virology*, 84, 153, 1978.
16. **Szybiak, U., Bouley, J. P., and Fritsch, C.,** Evidence for the existence of a coat protein messenger RNA associated with the top component of three tymoviruses, *Nucl. Acids Res.*, 5, 1821, 1978.
17. **Pleij, C. W. A., Neeleman, A., Van Vloten-Doting, L., and Bosch, L.,** Translation of turnip yellow mosaic virus RNA *in vitro:* a closed and open coat protein cistron, *Proc. Natl. Acad. Sci. U.S.A.*, 73, 4437, 1976.
18. **Briand, J. P., Keith, G., and Guilley, H.,** Nucleotide sequence at the 5′ extremity of turnip yellow mosaic virus genome RNA, *Proc. Natl. Acad. Sci. U.S.A.*, 75, 3168, 1978.
19. **Guilley, H. and Briand, J. P.,** Nucleotide sequence of turnip yellow virus coat protein mRNA, *Cell*, 15, 113, 1978.
20. **Florentz, C., Mengual, R., Briand, J. P., and Giege, R.,** Large-scale purification of the 3′-OH-terminal tRNA-like sequence (n = 159) of turnip-yellow-mosaic-virus RNA, *Eur. J. Biochem.*, 123, 89, 1982.
21. **Rietveld, K., Van Poelgeest, R., Pleij, C. W. A., Van Boom, J. H., and Bosch, L.,** The tRNA-like structure at the 3′ terminus of turnip yellow mosaic virus RNA. Differences and similarities with canonical tRNA, *Nucl. Acids Res.*, 56, 1929, 1982.
22. **Florentz, C., Briand, J. P., Romby, P., Hirth, L., Ebel, J. P., and Giege, R.,** The tRNA-like structure of turnip yellow mosaic virus RNA: structural organization of the last 159 nucleotides from the 3′ OH terminus, *EMBO J.*, 1, 269, 1982.
23. **Giege, R., Briand, J. P., Ebel, J. P., and Hirth, L.,** Valylation of the two RNA conmponents of turnip-yellow mosaic virus and specificity of the tRNA aminoacylation reaction, *Eur. J. Biochem.*, 84, 251, 1978.
24. **Hubert, E., Huez, G., and Marbaix, G.,** *In vivo* aminoacylation and processing of turnip yellow mosaic virus RNA in *Xenopus laevis* oocytes, *Nature (London)*, 275, 339, 1978.
25. **Pinck, M. and Hall, T. C.,** Aminoacylation of eggplant mosaic virus RNA, *Virology*, 88, 281, 1978.
26. **Joshi, S., Chapeville, F., and Haenni, A. L.,** Turnip yellow mosaic virus RNA is aminoacylated *in vivo* in chinese cabbage leaves, *EMBO J.*, 1, 935, 1982.
27. **Benicourt, C. and Haenni, A. L.,** *In vitro* synthesis of turnip yellow mosaic virus coat protein in a wheat germ cell-free system, *J. Virol.*, 20, 196, 1976.

28. **Klein, C., Fritsch, C., Briand, J. P., Richards, K. E., Jonard, G., and Hirth, L.,** Physical and functional heterogeneity in TYMV RNA; evidence for the existence of an independent messenger coding for coat protein, *Nucl. Acids Res.*, 3, 3043, 1976.
29. **Ricard, B., Barreau, C., Renaudin, H., Mouches, C., and Bove, J. -M.,** Messenger properties of TYMV-RNA, *Virology*, 79, 231, 1977.
30. **Pleij, C. W. A., Mellema, J. R., Noort, A., and Bosch, L.,** The occurrence of the coat protein messenger RNA in the minor components of turnip yellow mosaic virus, *FEBS Lett.*, 80, 19, 1977.
31. **Benicourt, C. and Haenni, A. -L.,** Differential translation of turnip yellow mosaic virus mRNAs *in vitro*, *Biochem. Biophy. Res. Commun.*, 84, 831, 1978.
32. **Benicourt, C., Pere, J. P., and Haenni, A. -L.,** Translation of TYMV RNA into high molecular weight proteins, *FEBS Lett.*, 86, 268, 1978.
33. **Mellema, J. -R., Benicourt, C., Haenni, A. -L., Noort, A., Pleij, C. W. A., and Bosch, L.,** Translational studies with turnip yellow mosaic virus RNAs isolated from major and minor virus particles, *Virology*, 96, 38, 1979.
34. **Morch, M. -D., Drugeon, G., and Benicourt, C.,** Analysis of the *in vitro* coding properties of the 3′ region of turnip yellow mosaic virus genomic RNA, *Virology*, 119, 193, 1982.
35. **Ricard, B., Renaudin, H., and Bove, J. M.,** Translation of eggplant mosaic virus RNA in wheat germ extracts and reticulocyte lysates, *Virology*, 91, 305, 1978.
36. **Renaudin, H., Ricard, B., Mouches, C., and Bove, J. M.,** Translation of eggplant mosaic virus RNA in wheat germ extracts, *Ann. Phytopathol.*, 10, 295, 1978.
37. **Hall, T. C., Pinck, M., Duranton, H. M., and German, T. L.,** Aminoacylation and messenger functions of eggplant mosaic virus RNA, *Virology*, 97, 354, 1979.
38. **Morch, M. -D. and Benicourt, C.,** Post-translational proteolytic cleavage of *in vitro*-synthesized turnip yellow mosaic virus RNA-coded high-molecular-weight proteins, *J. Virol.*, 34, 85, 1980.
39. **Morch, M. -D., Zagorski, W., and Haenni, A. -L.,** Proteolytic maturation of the turnip-yellow-mosaic-virus polyprotein coded *in vitro* occurs by internal catalysis, *Eur. J. Biochem.*, 127, 259, 1982.
40. **Zagorski, W., Morch, M. -D., and Haenni, A. -L.,** Comparison of three different cell-free systems for turnip yellow mosaic virus RNA translation, *Biochemistry*, 65, 127, 1983.
41. **Gepts, P., Angarita, A., and Kummert, J.,** Labeling of the replicative forms of three tymoviruses, *Parasitica*, 33, 95, 1977.
42. **Matthews, R. E. F.,** *Plant Virology*, 2nd ed., Academic Press, New York, 1981, 190.
43. **Ushiyama, R. and Matthews, R. E. F.,** The significance of chloroplast abnormalities with infection by turnip yellow mosaic virus, *Virology*, 42, 293, 1970.
44. **Garnier, M., Mamoun, R., and Bove, J. M.,** TYMV RNA replication in vivo: replicative intermediate is mainly single-stranded, *Virology*, 104, 357, 1980.
45. **Sugimura, Y., Given, N. K., and Matthews, R. E. F.,** TYMV replication in chinese cabbage protoplasts, in 5 Int. Congr. Virol., Abstract W25/01 Strasbourg, August 2 to 7, 1981.
46. **Hermodoson, M. A., Abad-Zapatero, C., Abdel-Meguid, S. S., Pundak, S., Rossmann, M. G., and Tremaine, J. H.,** Amino acid sequence of southern bean mosaic virus coat protein and its relationship to the three-dimensional structure of the virus, *Virology*, 119, 133, 1982.
47. **Abad-Zapatero, C., Abdel-Meguid, S. S., Johnson, J. E., Leslie, A. G. W., Rayment, I., Rossmann, M. G., Suck, D., and Tsukihara, T.,** Structure of southern bean mosaic virus at 2.8 Å resolution, *Nature (London)*, 286,, 33, 1980.
48. **Shepherd, R. J.,** Southern bean mosaic virus, *CMI/AAB Description of Plant Viruses*, No. 57, 1971.
48a. **Wu, S. -X., Hong, S. W., Mang, K. Q., and Kaesberg, P.,** Structure of the genome of southern bean mosaic virus, W27/26, in Am. Soc. Virol. 1983 Annu. Meet., East Lansing, July 10 to 14, 1983.
49. **Ghosh, A., Dasgupta, R., Salerno-Rife, T., Rutgers, T., and Kaesberg, P.,** Southern bean mosaic viral RNA has a 5′-linked protein but lacks 3′ terminal poly (A), *Nucl. Acids Res.*, 7, 2137, 1979.
50. **Veerisetty, V., El-Hassan, S. A., and Sehgal, O. P.,** Temperature induced structural stabilization of the encapsidated RNA and southern bean mosaic virus and its biological significance, *Virology*, 108, 286, 1981.
51. **Veerisetty, V. and Sehgal, O. P.,** Proteinase κ-sensitive factor essential for the infectivity of southern bean mosaic virus ribonucleic acid, *Phytopathology*, 70, 282, 1980.
52. **Salerno-Rife, T., Rutgers, T., and Kaesberg, P.,** Translation of southern bean mosaic virus RNA in wheat embryo and rabbit reticulocyte extracts, *J. Virol.*, 34, 51, 1980.
53. **Mang, K. Q., Ghosh, A., and Kaesberg, P.,** A comparative study of the cowpea and bean strains of southern bean mosaic virus, *Virology*, 116, 264, 1982.
54. **Rutgers, T., Salerno-Rife, T., and Kaesberg, P.,** Messenger RNA for the coat protein of southern bean mosaic virus, *Virology*, 104, 506, 1980.
55. **Ghosh, A., Rutgers, T., Ke-Qiang, M., and Kaesberg, P.,** Characterization of the coat protein mRNA of southern bean mosaic virus and its relationship to the genomic RNA, *J. Virol.*, 39, 87, 1981.

56. **Hull, R. and Aldous, D.,** The nucleic acid of southern bean mosaic and turnip rosette viruses, *John Innes Inst. 69th Annu. Rep.*, 106, 1979.
57. **Hull, R. and Morris-Krsinich, B. A. M.,** Protein covalently linked to the RNA of turnip rosette virus, *John Innes Inst. 70th Annu. Rep.*, 98, 1979.
58. **Morris-Krsinich, B. A. M. and Hull, R.,** Translation of turnip rosette virus RNA in rabbit reticulocyte lysates, *Virology*, 114, 98, 1981.
59. **Dawson, J. R. O.,** The *in vitro* translation of plant virus RNA using wheat germ cell-free extracts, *John Innes Inst. 70th Annu. Rep.*, 100, 1979.
60. **Wu, F. -S. and Murakishi, H. H.,** Infection and synthesis rate of southern bean mosaic virus in soybean callus cells under selected cultural conditions, *Phytopathology*, 68, 1389, 1978.
61. **Jarvis, N. P. and Murakishi, H. H.,** Infection of protoplasts from soybean cell culture with southern bean mosaic and cowpea mosaic viruses, *J. Gen. Virol.*, 48, 365, 1980.
62. **Morris-Krsinich, B. A. M., Hull, R., and Russo, M.,** Infection of turnip leaf protoplasts with turnip rosette virus, *J. Gen. Virol.*, 43, 339, 1979.
63. **Morris, T. J. and Dodds, J. A.,** Isolation and analysis of double-stranded RNA from virus-infected plant and fungal tissue, *Phytopathology*, 69, 854, 1979.
64. **Martelli, G. P.,** Tombusviruses, in *Handbook of Plant Virus Infections — Comparative Diagnosis*, Kurstak, E., Ed., Elsevier/North-Holland Biomedical Press, Amsterdam, 1981, chap. 4.
65. **Winkley, F. K., Schutt, C. E., Harrison, S. C., and Bricogne, G.,** Tomato bushy stunt virus at 5.5-Å resolution, *Nature (London)*, 256, 509, 1977.
66. **Harrison, S. C., Olson, A. J., Schutt, C. E., Winkler, F. K., and Bricogne, G.,** Tomato bushy stunt virus at 2.9 Å resolution, *Nature (London)*, 276, 368, 1978.
67. **Chauvin, C., Witz, J., and Jacrot, B.,** Structure of the tomato bushy stunt virus: a model for protein-RNA interaction, *J. Mol. Biol.*, 124, 641, 1978.
68. **Munowitz, M. G., Dobson, C. M., Griffin, R. G., and Harrison, S. C.,** On the rigidity of RNA in tomato bushy stunt virus, *J. Mol. Biol.*, 141, 317, 1980.
69. **Henriques, M. -I. C. and Morris, T. J.,** Evidence for different strategies in the plant tombusviruses, *Virology*, 99, 66, 1979.
70. **Reeves, J. T.,** *In Vitro* Translation of Tomato Bushy Stunt Virus RNA and Carnation Ringspot Virus RNA, Ph.D. thesis, Purdue University, West Layfayette, 1981.
71. **Dougherty, W. G. and Kaesberg, P.,** Turnip crinkle virus RNA and its translation in rabbit reticulocyte and wheat germ extracts, *Virology*, 115, 45, 1981.
72. **Altenbach, S. B. and Howell, S. H.,** Identification of a satellite RNA associated with turnip crinkle virus, *Virology*, 112, 25, 1981.
73. **Dougherty, W. G.,** unpublished data, 1981.
74. **Reeves, J. T., Larkin, B. A., and Jackson, A. O.,** *In vitro* synthesis of proteins directed by RNA from tomato bushy stunt virus, *Phytopathology*, 71, 251, 1981.
75. **Altenbach, S. B. and Howell, S. H.,** *In vitro* translation products of turnip crinkle virus RNA, *Virology*, 118, 128, 1982.
76. **Russo, M., DiFranco, A., and Martelli, G. P.,** Peroxisomes and multivesicular bodies in tombusvirus infections, in 4th Int. Conf. Comp. Virol., Banff, Canada, October 11 to 17, 1982.
77. **Russo, M. and Martelli, G. P.,** Ultrastructure of turnip crinkle- and saguaro cactus virus-infected tissues, *Virology*, 118, 109, 1982.
78. **Klein, W. H., Nolan, C., Lazar, J. M., and Clark, J. M.,** Translation of satellite tobacco necrosis virus ribonucleic acid. I. Characterization of *in vitro* prokaryotic and eukaryotic translation products, *Biochemistry*, 11, 2009, 1972.
79. **Rice, R. and Fraenkel-Conrat, H.,** Fidelity of translation of satellite tobacco necrosis virus ribonucleic acid in a cell-free *Escherichia coli* system., *Biochemistry*, 12, 181, 1973.
80. **Uyemoto, J. K.,** Tobacco necrosis and satellite viruses, in *Handbook of Plant Virus Infections — Comparative Diagnosis*, Kurstak, E., Ed., Elsevier/North-Holland Biomedical Press, Amsterdam, 1981, chap. 6.
81. **Henriksson, D., Tanis, R. J., Tashian, -R. E., and Nymam, P. O.,** Amino acid sequence of the coat protein subunit of satellite tobacco necrosis virus, *J. Mol. Biol.*, 152, 171, 1981.
82. **Unge, T., Liljas, L., Strandberg, B., Vaara, I., Kannan, K. K., Fridborg, K., Nordman, C. E., and Lentz, P. J., Jr.,** Satellite tobacco necrosis virus structure at 4.0 Å resolution, *Nature (London)* 285, 373, 1980.
83. **Liljas, L., Unge, T., Jones, T. A., Fridborg, K., Lovgren, S., Skoglung, U., and Standberg, B.,** Structure of satellite tobacco necrosis virus at 3.0 Å resolution, *J. Mol. Biol.*, 159, 93, 1982.
84. **Salvato, M. A. and Fraenkel-Conrat, H.,** Translation of tobacco necrosis virus and its satellite in a cell-free wheat germ system, *Proc. Natl. Acad. Sci. U.S.A.*, 74, 2288, 1977.

85. **Lesnaw, J. A. and Reichmann, M. E.,** Identify of the 5′ terminal RNA nucleotide sequence of the satellite tobacco necrosis virus and its helper virus: possible role of the 5′ terminus in recognition by virus-specific RNA replicase, *Proc. Natl. Acad. Sci. U.S.A.*, 66, 140, 1970.
86. **Wimmer, E., Change, A. Y., Clark, J. M., Jr., and Reichmann, M. E.,** Sequence studies of satellite tobacco necrosis virus RNA: isolation and characterization of a 5′-terminal trinucleotide, *J. Mol. Biol.*, 38, 59, 1968.
87. **Van Emmelo, J., Devos, R., Ysebaert, M., and Fiers, W.,** Construction and characterization of a plasmid containing a nearly full-size DNA copy of satellite tobacco necrosis virus RNA, *J. Mol. Biol.*, 143, 259, 1980.
88. **Ysebaert, M., van Emmelo, J., and Fiers, W.,** Total nucleotide sequence of a nearly full-size DNA copy of satellite tobacco necrosis virus RNA, *J. Mol. Biol.*, 143, 273, 1980.
89. **Kassanis, B. and Nixon, H. L.,** Activation of one tobacco necrosis virus by another, *J. Gen. Microbiol.*, 25, 459, 1961.
90. **Donis-Keller, H., Browning, K. S., and Clark, J. M.,** Sequence heterogeneity in satellite tobacco necrosis virus RNA, *Virology*, 110, 43, 1981.
91. **Condit, C. and Fraenkel-Conrat, H.,** Isolation of replicative forms of 3′ terminal subgenomic RNAs of tobacco necrosis virus, *Virology*, 97, 122, 1979.
92. **Steitz, J. A.,** Polypeptide chair initiation: nucleotide sequence of the three ribosomal binding sites in bacteriophage R17 RNA, *Nature (London)*, 224, 957, 1969.
93. **Fiers, W.,** Structure and function of RNA bacteriophages, Fraenkel-Conrat, H. and Wagner, R. R., Eds., Plenum Press, New York, 1979, chap. 3.
94. **Salomon, R., Bar Joseph, M., Soreq, H., Gozes, I., and Littauer, U. Z.,** Translation *in vitro* of carnation mottle virus RNA. Regulatory function of the 3′-region, *Virology*, 90, 288, 1978.
95. **Rowhani, A. and Stace-Smith, R.,** Purification and characterization of potato leafroll virus, *Virology*, 98, 45, 1979.
96. **Roberts, I. M., Tamada, T., and Harrison, B. C.,** Relationship of potato leafroll virus to luteoviruses: evidence from electron microscope serological tests, *J. Gen. Virol.*, 47, 209, 1980.
97. **Takanami, Y. and Kubo, S.,** Enzyme-assisted purification of two phloem-limited viruses: tobacco necrotic dwarf and potato leafroll, *J. Gen. Virol.*, 44, 153, 1979.
98. **Rochow, W. F. and Duffus, J. E.,** Luteoviruses and yellows disease, in *Handbook of Plant Virus Infections — Comparative Diagnosis*, Kurstak, E., Ed., Elsevier/North-Holland Biomedical Press, Amsterdam, 1981, chap. 7.
99. **Mayo, M. A., Barker, H., and Harrison, B. D.,** Evidence that potato leafroll virus RNA is positive-stranded, is linked to a small protein and does not contain polyadenylate, *J. Gen. Virol.*, 59, 163, 1982.
100. **Sarkar, S. and Blessing, J.,** DNA-like properties of the nucleic acid of potato leafroll virus, *Naturwissenschaften*, 60, 480, 1973.
101. **Sarkar, S.,** Potato leafroll virus contains a double-stranded DNA, *Virology*, 70, 265, 1976.
102. **Brakke, M. K. and Rochow, W. F.,** Ribonucleic acid of barley yellow dwarf virus, *Virology*, 61, 240, 1974.
103. **Takanami, Y. and Kubo, S.,** Nucleic acid of two phloem-limited viruses: tobacco necrotic dwarf and potato leafroll, *J. Gen. Virol.*, 44, 853, 1979.
104. **Kubo, S. and Takanami, Y.,** Infection of tobacco mesophyll protoplasts with tobacco necrotic dwarf virus, a phloem limited virus, *J. Gen. Virol.*, 42, 387, 1979.
105. **Van Regenmortel, M. H. V.,** Tobamoviruses, in *Handbook of Plant Virus Infections — Comparative Diagnosis*, Kurstak, E., Ed., Elsevier/North-Holland Biomedical Press, Amsterdam, 1981, chap. 19.
106. **Hirth, L. and Richards, K. E.,** Tobacco mosaic virus: model for structure and function of a simple virus, in *Advances in Virus Research*, Vol. 26, Smith, K. M. and Lauffer, M. A., Eds., Academic Press, New York, 1981, 145.
107. **Goelet, P., Lomonossoff, G. P., Butler, P. J. G., Akam, M. E., Gait, M. J., and Karn, J.,** Nucleotide sequence of tobacco mosaic virus RNA, *Proc. Natl. Acad. Sci. U.S.A.*, 79, 5818, 1982.
108. **Zimmern, D.,** The 5′ end group of tobacco mosaic virus is m^7Gppp^5Gp, *Nucl. Acids Res.*, 2, 1189, 1975.
109. **Knowland, J., Hunter, T., Hunt, T., and Zimmern, D.,** Translation of tobacco mosaic virus RNA and isolation of the messenger for TMV coat protein, *INSERM*, 47, 211, 1975.
110. **Beachy, R. N., Zaitlin, M., Bruening, G., and Israel, H. W.,** A genetic map for the cowpea strain of TMV, *Virology*, 73, 498, 1976.
111. **Pelham, H. R. B.,** Leaky UAG termination codon in tobacco mosaic virus RNA, *Nature (London)*, 272, 469, 1978.
112. **Beachy, R. N. and Zaitlin, M.,** Characterization and *in vitro* translation of the RNAs from less-than-full-length, virus related, nucleoprotein rods present in tobacco mosaic virus preparations, *Virology*, 81, 160, 1977.
113. **Zimmern, D. and Butler, P. J. G.,** The isolation of tobacco mosaic virus RNA fragments containing the origin for viral assembly, *Cell*, 11, 455, 1977.

114. **Zimmern, D.,** The nucleotide sequence at the origin for assembly on tobacco mosaic virus RNA, *Cell,* 11, 463, 1977.
115. **Jonard, G., Richards, R. E., Guilley, H., and Hirth, L.,** Sequence from the assembly nucleation region of TMV RNA, *Cell,* 11, 483, 1977.
116. **Guilley, H., Jonard, G., Kukla, B., and Richards, K. E.,** Sequence of 1000 nucleotides at the 3′ end of tobacco mosaic virus RNA, *Nucl. Acids Res.,* 6, 1287, 1979.
117. **Salomon, R., Sela, I., Soreq, H., Giveon, D., and Littauer, U. Z.,** Enzymatic acylation of histidine to tobacco mosaic virus RNA, *Virology,* 71, 74, 1976.
118. **Meshi, T., Ohno, T., and Okada, Y.,** Nucleotide sequence and its character of cistron coding for the 30K protein of tobacco mosaic virus (OM strain), *J. Biochem.,* 91, 1441, 1982.
119. **Meshi, T., Ohno, T., Iba, H., and Okada, Y.,** Nucleotide sequence of a cloned cDNA copy of TMV (cowpea strain) RNA, including the assembly origin, the coat protein cistron, and the 3′ noncoding region, *Mol. Gen. Genet.,* 184, 10, 1981.
120. **Meshi, T., Ohno, T., and Okada, Y.,** Nucleotide sequence of the 30K protein cistron of cowpea strain of tobacco mosaic virus, *Nucl. Acids Res.,* 10, 6111, 1982.
121. **Meshi, T., Kiyama, R., Ohno, T., and Okada, Y.,** Nucleotide sequence of the coat protein cistron and the 3′ noncoding region of cucumber green mottle mosaic virus (watermelon strain) RNA, *Virology,* 127, 54, 1983.
122. **Hunter, T., Jackson, R., and Zimmern, D.,** Multiple proteins and subgenomic mRNAs may be derived from a single open reading frame on tobacco mosaic virus RNA, *Nucl. Acids Res.,* 11, 801, 1983.
123. **Joshi, S., Pleij, C. W. A., Haenni, A. L., Chapeville, F., and Bosch, L.,** Properties of the tobacco mosaic virus intermediate length RNA -2 and its translation, *Virology,* 127, 100, 1983.
124. **Bruening, G., Beachy, R. N., Scalla, R., and Zaitlin, M.,** *In vitro* and *in vivo* translation of the ribonucleic acids of a cowpea strain of tobacco mosaic virus, *Virology,* 71, 498, 1976.
125. **Higgins, T. J. V., Goodwin, P. B., and Whitfield, P. R.,** Occurrence of short particles in beans infected with the cowpea strain of TMV. II.Evidence that short particles contain the cistron for coat protein, *Virology,* 71, 486, 1976.
126. **Hunter, T. R., Hunt, T., Knowland, J., and Zimmern, D.,** Messenger RNA for the coat protein of tobacco mosaic virus, *Nature (London),* 260, 759, 1976.
127. **Siegel, A., Hari, V., Montgomery, I., and Kolacz, K.,** A messenger RNA for capsid protein isolated from tobacco mosaic virus-infected tissue, *Virology,* 73, 363, 1976.
128. **Morris, T. J.,** Two nucleoprotein components associated with the cowpea strain of TMV, *Proc. Am. Phytopathol. Soc.,* 1, 83, 1974.
129. **Okada, Y., Fukuda, M., Takebe, I., and Osuki, Y.,** Initiation site of assembly of several strains of TMV and its relation to occurrence of the short particles in infected plants, *Biosystems,* 12, 257, 1980.
130. **Fukuda, M., Meshi, T., Okada, Y., Otsuki, Y., and Takebe, I.,** Correlation between particle multiplicity and location on virion RNA of the assembly initiation site for viruses of the tobacco mosaic virus group, *Proc. Natl. Acad. Sci. U.S.A.,* 78, 4231, 1981.
131. **Bienz, M. and Kubli, E.,** Wild type tRNA reads the TMV RNA stop codon but Q base-modified tRNA does not, *Nature (London),* 294, 188, 1981.
132. **Beier, H., Mundry, K. W., and Issinger, O. -G.,** *In vivo* and *in vitro* translation of the RNAs of four tobamoviruses, *Intervirology,* 14, 292, 1980.
133. **Wilson, T. M. A. and Gover, J. F.,** The origin of multiple polypeptides of molecular weight below 110000 encoded by tobacco mosaic virus RNA in the messenger-dependent rabbit reticulocyte lysate, *Biochim. Biophys. Acta,* 739, 35, 1983.
134. **Glover, J. F. and Wilson, T. M. A.,** Efficient translation of the coat protein cistron of tobacco mosaic virus in a cell-free system from *Escherichia coli, Eur. J. Biochem.,* 122, 485, 1982.
135. **Goelet, P. and Karn, J.,** Tobacco mosaic virus induces the synthesis of a family of 3′ coterminal messenger RNAs and their complements, *J. Mol. Biol.,* 154, 541, 1982.
136. **Siegel, A., Hari, V., and Kolacz, K.,** The effect of tobacco mosaic virus infection on host and virus-specific protein synthesis in protoplasts, *Virology,* 85, 494, 1978.
137. **Leonard, D. A. and Zaitlin, M.,** A temperature-sensitive strain of tobacco mosaic virus defective in cell-to-cell movement generates an altered viral-coded protein, *Virology,* 117, 416, 1982.
138. **Scalla, R., Romaine, P., Asselin, A., Rigaud, J., and Zaitlin, M.,** An *in vivo* study of a nonstructural polypeptide synthesized upon TMV infection and its identification with a polypeptide synthesized *in vitro* from TMV RNA, *Virology,* 91, 182, 1978.
139. **Roberts, B. E., Mathews, M. B., and Bruton, C. J.,** Tobacco mosaic virus RNA directs the synthesis of a coat protein peptide in a cell-free system from wheat, *J. Mol. Biol.,* 80, 733, 1973.
140. **Jackson, A. O., Mitchell, D. M., and Siegel, A.,** Replication of tobacco mosaic virus. I. Isolation and characterization of double-stranded forms of ribonucleic acid, *Virology,* 45, 182, 1971.
141. **Derrick, K. S.,** Double-stranded RNA is present in extracts of tobacco plants infected with tobacco mosaic virus, *Science,* 199, 538, 1978.

142. **Zelcer, A., Weaber, K. F., Balazs, E., and Zaitlin, M.,** The detection and characterization of viral-related double-stranded RNAs in tobacco mosaic virus-infected plants, *Virology*, 113, 417, 1981.
142a. **Dawson, W. O. and Dodds, J. A.,** Characterization of sub-genomic double-stranded RNAs from virus-infected plants, *Biochem. Biophy. Res. Commun.*, 107, 1230, 1982.
143. **Koenig, R. and Lesemann, D. -E.,** Potexvirus group, CMI/AAB Description of Plant Viruses, No. 200, 1978.
144. **Frowd, J. A. and Tremaine, J. H.,** Physical, chemical and serological properties of cymbidium mosaic virus, *Phytopathology*, 67, 43, 1977.
145. **Ricciardi, R. P., Goodman, R. M., and Gottlieb, D.,** Translation of PVX RNA *in vitro* by wheat germ. I. Characterization of the reaction and product size, *Virology*, 85, 310, 1978.
146. **Szczesna, E. and Filipowicz, W.,** Faithful and efficient translation of viral and cellular eukaryotic mRNAs in a cell-free S-27 extraction of *Saccharomyces Cerevisiase*. *Biochem. Biophys. Res. Commun.*, 92, 563, 1980.
147. **Wodnar-Filipowicz, A., Skrzeckowski, L. J., and Filipowicz, W.,** Translation of potato virus X RNA into high molecular weight proteins, *FEBS Lett.*, 109, 151, 1980.
148. **Abouhaidar, M. and Bancroft, J. B.,** The structure of the 5'-terminus of papaya mosaic virus RNA, *J. Gen. Virol.*, 39, 559, 1978.
149. **Sonenberg, N., Shatkin, A. J., Ricciardi, R. P., Rubin, M., and Goodman, R. M.,** Analysis of terminal structures of RNA from potato virus X, *Nucl. Acids Res.*, 5, 2501, 1978.
150. **Abouhaidar, M. G.,** The structure of the 5' and 3' ends of clover yellow mosaic virus RNA, *Can. J. Microbiol.*, 29, 151, 1983.
151. **Abouhaidar, M. G.,** Some specific RNA-protein interactions in the assembly of some helical RNA plant viruses, in 4th Int. Conf. Comp. Virol., Banff, Canada, October 11 to 17, 1982.
152. **Bendena, W. G., Abouhaidar, M., Mackie, G. A., and Bancroft, J. B.,** The *in vitro* translation of partially assembled papaya mosaic virus, in Proc. 11th Int. Congr. Biochem., Toronto, 1979, 134.
152a. **Bendena, W. G., Abouhaidar, M. B., and Mackie, G.,** Synthesis *in vitro* of the coat protein of papaya mosaic virus, personal communication, 1983.
152b. **Short, M. N. and Davies, J. W.,** Narcissus mosaic virus: a potexvirus with an encapsidated subgenomic messenger RNA for coat protein, *Bioscience Reports*, 3, 837, 1983.
153. **Abouhaidar, M. and Bancroft, J. B.,** The initiation of papaya mosaic virus assembly, *Virology*, 90, 54, 1978.
154. **Sanders, R. C. and Wood, K. R.,** Protein synthesis in tobacco leaves inoculated with potato virus X, in 4th Int. Congr. Virol., Abstract, The Hague, August 30 to September 6, 1978.
155. **Rao, D. V. and Hiruki, C.,** Infection of cowpea mesophyll protoplasts with a clover yellow mosaic virus, *J. Gen. Virol.*, 38, 303, 1978.
156. **Hollings, M. and Brunt, A. A.,** Potyviruses, in *Handbook of Plant Virus Infections — Comparative Diagnosis*, Kurstak, E., Ed., Elsevier/North-Holland Biomedical Press, Amsterdam, 1981, chap. 23.
157. **McDonald, J. G., Beveridge, T. J., and Bancroft, J. B.,** Self-assembly of protein from a flexuous virus, *Virology*, 69, 327, 1976.
158. **McDonald, J. G. and Bancroft, J. B.,** Assembly studies on potato virus Y and its coat protein, *J. Gen. Virol.*, 35, 251, 1977.
159. **Edwardson, J. G.,** *Some Properties of the Potato Virus Y-Group*, Florida Agricultural Experimental Station Monograph Series, No. 4, Gainesville, 1974, 398.
160. **Purcifull, D. E., Hiebert, E., and McDonald, J. G.,** Immunochemical specificity of cytoplasmic inclusions induced by viruses in the potato virus Y group, *Virology*, 55, 275, 1973.
161. **Knuhtsen, H., Hiebert, E., and Purcifull, D. E.,** Partial purification and some properties of tobacco etch virus induced intranuclear inclusions, *Virology*, 61, 200, 1974.
162. **Govier, D. A., Kassanis, B., and Pirone, T. P.,** Partial purification and characterization of the potato virus Y helper factors, *Virology*, 78, 306.
163. **Hill, J. H. and Benner, H. I.,** Properties of potyvirus RNAs: turnip mosaic, tobacco etch, and maize dwarf mosaic viruses, *Virology*, 75, 419, 1976.
164. **Hari, V.,** The RNA of tobacco etch virus: further characterization and detection of protein linked to RNA, *Virology*, 112, 391, 1981.
165. **Hari, V., Siegel, A., Rozek, C., and Timberlake, W. E.,** The RNA of tobacco etch virus contains poly(A), *Virology*, 92, 568, 1979.
166. **Dougherty, W. G. and Hiebert, E.,** Translation of potyvirus RNA in a rabbit reticulocyte lysate: reaction conditions and identification of capsid protein as one of the products of *in vitro* translation of tobacco etch and pepper mottle viral RNAs, *Virology*, 101, 466, 1980.
167. **Dougherty, W. G. and Hiebert, E.,** Translation of potyvirus RNA in a rabbit reticulocyte lysate: identification of nuclear inclusion proteins as products of tobacco etch virus RNA translation and cylindrical inclusion protein as a product of the potyvirus genome, *Virology*, 104, 174, 1980.

168. **Dougherty, W. G. and Hiebert, E.,** Translation of potyvirus RNA in a rabbit reticulocyte lysate: cell-free translation strategy and a genetic map of the potyviral genome, *Virology,* 104, 183, 1980.
169. **Hellman, G. M., Shaw, J. G., Lesnaw, J. A., Chu, L. -Y., Pirone, T. P., and Rhoads, R. E.,** Cell-free translation of tobacco vein mottling virus RNA, *Virology,* 106, 207, 1980.
170. **Hellman, G. M., Thornbury, D. W., Hiebert, E., Shaw, J. G., Pirone, T. P., and Rhoads, R. E.,** Cell-free translation of tobacco vein mottling virus RNA, II. Immunoprecipitation of products by antisera to cylindrical inclusion, nuclear inclusion, and helper component proteins, *Virology,* 124, 434, 1983.
171. **Thornbury, D. W. and Pirone, T. P.,** Helper components of two potyviruses are serologically distinct, *Virology,* 125, 487, 1983.
172. **Hiebert, E., Thornbury, D. W., and Pirone, T. P.,** Immunoprecipitation analysis of potyviral *in vitro* translation products using antisera to helper compenent of tobacco vein mottling virus and potato virus Y, *Virology,* 135, 1, 1984.
173. **de Mejia, M. V. G., Hiebert, E., and Purcifull, D. E.,** Identification of amorphous inclusion protein as a product of pepper mottle virus RNA (potyvirus), W28/323, in Am. Soc. Virol. 1983 Annu. Meet., East Lansing, July 10 to 14, 1983.
174. **Koziel, M. G., Hari, V., and Siegel, A.,** *In vitro* translation of tobacco etch virus RNA, *Virology,* 106, 177, 1980.
175. **Dougherty, W. G. and Hiebert, E.,** unpublished observations, 1983.
176. **Otal, T. and Hari, V.,** Detection and cell-free translation of subgenomic RNAs of tobacco etch virus, *Virology,* 125, 118, 1983.
177. **Dougherty, W. G.,** Analysis of viral RNA isolated from tobacco leaf tissue infected with tobacco etch virus, *Virology,* 131, 473, 1983.
178. **Hiebert, E. and Dougherty, W. G.,** A genetic map of the potyviral genome, *Phytopathology,* 71, 225, 1981.
179. **Lister, R. M. and Bar-Joseph, M.,** Closterviruses, in *Handbook of Plant Virus Infections — Comparative Diagnosis,* Kurstak, E., Ed., Elsevier/North-Holland Biomedical Press, Amsterdam, 1981, chap. 24.
180. **Bar-Joseph, M. and Murant, A. F.,** Closterovirus group, *CMI/AAB Description of Plant Viruses,* No. 260, 1982.
181. **Nagel, J., Hiebert, E., and Lee, R. F.,** Citrus tristeza virus RNA translated with a rabbit reticulocyte lysate: capsid protein identified as one of the products, *Phytopathology,* 72, 953, 1982.
182. **Bar-Joseph, M., Garnsey, S. M., and Gonsalves, D.,** The closteroviruses: a distinct group of elongated plant viruses, in *Advances in Virus Research,* Vol. 25, Smith, K. M. and Lauffer, M. A., Eds., Academic Press, New York, 1979, 93.
183. **Pelham, H. R. B.,** Translation of fragmented viral RNA *in vitro:* initiation at multiple sites, *FEBS Lett.,* 100, 195, 1979.
184. **Dodds, A. J. and Bar-Joseph, M.,** Double-stranded RNA from plants infected with closteroviruses, *Phytopathology,* 73, 419, 1983.
185. **Murant, A. F. and Mayo, M. A.,** Satellites of plant viruses, in *Annual Review of Phytopathology,* Vol. 20, Grogan, R. G., Zentmyer, G. A., and Cowling, E. B., Eds., Annual Reviews Inc., Palo Alto, 1982, 49.
186. **Mayo, M. A., Murant, A. F., Harrison, B. D., and Goold, R. A.,** Two protein and two RNA species in particles of strawberry latent ringspot virus, *J. Gen. Virol.,* 24, 29, 1974.
187. **Mayo, M. A., Barker, H., and Harrison, B. D.,** Polyadenylate in the RNA of five nepoviruses, *J. Gen. Virol.,* 43, 603, 1979.
188. **Harrison, B. D. and Barker, H.,** Protease-sensitive structure needed for infectivity of nepovirus RNA, *J. Gen. Virol.,* 40, 711, 1978.
189. **Mayo, M. A., Barker, H., and Harrison, B. D.,** Evidence for a protein covalently linked to tobacco ringspot virus RNA, *J. Gen. Virol.,* 43, 735, 1979.
190. **Koenig, I. and Fritsch, C.,** A protein linked at the 5′ end of satellite and genomic tomato black ring virus RNAs: study of *in vitro* translation after protease treatment, *J. Gen. Virol.,* 60, 343, 1982.
191. **Chu, P. W. G., Boccardo, G., and Francki, R. I. B.,** Requirement of a genome-associated protein of tobacco ringspot virus for infectivity but not for *in vitro* translation, *Virology,* 109, 428, 1981.
192. **Robinson, D. J., Barker, H., Harrison, B. D., and Mayo, M. A.,** Replication of RNA-1 of tomato black ring virus independently of RNA-2, *J. Gen. Virol.,* 51, 371, 1980.
193. **Mayo, M. A., Barker, H., and Robinson, D. J.,** Satellite RNA in particles of strawberry latent ringspot virus, *J. Gen. Virol.,* 63, 417, 1982.
194. **Mayo, M. A., Barker, H., and Harrison, B. D.,** Specificity and properties of the genome-linked proteins of nepoviruses, *J. Gen. Virol.,* 59, 149, 1982.
195. **Kiefer, M. C., Daubert, S. D., Schneider, I. R., and Bruening, G.,** Multimeric forms of satellite of tobacco ringspot virus RNA, *Virology,* 121, 262, 1982.
196. **Fritsch, C., Mayo, M. A., and Murant, A. F.,** Translation products of genome and satellite RNAs of tomato black ring virus, *J. Gen. Virol.,* 46, 381, 1980.

197. **Fritsch, C., Mayo, M. A., and Murant, A. F.,** Translation of the satellite RNA of tomato black ring virus *in vitro* and in tobacco protoplasts, *J. Gen. Virol.*, 40, 587, 1978.
198. **Owens, R. A. and Schneider, I. R.,** Satellite of tobacco ringspot virus RNA lacks detectable mRNA activity, *Virology*, 80, 222, 1977.
199. **Mayo, M. A. and Barker, H.,** Defective multiplication of a pseudorecombinant of tomato black ring virus in tobacco protoplasts, *J. Gen. Virol.*, 64, 649, 1983.
200. **Hull, R.,** Pea enation mosaic virus, in *Handbook of Plant Virus Infections — Comparative Diagnosis*, Kurstak, E., Ed., Elseiver/North-Holland Biomedical Press, Amsterdam, 1981, chap. 10.
201. **Gabriel, C. J. and de Zoeten, G. A.,** personal communication, 1983.
202. **Reisman, D. and de Zoeten, G. A.,** A covalently linked protein at the 5′ end of the genomic RNAs of pea enation mosaic virus, *J. Gen. Virol.*, 62, 187, 1982.
203. **German, T. L., de Zoeten, G. A., and Hall, T. C.,** Pea enation mosaic virus RNA contains no polyadenylate sequences and cannot be aminoacylated, *Intervirology*, 9, 226, 1978.
204. **Motoyoshi, F. and Hull, R.,** The infection of tobacco protoplasts with pea enation mosaic virus, *J. Gen. Virol.*, 24, 89, 1974.
205. **Adam, G. and Sander, E.,** Isolation and culture of aphid cells for the assay of insect-transmitted plant viruses, *Virology*, 70, 502, 1976.
206. **German, T. L. and de Zoeten, G. A.,** Purification and properties of the replicative forms and replicative intermediates of pea enation mosaic virus, *Virology*, 66, 172, 1975.
207. **Hull, R. and Lane, L. C.,** The unusual nature of the components of a strain of pea enation mosaic virus, *Virology*, 55, 1, 1973.
207a. **Matthews, R. E. F.,** The classification and nomenclature of viruses: summary of results of meetings of the international committee on taxonomy of viruses in Strasbourg, August 1981, *Intervirology*, 16, 53, 1981.
208. **Dodds, J. A., Tremaine, J. H., and Ronald, W. P.,** Some properties of carnation ringspot virus single- and double-stranded ribonucleic acid, *Virology*, 83, 322, 1977.
209. **Okuno, T., Hiruki, C., Rao, D. V., and Figueiredo, G. C.,** Genetic determinates distributed in two genomic RNAs of sweet clover necrotic mosaic, red clover mosaic and clover primary leaf necrosis viruses, *J. Gen. Virol.*, 64, 1907, 1983.
210. **Gould, A. R., Francki, R. I. B., Hatta, T., and Hollings, M.,** The bipartite genome of red clover necrotic mosaic virus, *Virology*, 108, 499, 1981.
211. **Morris-Krsinish, B. A. M., Forster, R. L. S., and Mossop, D. W.,** Translation of red clover necrotic mosaic virus RNA in rabbit reticulocyte lysate. Identification of the virus coat protein cistron on the larger RNA strand of the bipartite genome, *Virology*, 124, 349, 1983.
212. **Harrison, B. D. and Robinson, D. J.,** Tobraviruses, in *Handbook of Plant Virus Infections — Comparative Diagnosis*, Kurstak, E. E., Elsevier/North-Holland Biomedical Press, Amsterdam, 1981, chap. 18.
213. **Robinson, D. J.,** RNA species of tobacco rattle virus strains and their nucleotide sequence relationships, *J. Gen. Virol.*, 64, 657, 1983.
214. **Mayo, M. A. and Robinson, D. J.,** Revision of estimates of the molecular weights of tobravirus coat protein, *Intervirology*, 5, 313, 1975.
215. **Robinson, D. J., Mayo, M. A., Fritsch, C., Jones, A. T., and Raschke, J. H.,** Origin and messenger activity of two small RNA species found in particles of tobacco rattle virus strain SYM, *J. Gen. Virol.*, 64, 1591, 1983.
216. **Harrison, B. D. and Robinson, D. J.,** The tobraviruses, in *Advances in Virus Research*, Vol. 23, Smith, K. M. and Lauffler, M. K., Eds., Academic Press, New York, 1978, 25.
217. **Pelham, H. R.,** Translation of tobacco rattle virus RNAs *in vitro:* four proteins from three RNAs, *Virology*, 97, 256, 1979.
218. **Bisaro, D. M. and Siegel, A.,** A new viral RNA species in tobacco rattle virus-infected tissue, *Virology*, 107, 194, 1980.
219. **Ball, L. A., Minson, A. C., and Shih, D. S.,** Synthesis of plant virus coat proteins in an animal cell-free system, *Nature (New Biol.)*, 246, 206, 1973.
220. **Mayo, M. A., Fritsch, C., and Hirth, L.,** Translation of tobacco rattle virus RNA *in vitro* using wheat germ extracts, *Virology*, 69, 408, 1976.
221. **Fritsch, C., Mayo, M. A., and Hirth, L.,** Further studies on the translation products of tobacco rattle virus RNA *in vitro*, *Virology*, 77, 722, 1977.
222. **Harrison, B. D., Kubo, S., Robinson, D. J., and Hutcheson, A. M.,** The multiplication cycle of tobacco rattle virus in tobacco mesophyll protoplasts, *J. Gen. Virol.*, 33, 256, 1979.
223. **Mayo, M. A.,** Polypeptides induced by tobacco rattle virus during multiplication in tobacco protoplasts, *Intervirology*, 17, 240, 1982.
223a. **Bisaro, D. M. and Siegel, A.,** Subgenomic component of tobacco rattle virus in infected tissue, *Virology*, 118, 411, 1982.

224. **Matthews, R. E. F.,** Classification and nomenclature of viruses, fourth report of the International Committee on Taxonomy of Viruses, *Intervirology*, 17, 4, 1982.
225. **Van Regenmortel, M. H. V. and Pinck, L.,** Alfalfa mosaic virus, in *Handbook of Plant Virus Infections — Comparative Diagnosis*, Kurstak, E., Ed., Elsevier/North-Holland Biomedical Press, Amsterdam, 1981, chap. 14.
226. **Smit, C. H., Roosien, J., van Vloten-Doting, L., and Jaspars, E. M. J.,** Evidence that alfalfa mosaic virus infection starts with three RNA-protein complexes, *Virology*, 112, 169, 1981.
227. **Zuidema, P., Marti, F. A. B., Cornelissen, B. J. C., Bol, J. F., and Jaspars, E. M. J.,** Coat protein binding sites on RNA 1 of alfalfa mosaic virus, *Virology*, 125, 361, 1983.
228. **Bol, J. R., Brederode, F. T., Janze, G. C., and Rauh, D. K.,** Studies on sequence homology between the RNA's of alfalfa mosaic virus, *Virology*, 65, 1, 1975.
229. **Gould, A. R. and Symons, R. A.,** Alfalfa mosaic virus RNA. Determination of the sequence homology between the four RNA species and a comparison with the four RNA species of cucumber mosaic virus, *Eur. J. Biochem.*, 91, 269, 1978.
230. **Gunn, M. R. and Symons, R. H.,** Sequences homology at the 3′ termini of the four RNAs of alfalfa mosaic virus, *FEBS Lett.*, 109, 145, 1980.
231. **Pinck, L. and Pinck, M.,** Sequence homology at the 3′ ends of alfalfa mosaic virus RNAs, *FEBS Lett.*, 107, 61, 1979.
232. **Koper-Zwarthoff, E. C., Brederode, F. Th., Walstra, P., and Bol., J. F.,** Nucleotide sequence of the 3′ noncoding region of alfalfa mosaic virus RNA 4 and its homology with the genomic RNAs, *Nucl. Acids Res.*, 7, 1887, 1979.
233. **Cornelissen, B. J. C., Brederode, F. Th., Moormanna, R. J. M., and Bol., J. F.,** Complete nucleotide sequence of alfalfa mosaic virus RNA 1, *Nucl. Acids Res.*, 11, 1253, 1983.
234. **Cornelissen, B. J. C., Brederode, F. T., Veeneman, G. H., van Boom, J., and Bol., J. F.,** Complete nucleotide sequence of alfalfa mosaic virus RNA 2, *Nucl. Acids Res.*, 10, 3019, 1983.
235. **Barker, R. F., Jarvis, N. P., Thompson, D. V., Loesch-Fries, L. S., and Hall, T. C.,** Complete nucleotide sequence of alfalfa mosaic virus RNA 3, *Nucl. Acids Res.*, 11, 2881, 1983.
236. **Brederode, F. Th., Koper-Zwarthoff, E. C., and Bol, J. F.,** Complete nucleotide sequence of alfalfa mosaic virus RNA 4, *Nucl. Acids Res.*, 8, 2213, 1980.
237. **Koper-Zwarthoff, E. C., Brederode, F. Th., Veeneman, G., van Boom, J. H., and Bol, J. F.,** Nucleotide sequences at the 5′ termini of the alfalfa mosaic virus RNAs and the intercistronic junction in RNA 3, *Nucl. Acids Res.*, 8, 5635, 1980.
238. **Koper-Zwarthoff, E. C. and Bol, J. F.,** Nucleotide sequence of the putative recognition site for coat protein in the RNAs of alfalfa mosaic virus and tobacco streak virus, *Nucl. Acids Res.*, 8, 3307, 1980.
239. **Ravelonandro, M., Godefroy-Colburn, T., and Pinck, L.,** Structure of the 5′-terminal untranslated region of the genomic RNAs from two strains of alfalfa mosaic virus, *Nucl. Acids Res.*, 11, 2815, 1983.
240. **Swinkels, P. P. H. and Bol, J. F.,** Limited sequence variation in the leader sequence of RNA 4 from several strains of alfalfa mosaic virus, *Virology*, 106, 145, 1980.
241. **Van Tol, R. G. L. and van Vloten-Doting, L.,** Translation of alfalfa mosaic virus RNA 1 in the mRNA-dependent translation system from rabbit reticulocyte lysates, *Eur. J. Biochem.*, 93, 1979.
242. **Van Tol, R. G. L., van Gemeren, R., and van Vloten-Doting, L.,** Two leaky termination codons in AMV RNA 1, *FEBS Letts.*, 118, 67, 1980.
243. **Rutgers, A. S.,** *In Vitro* and *In Vivo* Translation of the RNAs of Alfalfa Mosaic Virus, Ph.D. thesis, University of Leiden, Leiden, 1977.
244. **Mohier, E., Hirth, L., LeMeur, M. A., and Gerlinger, P.,** Translation of alfalfa mosaic virus RNA's in mammalian cell-free systems, *Virology*, 68, 349, 1975.
245. **Gerlinger, P., Mohier, E., LeMeur, M. A., and Hirth, L.,** Monocistronic translation of alfalfa mosaic virus RNAs, *Nucl. Acids Res.*, 4, 813, 1977.
246. **Van Tol, R. G. L. and van Vloten-Doting, L.,** Lack of serological relationship between the 35K non-structural protein of alfalfa mosaic virus and the corresponding proteins of three other plant viruses with tripartite genomes, *Virology*, 109, 444, 1981.
247. **Mohier, E., Hirth, L., LeMeur, M., and Gerlinger, P.,** Analysis of alfalfa mosaic virus 17 S RNA translational products, *Virology*, 71, 615, 1976.
248. **Davies, J. W. and Kaesberg, P.,** Translation of virus mRNA protein synthesis directed by several virus RNAs in a cell-free extract from wheat germ, *J. Gen. Virol.*, 25, 11, 1974.
249. **Castel, A., Kraal, B., Konieczny, A., and Bosch, L.,** Translation by *Escherichia coli* ribosomes of alfalfa mosaic virus RNA 4 can be initiated at two sites on the monocistronic message, *Eur. J. Biochem.*, 101, 123, 1979.
250. **Van Ravenswaay Claasen, J. C., van Leeuwen, A. B. J., Duyts, G. A. H., and Bosch, L.,** *In vitro* translation of alfalfa mosaic virus RNA, *J. Mol. Biol.*, 23, 535, 1967.
251. **Lomonossoff, G. P., Shanks, M., Mathes, H. D., Singh, M., and Gait, M. J.,** The location of the first AUG in cowpea mosaic virus RNAs, *Nucl. Acids Res.*, 10, 4861, 1982.

252. **Nassuth, A., ten Bruggencate, G., and Bol, J. F.,** Time course of alfalfa mosaic virus RNA and coat protein synthesis in cowwpea protoplasts, *Virology*, 125, 75, 1983.
253. **Nassuth, A., Alblas, F., and Bol, J. F.,** Location of genetic information involved in the replication of alfalfa mosaic virus, *J. Gen. Virol.*, 53, 207, 1981.
254. **Nassuth, A. and Bol, J. F.,** Altered balance of the synthesis of plus- and minus- RNAs induced by RNAs 1 and 2 of alfalfa mosaic virus in the absence of RNA 3, *Virology*, 124, 75, 1983.
255. **Van Vloten-Doting, L., Hasrat, J. A., Oosterwijk, E., van T. Sant, P., Schoen, M. A., and Roosien, J.,** Description and complementation analysis of 13 temperature-sensitive mutants of alfalfa mosaic virus, *J. Gen. Virol.*, 46, 415, 1980.
256. **Sarachu, A. N., Nassuth, A., Rossien, J., van Vloten-Doting, L., and Bol, J. F.,** Replication of temperature sensitive mutants of alfalfa mosaic virus in protoplasts, *Virology*, 125, 64, 1983.
257. **Roosien, J. and van Vloten-Doting, L.,** Complementation and interference of ultraviolet-induced Mts mutants of alfalfa mosaic virus, *J. Gen. Virol.*, 63, 189, 1982.
258. **Van Vloten-Doting, L., Francki, R. I. B., Fulton, R. W., Kaper, J. M., and Lane, L. C.,** Tricornaviridea — a proposed family of plant viruses with tripartite, single-stranded RNA genomes, *Intervirology*, 15, 198, 1981.
259. **Fulton, R. W.,** Ilarviruses, in *Handbook of Plant Virus Infections — Comparative Diagnosis*, Kurstak, E., Ed., Elsevier/North-Holland Biomedical Press, Amsterdam, 1981, chap. 13.
260. **Van Vloten-Doting, L.,** Coat protein is required for infectivity of tobacco streak virus: biological equivalence of the coat proteins of tobacco streak and alfalfa mosaic viruses, *Virology*, 65, 215, 1975.
261. **Lister, R. M. and Saksena, K. N.,** Some properties of tulane apple mosaic and ILAR viruses suggesting grouping with tobacco streak virus, *Virology*, 70, 440, 1976.
262. **Gonsalves, D. and Garnsey, S. M.,** Infectivity of heterologous RNA-protein mixtures from alfalfa mosaic, citrus leaf rugose, citrus variegation, and tobacco streak viruses, *Virology*, 67, 319, 1975.
263. **Lane, L. C.,** Bromoviruses, in *Handbook of Plant Virus Infections — Comparative Diagnosis*, Kurstak, E., Ed., Elsevier/North-Holland Biomedical Press, Amsterdam, 1981, chap. 2.
264. **Lane, L. C.,** The bromoviruses, in *Advances in Virus Research*, Vol. 19, Smith, K. M. and Lauffer, M. A., Eds., Academic Press, New York, 1974, 151.
265. **Lane, L. C. and Kaesberg, P.,** Multiple genetic components in bromegrass mosaic virus, *Nature (New Biol.)*, 232, 40, 1971.
266. **Dasgupta, R., Harada, F., and Kaesberg, P.,** Blocked 5′ termini in brome mosaic virus RNA, *J. Virol.*, 18, 260, 1976.
267. **Loesch-Fries, L. S. and Hall, T. C.,** *In vitro* amino-acylation of brome mosaic and barley stripe mosaic virus RNAs, *Nature (London)*, 298, 771, 1982.
268. **Dasgupta, R. and Kaesberg, P.,** Sequence of an oligonucleotide derived from the 3′ end of each of the four brome mosaic viral RNAs, *Proc. Natl. Acad. Sci. U.S.A.*, 74, 4900, 1977.
269. **Ahlquist, P., Dasgupta, R., and Kaesberg, P.,** Near identity of 3′ RNA secondary structure in bromoviruses and cucumber mosaic virus, *Cell*, 23, 183, 1981.
270. **Dasgupta, R., Ahlquist, P., and Kaesberg, P.,** Sequence of the 3′ untranslated region of brome mosaic virus coat protein messenger RNA, *Virology*, 104, 339, 1980.
271. **Gunn, M. R. and Symons, R. H.,** The RNAs of bromoviruses: 3′ terminal sequences of four brome mosaic virus RNAs and comparison with cowpea chlorotic mottle virus RNA 4, *FEBS Lett.*, 115, 77, 1980.
272. **Ahlquist, P., Luckow, V., and Kaesberg, P.,** Complete nucleotide sequence of brome mosaic virus RNA 3, *J. Mol. Biol.*, 154, 23, 1981.
273. **Fowlks, E. R. and Lee, Y. F.,** Detection and sequence of an internal A-rich oligonucleotide series in brome mosaic viral RNA 3, *FEBS Lett.*, 130, 32, 1981.
274. **Dasgupta, R. and Kaesberg, P.,** Complete nucleotide sequence of the coat protein messenger RNAs of brome mosaic virus and cowpea chlorotic mottle virus, *Nucl. Acids Res.*, 10, 703, 1982.
275. **Shih, D. S. and Kaesberg, P.,** Translation of brome mosaic virus ribonucleic acid in a cell-free system derived from wheat embryo, *Proc. Natl. Acad. Sci. U.S.A.*, 70, 1799, 1973.
276. **Davies, J. W.,** The multipartite genome of brome mosaic virus: aspects of *in vitro* translation and RNA structure, *Ann. Microbiol. (Inst. Pasteur)*, 127A, 131, 1976.
277. **Shih, D. S. and Kaesberg, P.,** Translation of the RNAs of brome mosaic virus: the monocistronic nature of RNA 1 and RNA 2, *J. Mol. Biol.*, 103, 77, 1976.
278. **Payne, J. W. and Hall, T. C.,** Efficient ribosome binding of brome mosaic virus (BMV) RNA 4 contributes to its ability to out compete the other BMV RNAs for translation, *Intervirology*, 11, 23, 1979.
279. **Zagorski, W.,** Translational regulation of expression of the brome-mosaic virus RNA genome *in vitro*, *Eur. J. Biochem.*, 86, 465, 1978.
280. **Stubbs, J. D. and Kaesberg, P.,** Amino acid incorporation in an *Escherichia coli* cell-free system directed by brome grass mosaic virus ribonucleic acid, *Virology*, 33, 385, 1967.
281. **Guarino, L. A., Hruby, D. E., Ball, L. A., and Kaesberg, P.,** Translation of black beetle virus RNA and heterologous viral RNAs in cell-free lysates derived from *Drosophila melanogaster*, *J. Virol.*, 37, 500, 1981.

282. **Yamaguchi, K., Hidaka, S., and Miura, K. I.,** Relationship between structure of the 5′ noncoding region of viral mRNA and efficiency in the initiation step of protein synthesis in a eukaryotic system, *Proc. Natl. Acad. Sci. U.S.A.*, 79, 1012, 1982.
283. **Semancik, J. S., Conejero, V., and Gerhart, J.,** Citrus exocortis viroid survey of protein synthesis in *Xenopus laevis* oocytes following addition of viroid RNA, *Virology*, 80, 218, 1977.
284. **Davies, J. W. and Verduin, B. J. M.,** *In vitro* synthesis of cowpea chlorotic mottle virus polypeptides, *J. Gen. Virol.*, 44, 545, 1979.
285. **Evans, I. M. and Boulter, D.,** *In vitro* competition between cowpea chlorotic mottle virus RNA and pea mRNA, *Plant Sci. Lett.*, 27, 299, 1982.
286. **Dickerson, P. E., Malorni, M. C., Dawson, J. R. O., Morris-Krsinich, B. A. M., and Trim., A. R.,** An improved method for the preparation of four ribonucleic acids of cowpea chlorotic mottle virus, *Z. Naturforsch. Ser. C*, 36, 157, 1981.
287. **Rubicki, E. P. and von Wechmar, M. B.,** The serology of the bromoviruses. I. Serological interrelationships of the bromoviruses, *Virology*, 109, 391, 1981.
288. **Sakai, F., Dawson, J. R. O., and Watts, J. W.,** Synthesis of proteins in tobacco protoplasts infected with brome mosaic virus, *J. Gen. Virol.*, 42, 323, 1979.
289. **Watts, J. W. and Dawson, J. R. O.,** Double infection of tobacco protoplasts with brome mosaic virus and cowpea chlorotic mottle virus, *Virology*, 105, 501, 1980.
290. **Saki, F., Watts, J. W., Dawson, J. R. O., and Bancroft, J. B.,** Synthesis of proteins in tobacco protoplasts infected with cowpea chlorotic mottle virus, *J. Gen. Virol.*, 34, 385, 1977.
291. **Kiberstis, P. A., Loesch-Fries, L. S., and Hall, T. C.,** Viral protein synthesis in barley protoplasts inoculated with native and fractionated brome mosaic virus RNA, *Virology*, 112, 804, 1981.
292. **Loesch-Fries, L. S. and Hall, T. C.,** Synthesis, accumulation and encapsidation of individual brome mosaic virus RNA components in barley protoplasts, *J. Gen. Virol.*, 47, 323, 1980.
293. **Hardy, S. F., German, T. L., Loesch-Fries, L. S., and Hall, T. C.,** Highly active template-specific RNA-dependent RNA polymerase from barley leaves infected with brome mosaic virus, *Proc. Natl. Acad. Sci. U.S.A.*, 76, 4956, 1979.
294. **Bujarski, J. J., Hardy, S. F., Miller, W. A., and Hall, T. C.,** Use of dodecyl-β-D-maltoside in the purification and stabilization of RNA polymerase from brome mosaic virus-infected barley, *Virology*, 119, 465, 1982.
294a. **Ahlquist, P., Dasgupta, R., and Kaesberg, P. K.,** Structure of the brome mosaic virus genome, W27/253, in Am. Soc. Virol. 1983 Annu. Meet., East Lansing, July 10 to 14, 1983.
295. **Kaper, J. M. and Waterworth, H. E.,** Cucumoviruses, in *Handbook of Plant Virus Infections — Comparative Diagnosis*, Kurstak, E., Ed., Elsevier/North-Holland Biomedical Press, Amsterdam, 1981, chap. 11.
296. **Lot, H., Marchoux, G., Marrou, J., Kaper, J. M., West, C. K., van Vloten-Doting, L., and Hall, R.,** Evidence for three functional RNA species in several strains of cucumber mosaic virus, *J. Gen. Virol.*, 22, 81, 1974.
297. **Symons, R. H.,** Cucumber mosaic RNA contains 7-methylguanosine at the 5′ terminus of all four RNA species, *Mol. Biol. Rep.*, 2, 277, 1977.
298. **Kohl, R. J. and Hall, T. C.,** Amino acylation of RNA from several viruses: amino acid specificity and differential activity of plant, yeast and bacterial synthetases, *J. Gen. Virol.*, 25, 257, 1974.
299. **Symons, R.,** Extensive sequence homology at the 3′ termini of the four RNAs of cucumber mosaic virus, *Nucl. Acids Res.*, 7, 825, 1979.
300. **Wilson, P. A. and Symons, R. H.,** The RNAs of cucumoviruses: 3′-terminal sequence analysis of two strains of tomato aspermy virus, *Virology*, 112, 342, 1981.
301. **Gould, A. R. and Symons, R. H.,** Cucumber mosaic virus RNA 3. Determination of the nucleotide sequence provides the amino acid sequences of protein 3A and viral coat protein, *Eur. J. Biochem.*, 126, 217, 1982.
302. **Collmer, C. W., Tousignant, M. E., and Kaper, J. M.,** Cucumber mosaic virus-associated RNA 5. X. The complete nucleotide sequence of a CARNA 5 incapable of inducing tomato necrosis, *Virology*, 127, 230, 1983.
303. **Richards, K. E., Jonard, G., Jacquemond, M., and Lot, H.,** Nucleotide sequence of cucumber mosaic virus-associated RNAs, *Virology*, 89, 395, 1978.
304. **Gordon, K. H. J. and Symons, R. H.,** Satellite RNA of cucumber mosaic virus forms a secondary structure with partial 3′-terminal homology to genomal RNAs, *Nucl. Acids Res.*, 11, 947, 1983.
305. **Schwinghamer, M. W. and Symons, R. H.,** Fractionation of cucumber mosaic virus RNA and its translation in a wheat embryo cell-free systsm, *Virology*, 63, 252, 1975.
306. **Schwinghamer, M. W. and Symons, R. H.,** Translation of the four major RNA species of cucumber mosaic virus in plant and animal cell-free systems and in toad oocytes, *Virology*, 79, 88, 1977.
307. **Owens, R. A. and Kaper, J. M.,** Cucumber mosaic virus-associated RNA 5. II. *In vitro* translation in a wheat germ protein synthesis system, *Virology*, 80, 196, 1977.

308. **Gonda, J. J. and Symons, R. H.,** Cucumber mosaic virus replication in cowpea protoplasts: time course of virus, coat protein and RNA synthesis, *J. Gen. Virol.*, 45, 723, 1979.
309. **Robert, P. L. and Wood, K. R.,** Protein synthesis in cucumber mosaic virus-infected tobacco leaves, pre-infected cells and protoplasts, *Arch. Virol.*, 70, 115, 1981.
310. **Gordon, K. H. J., Gall, D. S., and Symons, R. H.,** Highly purified cucumber mosaic virus-induced RNA-dependent RNA polymerase does not contain any of the full length translation products of the genomic RNAs, *Virology*, 123, 284, 1982.
311. **Kaper, J. M.,** Perspective on CARNA5, cucumber mosaic virus-dependent replicating RNAs capable of modifying disease expression, *Plant Mol. Biol. Rep.*, 1, 49, 1983.
312. **Jackson, A. O. and Lane, L. C.,** Hordeiviruses, in *Handbook of Plant Virus Infections — Comparative Diagnosis*, Kurstak, E., Ed., Elsevier/North-Holland Biomedical Press, Amsterdam, 1981, chap. 20.
313. **Gustafson, G. D., Milner, J. J., McFarland, J. E., Pendersen, K., Larkin, B. A., and Jackson, A. D.,** Investigation of the complexity of barley stripe mosaic virus RNAs with recombinant DNA clones, *Virology*, 120, 182, 1982.
314. **Jackson, A. O., Dawson, J. R. O., Covey, S. N., Hull, R., Davies, J. W., McFarland, J. E., and Gustafson, G. D.,** Sequence relations and coding properties of a subgenomic RNA isolated from barley stripe mosaic virus, *Virology*, 127, 37, 1983.
315. **Agranovsky, A. A., Dolja, V. V., Kagramanova, V. K., and Atabekov, J. G.,** The presence of cap structure at the 5′ end of barley stripe mosaic virus RNA, *Virology*, 95, 208, 1979.
316. **Agranovsky, A. A., Dolja, V. V., and Atabekov, J. G.,** Structure of the 3′ extremity of barley stripe mosaic virus RNA: evidence for internal poly(A) and a 3′-terminal tRNA-like structure, *Virology*, 119, 51, 1982.
317. **McFarland, J. E., Brakke, M. K., and Jackson, A. O.,** Complexity of the Argentina mild strain of barley stripe mosaic virus, *Virology*, 130, 397, 1983.
318. **Dolja, V., Sokolova, N. A., Tjulkina, L. G., and Atabekov, J. G.,** A study of barley stripe mosaic virus (BSMV) genome. II. Translation of individual RNA species of two BSMV strains in a homologous cell-free system, *Mol. Gen. Genet.*, 175, 93, 1979.
319. **Dolja, V. V., Lunina, N. A., Leisie, R. M., Stanarius, T., Belzhelarskaya, S. N., Koxlov, Yu, V., and Atabekov, J. G.,** A comparative study on the *in vitro* translation products of individual RNAs from two-, three- and four-component strains of barley stripe mosaic virus, *Virology*, 127, 1, 1983.
320. **Gustafson, G. D., Larkins, B. A., and Jackson, A. O.,** Comparative analysis of polypeptides synthesized *in vivo* and *in vitro* by two strains of barley stripe mosaic virus, *Virology*, 111, 579, 1981.
321. **Jackson, A. O., McFarland, J. E., and Stanley, J.,** Complexity of the RNAs of strains of barley stripe mosaic virus, in *Current Communications in Molecular Biology*, Cold Spring Harbor Laboratory, New York, 1983, 94.
322. **White, J. L. and Brakke, M. K.,** Protein changes in wheat infected with wheat streak mosaic virus and in barley infected with barley stripe mosaic virus, *Physiol. Plant Pathol.*, 22, 87, 1983.
323. **Best, R. J.,** Tomato spotted wilt virus, in *Advances in Virus Research*, Vol. 13, Smith, K. M. and Lauffer, M. A., Eds., Academic Press, New York, 1968, 65.
324. **Francki, R. I. B. and Hatta, T.,** Tomato spotted wilt virus, in *Handbook of Plant Virus Infections — Comparative Diagnosis*, Kurstak, E., Ed., Elsevier/North-Holland Biomedical Press, Amsterdam, 1981, chap. 17.
325. **Verkleij, F. N.,** Characterization of a Defective Form of Tomato Spotted Wilt Virus, Ph.D. thesis, Agricultural College of Wageningen, Wageningen, 1982.
326. **Mohamed, N. A., Randles, J. W., and Francki, R. I. B.,** Protein composition of tomato spotted wilt virus, *Virology*, 56, 379, 1973.
327. **Tas, P. W. L. and Peters, D.,** The strucutral proteins of tomato spotted wilt virus, *J. Gen. Virol.*, 36, 267, 1977.
328. **Van den Hurk, J., Tas, P. W. L., and Peters, D.,** The ribonucleic acid of tomato spotted wilt virus, *J. Gen. Virol.*, 36, 81, 1977.
329. **Verkleij, F. N., de Vries, P., and Peters, D.,** Evidence that tomato spotted wilt virus RNA is a positive strand, *J. Gen. Virol.*, 58, 329, 1982.
330. **Mohamed, N. A,** Isolation and characterization of subviral structures from tomato spotted wilt virus, *J. Gen. Virol.*, 53, 197, 1981.
331. **Verkleij, F. N. and Peters, D.,** Characterization of a defective form of tomato spotted wilt virus, *J. Gen. virol.*, 64, 677, 1983.
332. **Shikata, E.,** Reoviruses, in *Handbook of Plant Virus Infections — Comparative Diagnosis*, Kurstak, E., Ed., Elsevier/North-Holland Biomedical Press, Amsterdam, 1981, chap. 15.
333. **Nuss, D. L.,** Molecular biology of wound tumor virus, in *advances in Virus Research*, Vol. 29, Smith, K. M. and Lauffer, M. A., Eds., Academic Press, New York, in press.
334. **Reddy, D. V. R. and Black, L. M.,** Electrophoretic separation of all components of the double-stranded RNA of wound tumor virus, *Virology*, 54, 557, 1973.

335. **Reddy, D. V. R. and MacLeod, R.,** Polypeptide component of wound tumor virus, *Virology*, 70, 274, 1976.
336. **Nuss, D. L. and Peterson, A. J.,** Expression of wound tumor virus gene products *in vivo* and *in vitro*, *J. Virol.*, 34, 532, 1980.
337. **Rhodes, D. P., Reddy, D. V. R., MacLeod, R., Black, L. M., and Banerjee, A. K.,** *In vitro* synthesis of RNA containing 5′-terminal structures mG(5′)ppp(5′)A ... by purified wound tumor virus, *Virology*, 76, 554, 1977.
338. **Nuss, D. L. and Peterson, A. J.,** *In vitro* synthesis and modification of mRNA by exvectorial isolates of would tumor virus, *J. Virol.*, 39, 954, 1981.
339. **Nuss, D. L.,** Characterization of subgenomic RNAs associated with exvectorial isolates of wound tumor virus, *in Current Communications in Molecular Biology*, Cold Spring Harbor Laboratories, New York, 1983, 111.
340. **Nuss, D. L.,** Molecular biology of wound tumor virus transmission: genome segment 5 and the loss of transmissibility, in *Double Stranded RNA Viruses*, Bishop, D. H. L. and Compans, R. W., Eds., Elsevier, New York, 1983, 204.

Chapter 3

STRUCTURE, REPLICATION, AND EXPRESSION OF THE BIPARTITE GENOME OF COWPEA MOSAIC VIRUS

Rob Goldbach and Albert van Kammen

TABLE OF CONTENTS

I. INTRODUCTION

A. Comoviruses

Cowpea mosaic virus (CPMV), type member of the comovirus group, has now been studied for more than 20 years and is one of the best characterized plant viruses. The comovirus group[1] contains viruses with a bipartite, single-stranded RNA genome separately encapsidated in icosahedral particles. All viruses are transmitted by beetles and have a rather narrow host range. Several review articles have been published dealing with the composition and genetics of comoviruses and the diagnosis and epidemiology of the diseases caused by these viruses (see References 2 to 5). Twelve plant viruses have been classified as comoviruses by the International Committee on Taxonomy of Viruses, and descriptions of the main characteristics of these viruses have been published.[6] Pea mild mosaic virus and glycine mosaic virus are probably also to be regarded as comoviruses[5,5a] and have therefore provisionally been added to the group of comoviruses listed in Table 1. Ten of these 14 viruses (BPMV, BRMV, BBSV, BBTMV, CPMV, CPSMV, GMV, PMMV, QPMV, and RCMV) have legumes as their natural hosts, with only a few hosts outside the Leguminosae.

B. Cowpea Mosaic Virus

CPMV causes mosaic diseases in cowpeas (*Vigna* spp.). The virus has been isolated from Surinam (Sb-isolate), Nigeria, and Kenya. In contrast to what has been described in the *CMI/AAB Description of Plant Viruses*,[7] its geographical distribution does probably not include Cuba and the U.S. In these latter countries only CPSMV, which was formerly regarded as a severe strain of CPMV but later recognized as a different comovirus,[8,9] has been unambiguously found. The Dg-isolate,[10,11] Vs-isolate,[12] and Ark-isolate,[9,13,14] formerly referred to as strains of CPMV, should now all be considered as strains of CPSMV. Actually, in the New World countries, CPMV has only been found with certainty in Surinam. The viruses isolated from Nigeria, Kenya, and Surinam have in the past been described as different isolates of the "yellow" strain of CPMV to draw a distinction with the "severe" strain. Now that the former yellow and severe strain of CPMV are proven to be different viruses, these specifications should not longer be used with CPMV.

CPMV multiplies rapidly in its natural host, *Vigna unguiculata*. If primary leaves of *Vigna* plants are inoculated with CPMV and grown at 30°C, up to 15 μg virus per milligram of leaf tissue can be isolated from these leaves after 4 days.[15] Yields of virus may reach 2 g/kg leaf tissue from whole *Vigna* plants grown at 30°C. When isolated cowpea mesophyll protoplasts are inoculated with CPMV, yields of virus have been estimated to amount to 20 to 30 μg per 10^6 protoplasts at 41 hr after inoculation.[16] This intensive multiplication offers a good starting point for studying the replication of this virus.

C. Molecular Structure of Cowpea Mosaic Virus

CPMV, like the other comoviruses, produces in infected plants three classes of icosahedral particles, each approximately 28 nm in diameter: B (bottom) and M (middle) components contain each a different RNA molecule, whereas T (top) components have no RNA, but consist of empty particles (Figure 1). Both B and M components, or their RNAs, are necessary for virus multiplication.[17,18] B components, unlike M components, show some degree of

Table 1
THE COMOVIRUS GROUP

Virus	Abbreviation
Andean potato mottle virus	APMV
Bean pod mottle virus	BPMV
Bean rugose mosaic virus	BRMV
Broad bean stain virus	BBSV
Broad bean true mosaic virus	BBTMV
Cowpea mosaic virus	CPMV
Cowpea severe mosaic virus	CPSMV
Glycine mosaic virus	GMV
Pea mild mosaic virus[a]	PMMV
Quail pea mosaic virus	QPMV
Radish mosaic virus	RaMV
Red clover mottle virus	RCMV
Squash mosaic virus	SqMV
Ullucus virus C[b,171]	

[a] Added to the list on basis of the report of Stace-Smith,[5] but not yet considered by the International Committee on Taxonomy of Plant Viruses.

[b] Added to the list in agreement with the provisional grouping by the editors of *CMI/AAB Descriptions of Plant Viruses*, but not yet considered by the International Committee on Taxonomy of Plant Viruses.

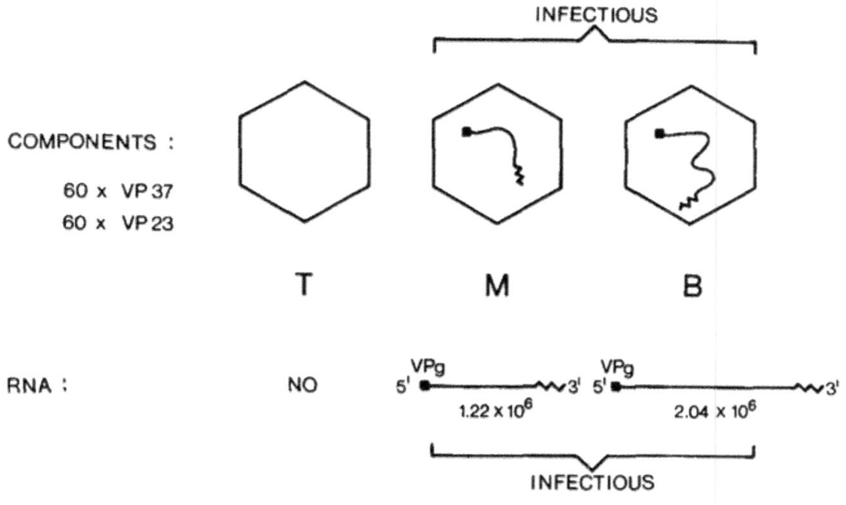

FIGURE 1. Composition of CPMV. Purified preparations of CPMV consists of three components, all isometric particles approximately 28 nm in diameter, with capsids composed of 60 copies of each of VP37 (molecular weight 37,000) and VP23 (molecular weight 23,000). Components T are devoid of RNA, whereas M and B components contain different amounts of RNA. Both B and M components, or their RNAs, are required for virus multiplication. Both B-RNA and M-RNA carry a small protein, VPg (molecular weight ~ 4000) at their 5'-termini and a poly(A) tail at their 3'-ends. The indicated molecular weights of the two RNAs have been calculated from their primary structure (see Figure 6 and 7).

Table 2
ESTIMATES OF THE MOLECULAR WEIGHTS ($\times 10^{-6}$) OF CPMV RNAs

Estimate	B-RNA	M-RNA
From electrophoresis in polyacrylamide at 60°C in 6*M* urea[49]	2.02	1.37
From electrophoresis in agarose after glyoxylation[50]	2.39	1.39
From primary structure	2.04	1.22

independence in that inoculation of cowpea protoplasts with these components leads to independent replication and expression of its (B-) RNA (Reference 19; see, for a more detailed discussion on this property, Section II.C). The capsids of T, M, and B components have an identical structure and are built up of 60 copies of each of 2 different proteins, referred to as VP37 (M_r 37,000) and VP23 (M_r 23,000)[20,22] At least one of these proteins is partly glycosylated.[23] The two virion RNAs of CPMV, M-RNA, and B-RNA, are both "plus" type single-stranded RNAs and are polyadenylated at their 3′ terminus.[24-26] Both RNAs also possess a small protein of approximately 4000 M_r, called VPg (viral protein genome-bound), covalently bound to their 5′ terminus.[27-28] Removal of this protein with proteinase K does not lead to loss of infectivity.[28] The presence of both a poly(A) tail and a VPg has been established with the genome of all comoviruses tested so far.[29] Such terminal structures also occur at the ends of the viral RNAs of the bipartite nepoviruses,[30-34] the bipartite pea enation mosaic virus (PEMV),[35] and at the ends of the single RNA of potyviruses.[36,37] As found for CPMV, removal of VPg from the RNA has no effect on the infectivity of PEMV[35] but for nepoviruses, on the contrary, removal of VPg results in a drastic decrease of infectivity.[34]

The combination of a VPg at the 5′ end of the viral RNA and a poly(A) tail at the 3′ end is also known to occur with the animal picornaviruses, like poliovirus, encephalomyocarditis virus, and foot-and-mouth disease virus,[38-43] and caliciviruses.[44] Other plant viruses, like the monopartite sobemoviruses,[45-47] and luteoviruses[48] carry a VPg at the 5′ ends of their RNA, but lack a poly(A) tail at the 3′ ends.

The molecular weights of B- and M-RNA (sodium salts) from CPMV were first estimated by gel electrophoresis under denaturing conditions by Reijnders et al.[49] and later by Murant et al.[50] Their estimates are compiled in Table 2 and are compared with the molecular weights (sodium salts) as deduced from the primary structure of both RNAs[51,52] (see also section III). Compared to the molecular weight as deduced from the nucleotide sequence, the estimates for the molecular weight of M-RNA made by both groups are too high. The experimental estimate for the molecular weight of the B-RNA by Reijnders et al.[49] agrees exceptionally well with the molecular weight as calculated from the nucleotide sequence of this RNA, whereas there is rather a discrepancy with the experimental data of Murant et al.[50] Figure 1 summarizes schematically the molecular organization of CPMV particles.

II. PROTEINS SYNTHESIZED BY CPMV

A. Translation of CPMV RNAs in Cell-Free Extracts

Before it emerged that CPMV produces considerable quantities of both structural and nonstructural proteins in cowpea leaves and isolated cowpea protoplasts, the translation strategy of the CPMV RNAs was studied by translation in vitro. Using both messenger-dependent rabbit reticulocyte lysates and extracts from wheat germ, it was found that the virion RNAs of CPMV are messenger-sense and translated into large polypeptides of specific size.[53,54] Actually, these in vitro translation studies provided the first evidence that the expression of CPMV RNAs involves proteolytic cleavage of large primary translation products rather than the production of subgenomic mRNAs. Translation of the CPMV RNAs in reticulocyte lysates and in wheat germ extracts yields, in principle, the same polypeptides. B-RNA produces in vitro a primary translation product of approximately 200 kilodaltons (K),[53-56] in good agreement with the length of the unique, long open reading frame of this RNA (see Figure 2 and Section III). M-RNA produces in vitro two polypeptides with sizes of around 95K and 105K[53-58] (Figure 2). It should be noticed that the reported sizes of both the B-RNA-encoded and M-RNA-encoded polypeptides vary significantly, but these variations appear to be completely due to the use of different gel electrophoresis systems and different sets of molecular weight markers. Thus it has been observed that in some polyacrylamide gels the larger of the two M-RNA-encoded polypeptides migrates slower than *E. coli* β-galactosidase (mol wt 116,353[59]), whereas in other gel systems the same polypeptide migrates significantly faster than this marker protein.[60] This observation once again demonstrates that by determining the electrophoretic mobilities of proteins, only apparent molecular weights are obtained which are at best good approximations of the real sizes of the proteins. While in the first in vitro translation studies with B-RNA only the production of a 200K-polypeptide was reported, in later experiments using rabbit reticulocyte lysates Pelham[55] showed that under proper conditions this 200K polypeptide is cleaved, probably by autocatalytic proteolysis, into 32K and 170K polypeptides. This cleavage requires ATP and dithiothreitol (2 m*M*), and inhibition studies further indicated that the protease involved is of the thiol type.[55] The two in vitro cleavage products turned out to be similar to virus-specific 32K and 170K polypeptides found in infected cowpea protoplasts[22,57] (see Section II.B), indicating the faithfulness of the rabbit reticulocyte system. Very recently,[61,62] further cleavage of the 170K polypeptide has been achieved in this in vitro system by proceeded incubation (10 to 18 hr) (Figure 3) at 30°C. During this period, the 170K polypeptide is slowly converted into the same 110K, 87K, 84K, and 60K virus-specific polypeptides as are found in infected cells (Figure 2 and Section II.B), together with a polypeptide of approximately 28K not detected in vivo so far. Since the proteolytic cleavage in the rabbit reticulocyte lysate proceeds similarly to that in the cowpea plant (see Section II.B), it appears reasonable to assume that the processing activity is encoded by B-RNA itself. In Section II.E, evidence will be discussed that the relevant protease activity resides in the 28K polypeptide. Although for wheat germ extracts only the production of the 200K polypeptide from B-RNA was reported,[54,56] more recent analyses have shown that in this system at least the primary cleavage of this polypeptide can also be achieved.[63] While in vitro translation of B-RNA produces polypeptides which are also obtained from this RNA in vivo (cf. Section II.B) the 105K and 95K polypeptides translated in vitro from M-RNA have so far never been detected in infected cowpea leaves or protoplasts. These two polypeptides have overlapping amino acid sequences and are produced from two different initiation sites on M-RNA.[55-58] Whether both initiation sites are also used in vivo remains to be ascertained. Pelham[55] first showed that the 95K and 105K polypeptides are specifically cleaved by a proteolytic activity in the in vitro translation product of B-RNA. Later, Franssen et al.[58] were able to demonstrate that this activity is also produced in vivo and that it resides in the 32K polypeptide.[64] The proteolytic cleavage products generated have sizes of 60K, 58K, and 48K (see Figure 2). The 60K polypeptide reacts both with antibodies against VP37 and antibodies against VP23 and represents presumably a specific, common precursor to the two capsid proteins.[58]

Thus, two different proteases appear to be encoded by B-RNA, involved in the processing of the M- and B-RNA-encoded polyproteins, respectively. In Section II.E these different proteases will be further discussed.

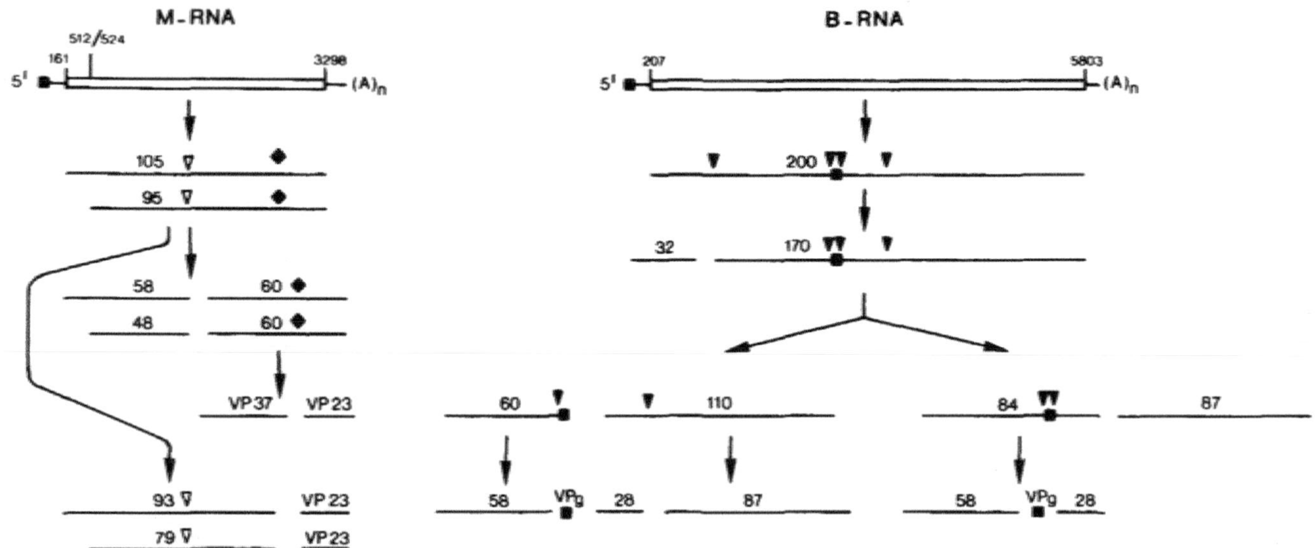

FIGURE 2. Model for the translation of the bipartite genome of CPMV. The proteolytic cleavage pathways are based on both in vivo and in vitro translation studies. M-RNA encodes both capsids proteins VP37 and VP23, and B-RNA encodes VPg (4K), a protease (32K) involved in the cleavage of the M-RNA-coded polyproteins, a second protease (28K) for the cleavage of the B-RNA-coded polyprotein, and the viral replicase (110K) (see also Sections II.E and IV). The open reading frames in both RNAs are indicated with double-lined bars. The positions of the start and stop-codons in M- and B-RNA are taken from Van Wezenbeek et al.[51] and Lomonossoff and Shanks,[52] respectively. The B-RNA-encoded 130K and 112K polypeptides, sometimes found in small amounts, are not included in the scheme. Proteolytic cleavage sites are indicated as follows: ▽, glutamine-methionine; ◆, glutamine-glycine; ▼, glutamine-serine. The various cleavage sites are discussed in Sections III.A.3 and III.B.2.

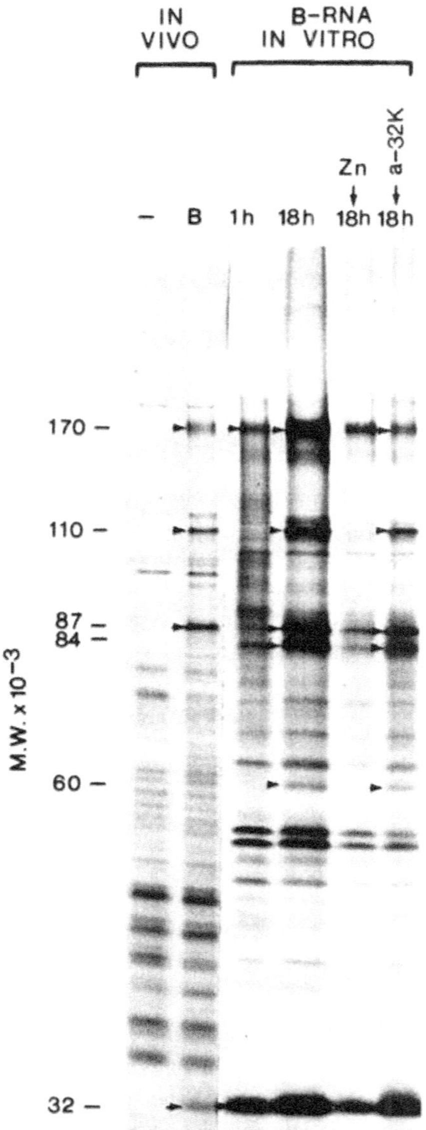

FIGURE 3. In vitro translation of B-RNA in the rabbit reticulocyte lysate. After 1 hr of translation (lane: 1 hr) predominantly 170K and 32K polypeptides are found, which are generated by a single cleavage of the 200K primary translation products. After prolonged incubation for 18 hr the 170K polypeptide is slowly further cleaved into 110K, 87K, 84K, 60K, and 28K polypeptides (lane: 18hr). These additional cleavages are just like the primary cleavage significantly inhibited by 1 m*M* Zn^{2+} (lane: Zn), but not by antiserum raised against the 32K polypeptide (lane: a-32K). The left two lanes (in vivo) contain S30-extract from cowpea protoplasts, either uninoculated (–) or inoculated with B-components. (From Franssen, H., Goldbach, R., and van Kammen, A., in *Virus Research*, Vol. 1 (No. 1), 1984, 44. With permission.)

An additional proteolytic activity present in infected cowpea leaves is capable of releasing the small capsid protein VP23 from the 105 and 95K polypeptides but achieves no further cleavage to generate the other capsid protein VP37 (reference 58 and Figure 2).

The analyses of Franssen et al.[58] provide direct evidence that both capsid proteins of CPMV are coded by M-RNA. The location of the capsid protein genes on M-RNA is in full agreement with the finding[19] that upon inoculation of cowpea protoplasts with only B components all viral nonstructural proteins are synthesized but not the capsid proteins (see Figure 2). Previous genetical complementation experiments with separated components of wild-type and mutant G3 of CPMV (Nigerian isolate) had already indicated that at least the small capsid protein must be encoded by M-RNA.[65]

B. Viral Polypeptides in Cowpea Mesophyll Protoplasts

The development of plant protoplast systems for studying virus multiplication has greatly enhanced the possibilities of examining the expression of plant viral genomes in vivo. In leaf extracts of virus-infected plants, the only detectable virus-specific proteins are usually the structural proteins. However, the efficient, approximately synchronous infection of isolated mesophyll protoplasts has enabled the detection of virus-specific nonstructural proteins with several plant viruses (see Reference 66 and 67 and the chapter in this volume by Maule for review). Thus the development of a procedure for the isolation of cowpea mesophyll protoplasts[68,69] marked a real breakthrough in our understanding of the expression mechanism of CPMV. Even though some noncapsid proteins are detectable in extracts from *Vigna* leaves infected with CPMV, full information has only been obtained by analyzing infected mesophyll protoplasts. From about 16 hr after inoculation of a protoplast suspension, virus-encoded polypeptides can be detected by labeling with [^{35}S] methionine. Besides the capsid proteins, VP37 and VP23, virus-specific polypeptides of approximately 170K, 110K, 87K, 84K, 60K, 58K, and 32K can be recognized, all encoded by B-RNA.[16,19,22,57,70,71] In addition, minor amounts of virus-induced polypeptides of approximately 130K and 112K have sometimes been detected.[16,57,72]

Figure 4 shows the autoradiograph of an SDS-polyacrylamide gel with the [^{35}S]-methionine-labeled proteins from the soluble fraction (31,000 $\times$ *g* supernatant) of CPMV-infected protoplasts (lane M + B). The 60K and 58K polypeptides are not visible in this gel since these polypeptides occur mainly membrane-bound and are predominantly found in the 31,000 $\times$ *g* pellet fraction.[71,73] Moreover, in the soluble fraction they are masked by host proteins of approximately similar sizes. Figure 4 shows, furthermore, that no shut-off of host protein synthesis occurs after infection.

C. The Independence of the Bottom Component

Although both B and M components, or their RNAs, are necessary for virus multiplication,[17,18] inoculation of cowpea protoplasts with purified B components has been shown to result in replication of B-RNA and synthesis of all known B-RNA-encoded polypeptides (Reference 19 and Figure 4, lane B). Furthermore, infection with B components induces the formation of cytopathic structures in protoplasts,[74] indistinguishable from those found in cells infected with the complete virus.[68,75] On the other hand, M-RNA is not replicated in the absence of B-RNA, which suggests that the B-RNA-encoded polypeptides are necessary for virus RNA replication. B-RNA-encoded polypeptides are even produced in larger amounts in cells infected with B components alone, then in cells infected by both M and B components, as can be inferred from Figure 4 (cf. lanes designated B and M + B). This may be due to the fact that no capsid proteins are synthesized and consequently the progeny B-RNA molecules are not encapsidated but remain available for translation.

For two other two-component viruses, tobacco rattle virus (TRV, type member of the tobravirus group) and tomato black ring virus (TBRV, member of the nepovirus group), it has also been shown that the larger RNA (RNA-1) is capable of replicating in the absence of the smaller RNA (RNA-2).[76-80] In both cases the smaller RNAs specify the capsid proteins just as M-RNA of CPMV.[77,81-86] Unlike RNA-1 of TRV, B-RNA of CPMV and RNA-1 of

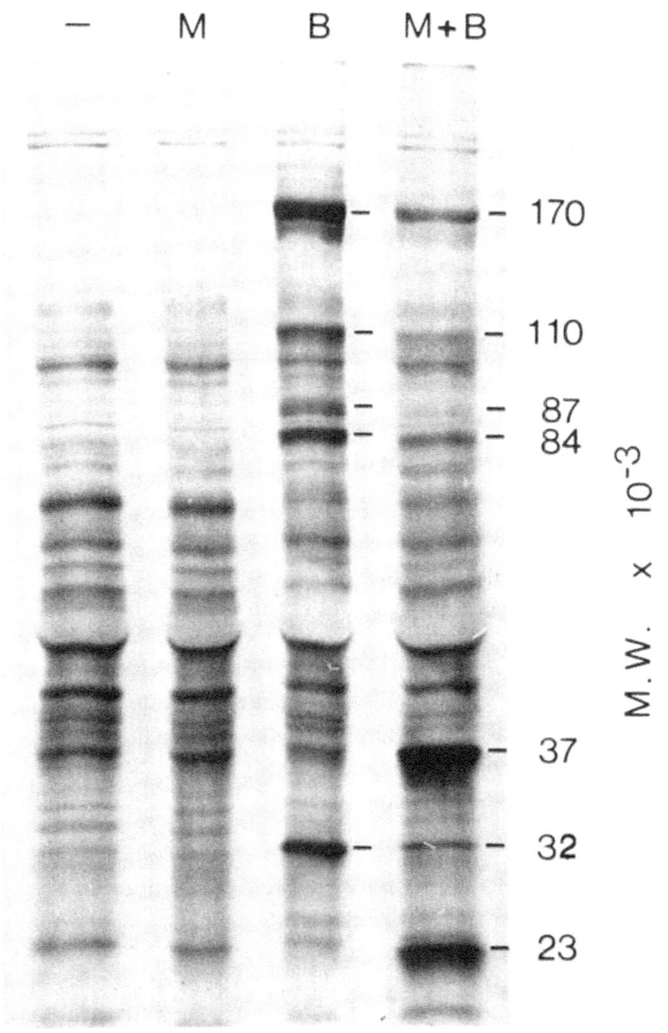

FIGURE 4. Analysis of the ^{35}S-methionine-labeled proteins from S30-fractions of uninfected cowpea protoplasts (−) and protoplasts inoculated with M, B, or M + B components. (Reprinted with permission from Goldbach, R., Rezelman, G., and Van Kammen, A., *Nature (London)*, 286, 297, 1980. Copyright (c) 1980 Macmillan Journals Limited.)

TBRV do not produce symptoms in inoculated plants. For CPMV it has been shown that this is caused by the inability of B-RNA to spread through the leaf in the absence of M-component.[74] It has not yet been resolved whether encapsidation is essential for spreading of CPMV RNAs to surrounding cells or whether M-RNA encodes another function involved in virus transport beside the two capsid proteins. The 58K and/or 48K polypeptides which arise by processing of the M-RNA primary translation products (see Figure 2) may be considered for such function. The independence of B-RNA is thus confined to the cell level as it is replicated and expressed only in cells it has entered upon inoculation. Isolation and analysis of B-RNA synthesized in the absence of M-RNA showed that it contains VPg. This finding provided evidence that VPg is encoded by B-RNA[87] and strengthened the suggestion that B-RNA-encoded proteins are involved in viral RNA replication.

D. The Translation Strategy of the Bipartite Genome of CPMV

The relationship between the virus-specific polypeptides produced in B component-inoculated protoplasts has been elucidated by both comparison of the proteolytic digestion patterns of the separate proteins and immunological analysis.[56,57,71,73,88] It was demonstrated that the 170K and 32K polypeptides found in CPMV-infected protoplasts are identical to the 170K and 32K polypeptides obtained upon in vitro translation of B-RNA and are therefore indeed encoded by B-RNA.[57] The peptide patterns of the various polypeptides as obtained after digestion with *Staphylococcus aureus* protease V8 indicated that the 200K primary translation product of B-RNA is proteolytically cleaved to give the smaller B-RNA-encoded polypeptides as depicted in the scheme of Figure 2. The 170K polypeptide generated by the primary cleavage is further cleaved either into 60K and 110K polypeptides or, alternatively, into 84K and 87K polypeptides. Additional cleavages produce 87K, 58K, and 28K polypeptides and VPg as the final products. The 28K polypeptide included in the model of Figure 2 has never been detected with certainty in vivo, but has been found upon in vitro processing of B-RNA-encoded translation products (see Figure 3, Section II.A). The 60K polypeptide has been shown to be a predominantly membrane-bound precursor to VPg,[73] which is cleaved into VPg and a 58K polypeptide (Reference 71 and Figure 2). So far, VPg has not been found in free form but only covalently linked to the viral RNAs. Indeed, if VPg acts as a primer for initiation of RNA replication (see, for discussion on this point, Section IV) it may directly be linked to nascent RNA-chains and not occur as a free protein.

The orientation of the polypeptides within the cleavage map has been determined by introducing, during in vitro translation of B-RNA in reticulocyte lysates, gradients of specific radioactivity into the 200K and 170K precursor polypeptides.[88] The relationship between the B-RNA-encoded polypeptides as established by comparison of their proteolytic peptide patterns has been fully confirmed by immunological analyses using antisera raised against the 170K, 110K, 87K, and 32K polypeptides, and against VPg.[71,73] Some results obtained with these antisera are shown in Figure 5. Under standard conditions for immunoprecipitation (i.e., in buffer containing detergents; Figure 5 lanes indicated +) anti-VPg serum specifically precipitates the VPg-containing 60K, 84K, and 170K polypeptides from an extract of B-component-inoculated protoplasts. The amount of the 200K primary translation product in protoplasts is mostly too low to allow its detection. Anti-32K serum precipitates in accordance with the expectations only the 32K polypeptide, and anti-170K serum reacts specifically with the 170K polypeptide and with the 110K and 87K polypeptide which are both derived from the carboxy-terminal part of the 170K polypeptide.[88] It is unclear why anti-170K serum does not precipitate the 84K and 60K polypeptides derived from the amino-terminal part of the 170K polypeptide. Perhaps these polypeptides are more hydrophobic than the 87K and 110K sequences and therefore less antigenic. Indeed the 60K polypeptide appears to be predominantly membrane-bound.[73] Figure 5 includes, furthermore, immunoprecipitations obtained from extracts from B-component-inoculated protoplasts in the absence of detergents (lanes indicated −). Under such conditions possible complexes of viral and host proteins remain stable. In particular, the immunoprecipitation with anti-32K serum demonstrates here that part of the viral polypeptides are contained in specific complexes. Apparently, 32K polypeptides do not occur as free proteins but seem always complexed to 58K, 60K, 84K, and/or 170K polypeptides.

The 130K and 112K polypeptides, sometimes found in small amounts, are not included in the translation map of Figure 2. The 130K polypeptide is probably a more-or-less specific, nonfunctional breakdown product of the 170K polypeptide.[89] On the other hand, the 112K polypeptide might result from a potential third cleavage pathway of the 170K polypeptide, by which the 58K polypeptide is directly released together with a 112K polypeptide containing the amino acid sequence of the 110K polypeptide and VPg.[72]

The translation model for M-RNA as shown in Figure 2 has been deduced from in vitro translation and processing studies[58] and remains to be verified in vivo. Until now, M-RNA-encoded polypeptides other than the two mature capsid proteins VP37 and VP23 have been

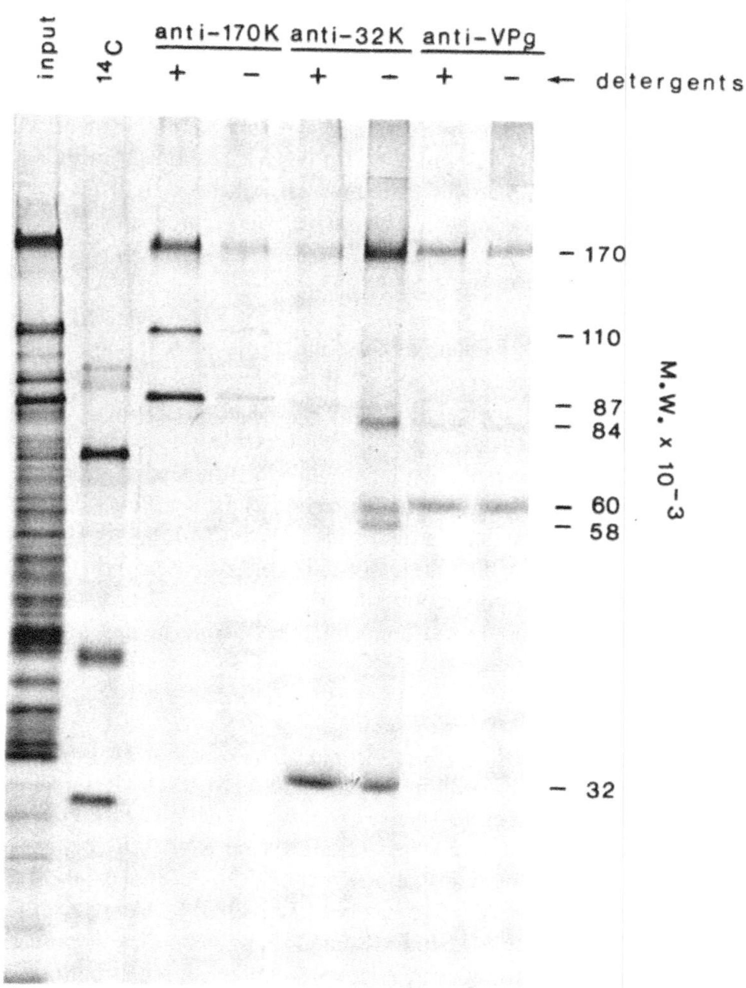

FIGURE 5. Analysis of B-RNA-encoded polypeptides with various antisera. Antiserum raised against the 170K polypeptide, 32K polypeptide, or VPg was added to extracts from B-component-inoculated protoplasts, either in the presence (+) or absence (−) of detergents, and the immunoprecipitates were analyzed by gel electrophoresis. The lane designated "input" contains the unfractionated extract used for the immunoprecipitations; the lane designated ^{14}C contains ^{14}C-methylated marker proteins: phosphorylase b, 100K and 92.5K; bovine serum albumin, 68K; ovalbumin, 46K; and carbonic anhydrase, 30K. (From Franssen, H., Moerman, M., Rezelman, G., and Goldbach, R., *J. Virol.*, 1984, 50, 183, 1984. With permission.)

detected neither in infected protoplasts nor in leaves. Using high titer antisera raised against the electrophoretically separated VP37 and VP23 proteins, no precursors to these proteins could be detected in lysates of CPMV-infected protoplasts,[58] suggesting that the primary product(s) from M-RNA are already processed before their synthesis is completed. Such a conclusion is in accordance with the observation that the B-RNA-encoded 32K protease responsible for processing (see Section II.E) is capable of cleaving the nascent in vitro products of M RNA.[58]

E. Functions of the Proteins Encoded by CPMV

With the proteins encoded by CPMV identified, a remaining task is to resolve the functions of the various proteins. Besides the capsid proteins VP37 and VP23, the genome of CPMV encodes a number of nonstructural polypeptides which may be involved in processes like induction of cytopathic membrane structures, RNA replication, proteolytic cleavage of primary translation products, encapsidation, and transport of RNA. Since B-RNA is capable of self-replication, it has been proposed that this RNA encodes one or more polypeptides essential for viral RNA replication. In Section IV the available evidence for a role of the B-encoded 110K polypeptide in this process will be discussed and, in the same section a possible role for VPg as primer in the initiation of replication will be considered. Here special attention will be given to other processes in which virus-specific polypeptides are involved.

1. Encapsidation

In Section II.A evidence was presented that both capsid proteins of CPMV are encoded by M-RNA (Figure 2). The location of the capsid protein genes on M-RNA probably holds good for all comoviruses. Complementation experiments with separated components of the Puerto Rico and Arkansas isolates of CPSMV have provided evidence that the capsid proteins are specified by M-RNA.[90] The reactivity of the in vitro translation products from SqMV M-RNA towards antisera raised against the separated capsid proteins clearly demonstrate that with SqMV the capsid proteins are also coded for by M-RNA.[91] The capsid proteins can assemble to T components which are capsids devoid of RNA. Whether for the encapsidation of comoviral RNA other polypeptides are needed besides the capsid proteins remains unknown. For example, the possibility that VPg plays a role in this process may not be excluded.

2. Transport of RNA

Although B-RNA alone is efficiently replicated and expressed in inoculated cells, it is unable to transfer to surrounding cells in the absence of M-RNA (References 19 and 74, and Section II.C). M-RNA therefore seems to encode functions essential for cell-to-cell transport of virus RNA. Obviously, encapsidation of viral RNA may be a prerequisite for its transport, but it is equally possible that an other M-RNA-encoded function is required. As with TMV, the viral 30K polypeptide has been suggested to have a role in virus transport,[92,93] so with CPMV the M-RNA-coded 58K and 48K polypeptides are possible candidates for such function. This remains, however, pure speculation as long as both polypeptides are not detectable in vivo, and inquiry into their possible functions is precluded.

3. Membrane Attachment

A major part of the viral multiplication process seems to take place at membranes. Soon after infection of cowpea cells, cytopathic structures are induced which contain large numbers of membranous vesicles.[74,75,94] Viral RNA synthesis takes place in close association with the vesicular membranes and recently, the viral RNA replication complex has been shown to be strongly membrane bound, being only solubilized by treatment with 1% Triton® X-100.[95,96] The crude membrane fraction from CPMV-infected leaves contains considerable amounts of the 170K, 112K, 110K, 87K, 84K, and 60K B-RNA-encoded polypeptides. The bulk of these viral proteins is released by washing the crude membrane fraction with Mg^{2+}-deficient buffer but some of these polypeptides remain firmly bound and are only solubilized by treating the washed membrane fraction with nonionic detergents.[96] In order to anchor the various viral proteins to membranes it seems not unreasonable to assume that one or more of the viral products bear the function of membrane attachment. From the viral polypeptides only the (overlapping) B-RNA-coded 60K and 58K polypeptides are distinct membrane proteins[57,71,73] and may therefore be candidates for such a function. The amino acid sequences of the 60K and 58K polypeptides are also contained in the 170K and 84K polypeptides and may provide for membrane association of these polypeptides, whereas other polypeptides are more peripherally bound.

4. Viral Protease(s)

The processing by proteolytic cleavages of the primary translation products obtained from B-RNA and M-RNA is largely, if not completely, accomplished by B-RNA-encoded functions and appears to involve at least two distinct activities.

The M-RNA-encoded (105K and 95K) primary translation products are cleaved at two different sites: a glutamine-methionine peptide bound is split to release 60K polypeptides. The two capsid proteins VP37 and VP23 are liberated from this cleavage product by subsequent cleavage of a glutamine-glycine bond.[51] The activity which cleaves the glutamine-methionine bond is clearly coded for by B-RNA since in vitro translation products from B-RNA, as well as extracts from protoplasts infected with B components, contain the capacity of releasing the 60K capsid protein precursor.[55,58] Recently, evidence has been obtained that the B-RNA-encoded 32K polypeptide is the protease responsible for cleaving the glutamine-methionine bond.[64] Antiserum raised against 32K polypeptide, but not antiserum against VPg or 170K polypeptide, inhibits the release of the 60K polypeptide from the M-RNA-encoded 105K and 95K translation products. When the soluble fraction from B-inoculated protoplasts containing this cleavage activity was analyzed by glycerol gradient centrifugation, particularly 32K polypeptides complexed with 170K polypeptide seemed active. The activity of the 32K polypeptide was significantly less if it occurred in complexes with the smaller B-RNA-encoded polypeptides.[64] These observations suggest that the proteolytic activity of the 32K polypeptide is rather unstable as it predominantly resides in newly synthesized molecules, still associated with 170K polypeptides just after cleavage of the 200K primary translation product.

The activity involved in producing the two capsid proteins from the 60K precursor has not yet been identified. A proteolytic activity capable of cleaving VP23 from the 105 and 95K translation product from B-RNA, has been found in CPMV-infected cowpea leaves.[58] It has not yet been possible to determine whether this activity is host or virus encoded.

In addition, B-RNA encodes at least one other protease, involved in the processing of the 200K polyprotein specified by B-RNA itself. As described in Section II.A, in rabbit reticulocyte lysates, faithful processing of the B-RNA-coded primary translation product can be obtained, generating 110K, 87K, 84K, and 60K polypeptides identical to the B-RNA-encoded polypeptides found in vivo. Such results suggest that the responsible protease is coded by B-RNA. Since glutamine-methionine or glutamine-glycine pairs do not occur at or around most of the expected cleavage positions in the B-RNA-encoded polyprotein (see Figure 7 in Section III for the complete sequence of this polyprotein), this protease is probably different from the 32K polypeptide. Recently, circumstantial evidence has been obtained that the 28K polypeptide located between the 60K and 87K polypeptide sequences within 170K (Figure 2) represents this protease.[61] This idea has been strengthened by the recent discovery[97] that part of this protein exhibits significant amino acid sequence homology to the protease (protein P3-7c) from poliovirus (see also Section V). Thus, after translation of both CPMV RNAs a cascade-like action of proteases seems to occur: first the 28K protease releases the 32K protease, which, in turn, becomes active to cleave both M-RNA-encoded polyproteins.

F. Proteins Synthesized by Other Comoviruses

CPSMV is the only other comovirus for which information on its expression in vivo is available. The proteins synthesized by the Dg strain of CPSMV have been studied by Beier et al.[11] by labeling infected cowpea protoplasts with ^{35}S-methionine. Four virus-specific polypeptides with sizes of 125K, 98K, 86K, and 65K could be detected in addition to 39K and 22K capsid proteins. At least the 125K, 86K, and 65K polypeptides are specified by the B-RNA, since they were also formed upon inoculation of protoplasts with B-components alone.[11] No polypeptides corresponding with the CPMV-specific 170K and 32K proteins

Table 3
POLYPEPTIDES TRANSLATED IN VITRO FROM M-RNAs OF VARIOUS COMOVIRUSES

Virus (strain)	Polypeptides (in kilodaltons)	Ref.
CPSMV (Ark)	95 + 91	99
CPSMV (Dg)	93 + 100	11,99
RCMV	96 + 102	99
CPMV	95 + 105	See Section II.A
BPMV	97 + 107	100
SqMV	100 + 107	91,99

Note: The in vitro translation of different M-RNAs was done in rabbit reticulocyte lysates[11,91,99,100] and wheat germ extract.[99]

were found, suggesting that the translation strategy of CPSMV is distinct from that followed by CPMV. At any rate, the CPSMV polypeptides have little homology to the CPMV-specified proteins as antisera raised against the 170K, 110K, 32K polypeptides or VPg from CPMV do not precipitate any of the CPSMV polypeptides.[98] Only antiserum raised against the 87K polypeptide of CPMV showed a slight reaction with the CPSMV-specific 86K polypeptide.[98] However, in vitro translation studies seem to point out that CPMV and CPSMV follow in principle the same translation strategy: M-RNA of CPSMV — just as M-RNA of CPMV — is translated into two overlapping polypeptides[11,99] and B-RNA of CPSMV, comparable to B-RNA of CPMV, is translated into a polypeptide of approximately 200K which is cleaved in 170K and 35K polypeptides.[98] While not detected in vivo, these 170K and 35K polypeptides produced in vitro seem equivalent to the 170K and 32K polypeptides of CPMV. The observed discrepancies between the in vitro and in vivo translation studies on CPSMV make further studies on the expression of this virus desirable.

For three other comoviruses, BPMV, RCMV, and SqMV, the expression of the RNAs has only been studied by in vitro translation. The available evidence indicates that with all three viruses large polyproteins are produced as primary translation products suggesting that the same translation strategy is followed as used by CPMV. B-RNA of BPMV produces in rabbit reticulocytes a polypeptide of 181K[100] and B-RNA of SqMV a polypeptide of 190K.[91] There are no reports on the in vitro translation of RCMV B-RNA.

M-RNAs of BPMV, SqMV, and RCMV direct upon in vitro translation, just as with CPMV and CPSMV, the synthesis of two polypeptides with sizes of approximately 100K[91,99,100] (see Table 3). Translation into two large-sized proteins is apparently a common feature of comoviral M-RNAs, which is the more remarkable as the sequence homology between CPMV-M RNA and the M-RNAs of the four other viruses is less than 10%.[98] The absence of extensive sequence homology among M-RNAs of different comoviruses is further indicated by the lack of substantial cross-reactivity of antisera against individual viruses of the group.[13] Likewise, the proteases involved in the cleavage of these polypeptides are virus specific and not interchangeable. For example, the 32K protease of CPMV is not capable of cleaving the polypeptides from M-RNA of any of the other viruses.[99] Table 3 summarizes the sizes determined for the double polypeptides specified by the M-RNAs of comoviruses tested so far.

III. THE PRIMARY STRUCTURE OF THE CPMV GENOME

CPMV, together with TMV[101] and AMV,[102-105] is one of the first plant RNA viruses of which the complete nucleotide sequence of the genome has been elucidated.[51,52] Such information allows more detailed exploration of the organization of the viral genome on the molecular level. As with other RNA viruses, the 3′- and 5′-ends of B- and M-RNA were the first parts of the CPMV genome of which the sequence was elucidated.[106-109] Subsequently, by molecular cloning of double-stranded cDNA of viral RNAs sequencing of the entire genome became possible. The following sections deal with the primary structure of both CPMV RNAs and summarize the information which can be derived from them.

A. The Primary Structure of M-RNA

1. General Description

The nucleotide sequence of M-RNA was obtained by sequencing different cDNA inserts subcloned in M13 phage derivatives[110] using the dideoxy chain-termination method.[111] Exclusive of the poly(A) tail, M-RNA has a sequence of 3481 nucleotides, containing a unique open reading frame starting at nucleotide 161 and ending with a UAA stopcodon at nucleotide 3298 (Figure 6). Including the poly(A) tail, with an average length of 167 residues,[26] a molecular weight of 1.22×10^6 is calculated for M-RNA (sodium salt). The poly(A) track is not preceded by an AAUAAA polyadenylation signal.[106-109]

The open reading frame within the sequence corresponds with a primary translation product of 116,085 daltons,[51] which is larger than the size estimated for the largest of the two overlapping in vitro translation products of M-RNA (105,000 daltons, see Section II.A.). However, as mentioned earlier, sizes of proteins determined by polyacrylamide gel electrophoresis (PAGE) strongly depend on the choice of the gel system and the molecular weight markers. Similar discrepancies between calculated and experimental values for sizes of proteins have been found with other large viral proteins. For example, for the two largest translation products of TMV molecular weights of 183,253 and 125,941 were calculated from the primary structure,[101] whereas previously molecular weights of 165,000 and 130,000 had been reported based on electrophoretic mobility.[112,113] Similarly a molecular weight of 246,530 for the polyprotein of poliovirus was deduced from the primary structure[114] vs. 220,000 as derived from electrophoretic mobility.[115] The sum of the two cleavage products of the 105K polypeptide (60K + 58K = 118K; see Figure 2) fits better with the calculated size, which is in agreement with the general experience that molecular weight determinations by gel electrophoresis are more precise for smaller polypeptides.

As suggested by Van Wezenbeek et al.,[51] the internal AUG-codons at positions 512 or 524 may act as startcodons for the synthesis of the smaller (95K) translation product from M-RNA. According to the ribosome scanning model of Kozak,[116] an AUG startcodon should be preceded by a purine at position -3 and/or a guanine at position $+4$. The first condition is fulfilled with the three AUG codons at positions 161, 512, and 524, respectively, but only the AUG-codon at position 512 meets the second requirement (Figure 6). Therefore, the synthesis of the 95K polypeptide will most likely start at this latter codon.

2. Mapping of the Coding Region of VP37 and VP23

As described in Section II.A, the results of in vitro translation and processing experiments[58] indicated that both capsid proteins VP37 and VP23 are located in the carboxy-terminal half of the primary translation products of M-RNA, with VP37 preceding VP23. This location is fully confirmed by the nucleotide sequence data. The sequence of the first 11 amino acids from the N-terminal end of VP23, determined by protein sequencing, indicated that the coding region of this protein starts at position 2660 and proceeds to the end of the open reading frame (see Figure 6). The amino acid composition calculated from the nucleotide sequence of the coding region resembles the amino acid composition of VP23 previously determined.[21]

The N-terminal amino acid of VP37 is an acetylated methionine residue.[117] The coding

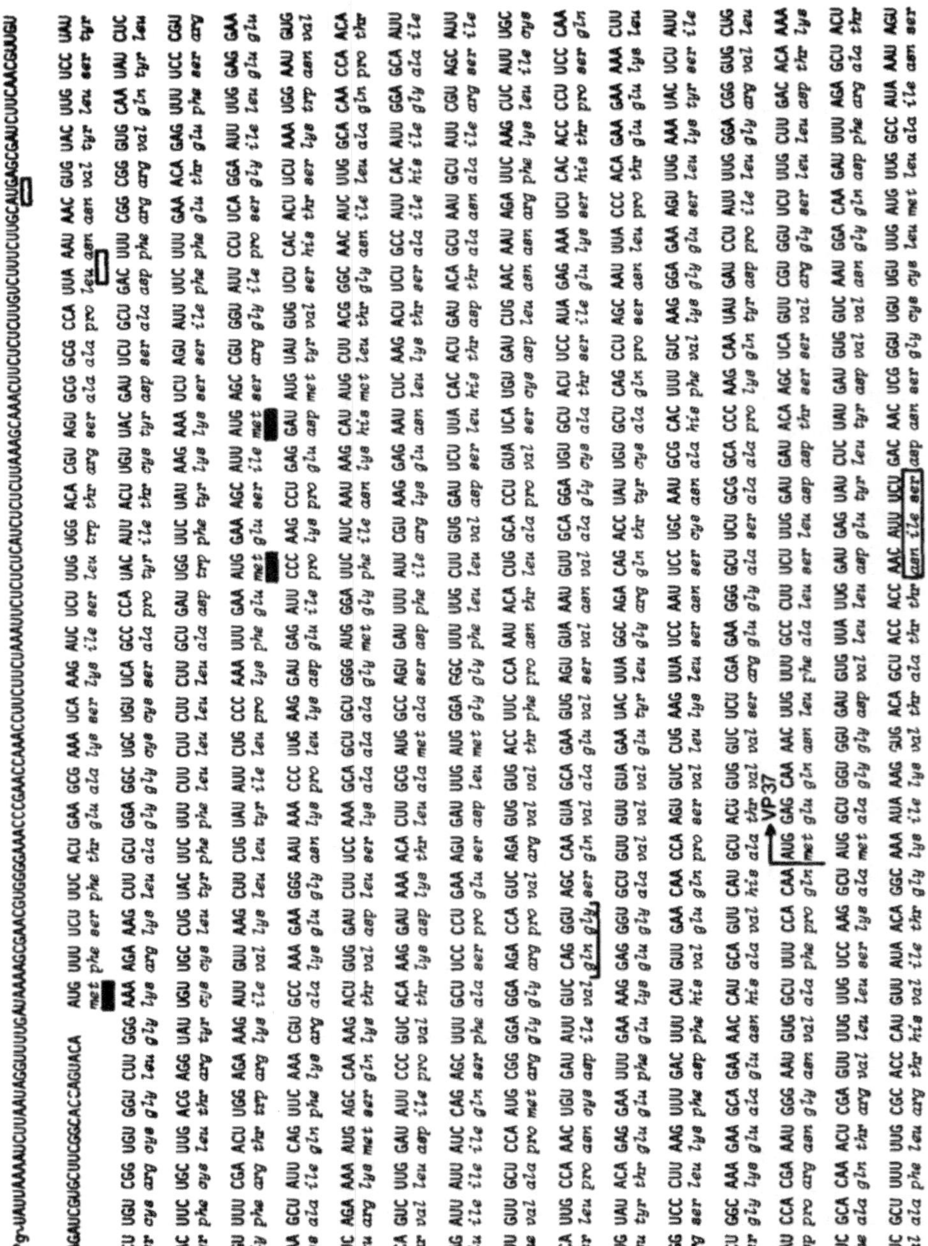

FIGURE 6

```
1820  GGU GUG AGG GGU AAG UAU AGU ACU GAU GUU UAU ACU AUC UGC UCU CAA GAC UCC AUG ACG UGG AAC CCA GGG UGC AAA AAG AAC UUC UCG UUC ACA UUU AAU CCA
      gly val arg gly lys tyr ser thr asp val tyr thr ile cys ser gln asp ser met thr trp asn pro gly cys lys lys asn phe ser phe thr phe asn pro

1925  AAC CCU UGU GGG GAU UCU UGG UCU GCU GAG AUG AUA AGU CGA AGC AGA GLU AGG AUG ACA GUU AUU UGU GUU UCG GGA UGG ACC UUA UCU CCU ACC ACA GAU GUG
      asn pro cys gly asp ser trp ser ala glu met ile ser arg ser arg val arg met thr val ile cys val ser gly trp thr leu ser pro thr thr asp val

2030  AUU GCC AAG CUA GAC UGG UCA AUU GUC AAU GAG AAA UGU GAG CCC ACC AUU UAC CAC UUG GCU GAU UGU CAG AAU UGG UUA CCC CUU AAU CGU UGG AUG GGA AAA
      ile ala lys leu asp trp ser ile val asn glu lys cys glu pro thr ile tyr his leu ala asp cys gln asn trp leu pro leu asn arg trp met gly lys

2135  UUG ACU UUU CCC CAG GGU GUG ACA AGU GAG GUU CGA AGG AUG CCU CUU UCU AUA GGA GGC GGU GCU GGU GCG ACU CAA GCU UUC UUG GCC AAU AUG CCC AAU UCA
      leu thr phe pro gln gly val thr ser glu val arg arg met pro leu ser ile gly gly gly ala gly ala thr gln ala phe leu ala asn met pro asn ser

2240  UGG AUA UCA AUG UGG AGA UAU UUU AGA GGU GAA CUU CAC UUU GAA GUU ACU AAA AUG AGC UCU CCA UAU AUU AAA GCC ACU GUU ACA UUU CUC AUA GCU UUU GGU
      trp ile ser met trp arg tyr phe arg gly glu leu his phe glu val thr lys met ser ser pro tyr ile lys ala thr val thr phe leu ile ala phe gly

2345  AAU CUU AGU GAU GCC UUU GGU UUU UAU GAG AGU UUU CCU CAU AGA AUU GUU CAA UUU GCU GAG GUU GAG GAA AAA UGU ACU UUG GUU UUC UCC CAA CAA GAG UUU
      asn leu ser asp ala phe gly phe tyr glu ser phe pro his arg ile val gln phe ala glu val glu glu lys cys thr leu val phe ser gln gln glu phe

2450  GUC ACU GCU UGG UCA ACA CAA GUA AAC CCC AGA ACC ACA CUU GAA GCA GAU GGU UGU CCC UAC CUA UAU GCA AUU AUU CAU GAU AGU ACA ACA GGU ACA AUC UCC
      val thr ala trp ser thr gln val asn pro arg thr thr leu glu ala asp gly cys pro tyr leu tyr ala ile ile his asp ser thr thr gly thr ile ser
                                                                                                                                      VP37←
2555  GGA GAU UUU AAU CUU GGG GUC AAG CUU GUU GGC AUU AAG GAU UUU UGU GGU AUA GGU UCU AAU CCG GGU AUU GAU GGU UCC CGC UUG CUU GGA GCU AUA GCA CAA
      gly asp phe asn leu gly val lys leu val gly ile lys asp phe cys gly ile gly ser asn pro gly ile asp gly ser arg leu leu gly ala ile ala gln
      →VP23
2660  GGA CCU GUU UGU GCU GAA GCC UCA GAU GUG UAU AGC CCA UGU AUG AUA GCU AGC ACU CCU CCU GCU CCA UUU UCA GAC GUU ACA GCA GUA ACU UUU GAC UUA AUC
      gly pro val cys ala glu ala ser asp val tyr ser pro cys met ile ala ser thr pro pro ala pro phe ser asp val thr ala val thr phe asp leu ile

2765  AAC GGC AAA AUA ACU CCU GUU GGU GAU GAC AAU UGG AAU ACG CAC AUU UAU AAU CCU CCA AUU AUG AAU GUC UUG CGU ACU GCU GCU UGG AAA UCU GGA ACU AUU
      asn gly lys ile thr pro val gly asp asp asn trp asn thr his ile tyr asn pro pro ile met asn val leu arg thr ala ala trp lys ser gly thr ile

2870  CAU GUU CAA CUU AAU GUU AGG GGU GCU GGU GUC AAA AGA GCA GAU UGG GAU GGU CAA GUC UUU GUU UAC CUG CGC CAG UCC AUG AAC CCU GAA AGU UAU GAU GCG
      his val gln leu asn val arg gly ala gly val lys arg ala asp trp asp gly gln val phe val tyr leu arg gln ser met asn pro glu ser tyr asp ala

2975  CGG ACA UUU GUG AUC UCA CAA CCU GGU UCU GCC AUG UUG AAC UUC UCU UUU GAU AUC AUA GGG CCG AAU AGC GGA UUU GAA UUU GCC GAA AGC CCA UGG GCC AAU
      arg thr phe val ile ser gln pro gly ser ala met leu asn phe ser phe asp ile ile gly pro asn ser gly phe glu phe ala glu ser pro trp ala asn

3080  CAG ACC ACC UGG UAU CUU GAA UGU GUU GCU ACC AAU CCC AGA CAA AUA CAG CAA UUU GAG GUC AAC AUG CGC UUC GAU CCU AAU UUC AGG GUU GCC GGC AAU AUC
      gln thr thr trp tyr leu glu cys val ala thr asn pro arg gln ile gln gln phe glu val asn met arg phe asp pro asn phe arg val ala gly asn ile

3185  CUG AUG CCC CCA UUU CCA CUG UCA ACG GAA ACU CCA CCG UUA UUA AAG UUU AGG UUU CGG GAU AUU GAA CGC UCC AAG CGU AGU GUU AUG GUU GGA CAC ACU GCU
      leu met pro pro phe pro leu ser thr glu thr pro pro leu leu lys phe arg phe arg asp ile glu arg ser lys arg ser val met val gly his thr ala

3290  ACU GCU GCU UAA  CUCUGGUUUCAUUAAAUUUUCUUUAGUUUGAAUUUACUGUUAUUUGGUGUGCAUUUCUAUGUUUGGUGAGCGGUUUUCUGUGCUCAGAGUGUGUUUAUUUUAUGUAAUUUAAUUUCUUUGUG
      thr ala ala

3425  AGCUCCUGUUUAGCAGGUCGUCCCUUCAGCAAGGACACAAAAAGAUUUUAAUUUUAUU-poly(A)
```

FIGURE 6. Nucleotide sequence of CPMV M-RNA and the amino acid sequence (three-letter code) deduced from the long open reading frame. Open boxes indicate the start and stop-codons of the first possible open reading frame. Black boxes indicate possible start codons within the long open reading frame. Underlined amino acids of VP23 have been sequenced by Edman degradation.[118] The boxed tripeptide sequences are possible glycosylation sites. (Reprinted with permission from Van Wezenbeek, P., Verver, J., Harmsen, J., Vos, P., and van Kammen, A., *EMBO J.*, 2, 941, 1983. Includes some additional specifications.)

region of this polypeptide most probably starts with the AUG-codon at position 1538 and continues to the CAA codon for glutamine at position 2657, adjoining the region coding for VP23. The amino acid composition deduced from this region agrees with the experimentally determined composition of VP37.[21] Moreover, the amino acid composition of the chymotryptic peptide derived from the N-terminal end of VP37 corresponds with the composition deduced from the nucleotide sequence of the first six triplets of the proposed (see Figure 6) coding region of VP37.[118]

3. Proteolytic Cleavage Sites within the Primary Translation Products

If the VP37 coding region indeed starts at position 1538 on the M-RNA sequence, it follows that the proteolytic cleavage by the B-RNA-encoded 32K polypeptide to release the 60K capsid protein precursor from the primary M-RNA-translation products, is effected at a glutaminyl-methionine pair (see Figure 6). On the other hand, the cleavage of the 60K precursor into VP37 and VP23 (see Section II.A), is achieved at a glutaminyl-glycine pair (Figure 6). Two proteolytic cleavages at different amino acid pairs for generating VP37 and VP23 is in agreement with the results from in vitro translation studies, which suggested that two different virus-induced proteolytic activities are involved.[58]

4. Glycosylation Sites in the Primary Translation Products

Partridge et al.[23] found that the capsid proteins of CPMV contain some carbohydrate residues, and this gives a good reason to screen the amino acid sequence of the M-RNA primary translation products, deduced from the nucleotide sequence, for the presence of possible glycosylation sites. Two types of glycosylation are common: first, glycosylation at hydroxyl groups of serine and threonine residues, and second, glycosylation through the free amino group of asparagine residues occurring in the sequences asn-x-thr or asn-x-ser, where x may be any amino acid. With regard to the first type of glycosylation, no predictions can be derived from the amino acid sequences of the primary translation products. As for the second type, it is interesting to note that asn-x-ser and asn-x-thr sequences are only found within the VP37 (three times) and VP23 (twice) (Figure 6, boxed amino acid sequences). These sequences may be used for glycosylation. Similar potential glycosylation sites are lacking in the preceding part of the primary translation product, from which the overlapping 58K and 48K polypeptides are derived (see Figure 2).

B. The Primary Structure of B-RNA

1. General Description

The nucleotide sequence of B-RNA as determined by Lomonossoff and Shanks[52] is shown in Figure 7. The sequence exclusive of the poly(A) tail runs to 5889 nucleotides. The poly(A) tail of B-RNA, just as with M-RNA, is not preceded by a AAUAAA polyadenylation signal. From the nucleotide sequence, a molecular weight of 2.04×10^6 can be calculated for B-RNA (sodium salt) including a poly(A) tail with an average length of 87 residues.[26] Only one open reading frame is found in the sequence, starting with the AUG codon at position 207, and extending to a UAG stopcodon at position 5804. The molecular weight of the predicted protein encoded by the reading frame is 207,760 which is in good agreement with the experimentally determined size of this polyprotein (200,000). The AUG codon at position 207 is surrounded by an A residue at position -3 and a G residue at position $+4$ and therefore meets the requirements for efficient initiation.[116]

2. Mapping of the Coding Region of VPg and Possible Proteolytic Cleavage Sites

Zabel et al.[119] have recently determined a partial amino acid sequence of the genome-linked protein VPg. Straightforward Edman degradation of unlabeled VPg revealed serine and arginine residues on positions 1 and 2, respectively. Upon radiolabeling with chloramine

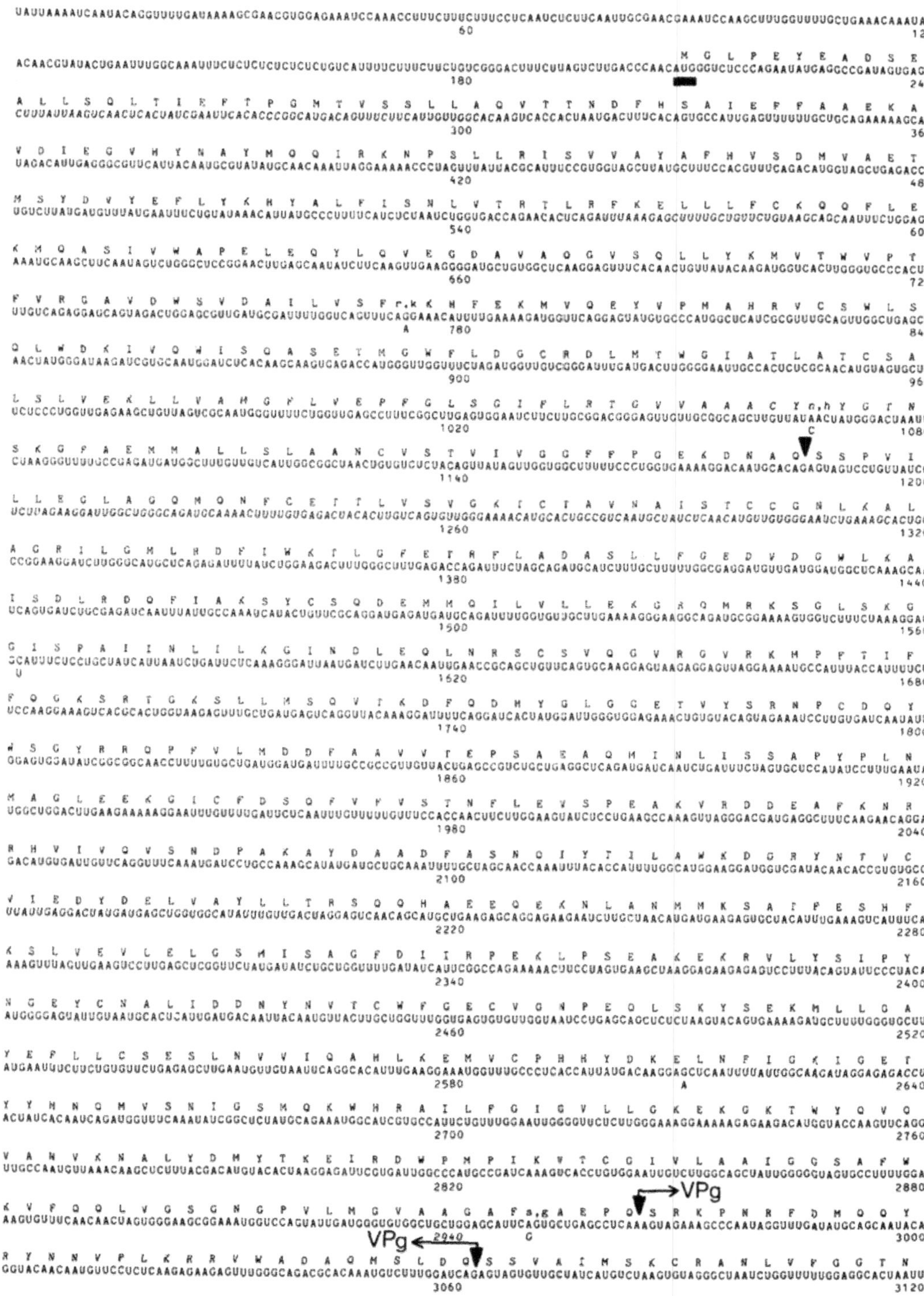

FIGURE 7. Nucleotide sequence of CPMV B-RNA and the amino acid sequence (one-letter code) deduced from the long open-reading frame. Start and stop codons of the open-reading frame are indicated with black boxes. The indicated position of VPg within the amino acid sequence of the open reading frame, is based on the protein sequencing data from purified VPg as obtained by Zabel et al.[119] Figure continued on next page. (Modified and reprinted with permission from Lomonossoff, G. P. and Shanks, M. S., *EMBO J.*, 2, 2153, 1983.)

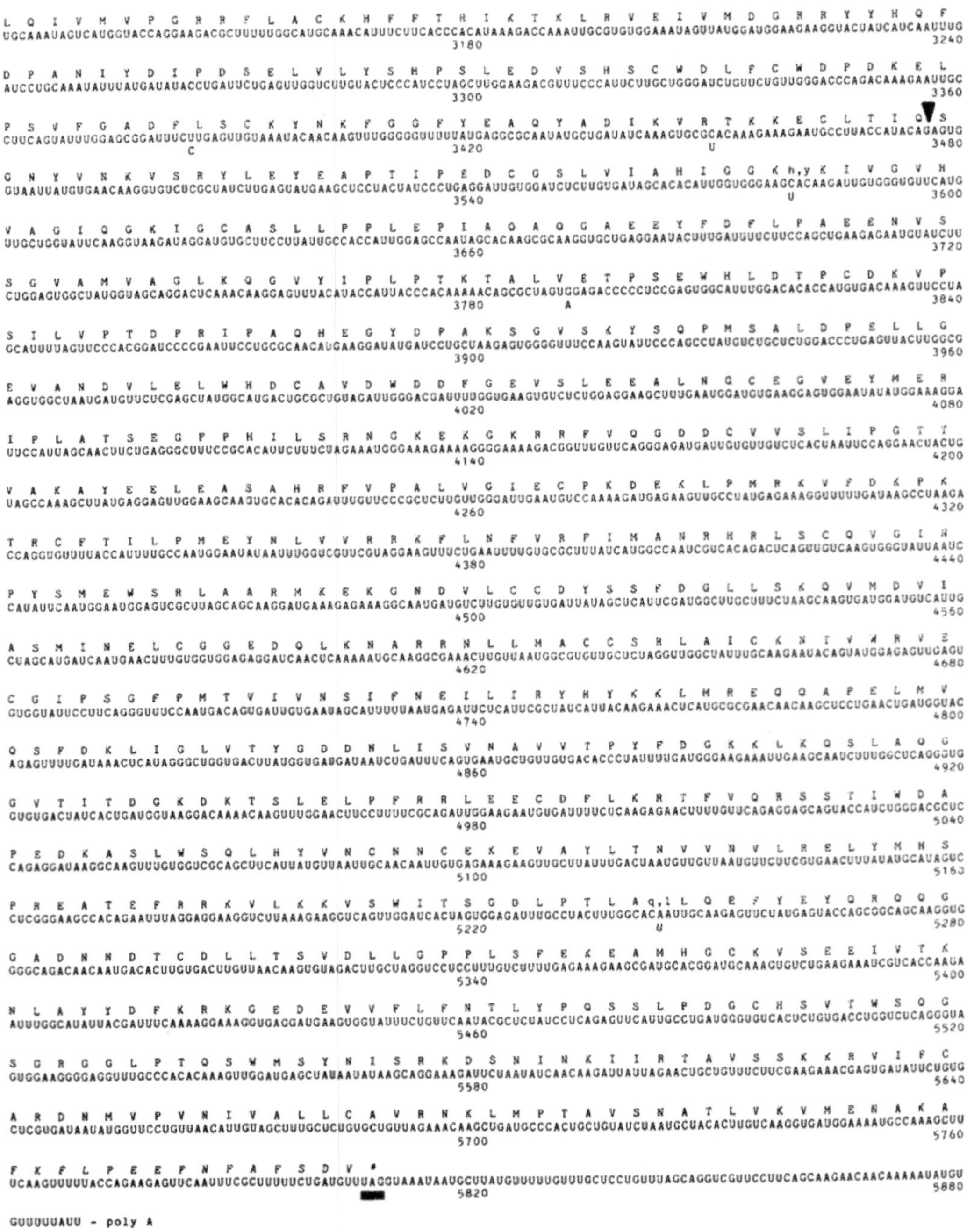

FIGURE 7 (continued).

T and Bolton-Hunter reagent the presence of two tyrosines on positions 12 and 14 and of a lysine residue on position 3 could be determined. The data obtained were sufficient to map the sequence coding for VPg in the region from nucleotide 2964 downstream (Figure 7). This location of the VPg gene is in good agreement with the immunological mapping studies identifying VPg as the C-terminal end of its 60K precursor (Figure 2 and References 71, 73). The position of VPg within the sequence of the 200K polyprotein indicates that the site used for cleaving the N-terminal of VPg from the 60K precursor is a glutamine-serine sequence (cf. Figure 7). Since evidence has been obtained suggesting that most, if not all, cleavage sites utilized within the B polyprotein are identical,[61] Zabel et al.[119] inferred that a similar cleavage site, found at nucleotide positions 3060 to 3065 (see Figure 7), might be used to release VPg at its C-terminal. The sequence of VPg would then consist of 33 amino acids, in good agreement with its apparent size (4 to 5K) in polyacrylamide gels. As glutamine-serine pairs are found at all positions within the primary structure of the B polyprotein where cleavages are expected, it is tempting to assume that the protease involved in the processing of the B-RNA-encoded polyprotein specifically cleaves the bonds between glutamine-serine pairs. As already discussed in Section II.E, the 28K polypeptide, situated in the central part of the 200K polyprotein, is the most plausible candidate for this proteolytic activity. If the 28K polypeptide were capable of cleaving itself out of the 170K chain, first left and then right, or the other way around, then the 110K, 87K, 84K, and 60K polypeptides are all four direct cleavage products (Figure 2).

3. Homologous Sequences in M- and B-RNA

In addition to coding regions, B- and M-RNA contain noncoding sequences which may be involved in RNA replication, ribosome binding or various interactions with different viral and host proteins. Among the proteins anticipated to interact with the viral RNAs will be the viral replicase (see Section IV), the capsid proteins, and ribosomal proteins. Since such interactions will occur with both viral RNAs, one might expect some analogy in the primary and/or secondary structure of the regions concerned. Comparison of the noncoding regions at the 5′-terminal ends of B- and M-RNA shows considerable homology in the sequence of the first 44 nucleotides of both RNAs with differences in only 6 nucleotides, respectively, at positions 11 to 15 and 40.[107,109] In addition homologous sequences are found beyond nucleotide 44, including an alternating C-U stretch around positions 75 in M-RNA, and around position 150 in B-RNA.[109] The 3′-terminal sequences of the two RNAs are also rather similar as the first 65 bases of each RNA are approximately 80% homologous[106] (Figures 6 and 7). Furthermore, some regions near the 5′ and 3′ termini of each RNA show to a certain extent complementarity which might enable the RNAs to circularize. For example, the sequences of nucleotides 19 to 29 of both RNAs are complementary to the first 4 A-residues of the poly(A) tail and the 7 adjacent nucleotides. Such common structural features in the noncoding regions of both RNAs may play a role in some stage of virus multiplication. However, it is not possible to correlate these regions to a particular process like replication or translation of viral RNA.

IV. REPLICATION OF CPMV RNA

A. Involvement of the Host Cell

Infection of cells with CPMV is accompanied by the appearance of characteristic cytopathic structures which consist of arrays of vesicles forming a kind of reticulum, the spaces between the vesicles being filled with electron-dense material.[75,94] The replication of viral RNA is associated with the membranes of the vesicles in these strucutres as shown by fractionation of virus-infected cells and autoradiography.[75,94] Similar cytopathic structures are found in cowpea mesophyll protoplasts upon infection with CPMV[68] and, moreover, in protoplasts inoculated with alone B components.[74] Since, in cowpea protoplasts infected with B components B-RNA is readily expressed and replicated[19] (see Section II.C) the development of cytopathic structures in B-inoculated protoplasts confirms that such structures are linked to viral RNA replication rather than to the accumulation of virus particles.

During the earliest stage of infection, CPMV multiplication is somehow dependent on host-DNA-specified RNA synthesis as can be concluded from experiments with actinomycin

D. Lockhart and Semancik[120,121] reported that the replication of CPMV and BPMV in cowpea hypocotyl tissue was inhibited by actinomycin D provided that the antibiotic was administered shortly after inoculation. Similar results were obtained by Rottier et al.[22] for cowpea mesophyll protoplasts infected with CPMV. Actinomycin D at a concentration of 10 μg/mℓ prevented the production of virus particles if the drug was added at the time of inoculation or immediately thereafter. The inhibitory effect of actinomycin D rapidly decreased if added after inoculation. At 8 hr after inoculation, virus replication had become completely insensitive to the drug. Similar observations have also been made on the replication of several other plant viruses such as potato virus X in tobacco protoplasts,[122] alfalfa mosaic virus in cowpea protoplasts,[123] tobacco mosaic virus in tobacco leaves,[124,125] turnip yellow mosaic virus in cabbage protoplasts,[125a] and turnip rossette virus in turnip protoplasts.[126,127]

DNA-dependent RNA synthesis as a prerequisite for CPMV RNA replication suggests the involvement of some essential virus-induced but host-specified component. One possibility might be that a host-encoded factor is an essential part of a functional virus replicating complex.[125b] Another possibility is that initiation of the formation of virus-specific cytopathic structures as sites for virus RNA replication will require host-directed protein synthesis. Inhibition of CPMV multiplication by actinomycin D is probably not due to interference with the process of viral RNA replication per se, since addition of the antibiotic during the period of active RNA replication does not affect virus synthesis.

B. Identification of a CPMV RNA Replication Complex

Zabel et al.[15] demonstrated that the crude membrane fraction of CPMV-infected cowpea leaves contains RNA-dependent RNA polymerase activity capable of synthesizing double-stranded and possibly some single-stranded viral RNAs in vitro on endogenous template RNA. Since such results indicated that the crude membrane fraction of infected leaves holds the viral replication complex, this fraction has been used as starting material for purification of the presumptive viral RNA replicase.[15,128]

Characterization of RNA products synthesized in vitro by the RNA-dependent RNA polymerase in the crude membrane fraction of infected leaves revealed the occurrence of two functionally different enzyme-template complexes.[95] One of the activities constitutes the CPMV RNA replication complex, defined by its capacity of elongating in vitro nascent chains initiated in vivo, to full-length plus-type viral RNA mainly found in double-stranded form (RF_B and RF_M). The other activity which is host specified transcribes endogenous plant RNAs and viral RNAs into small minus-type RNA molecules. This activity can also be detected in the membrane fraction of uninfected leaves but is greatly enhanced by infection with CPMV. The essential first step in separating and purifying the two activities is extraction of the membrane fraction with a Mg^{2+} deficient buffer. The polymerase synthesizing viral double-stranded RNAs remains membrane bound, even after extensive washings. The other RNA polymerase activity is solubilized by washing in Mg^{2+}-deficient buffer and retains full activity. Both activities have been purified to determine the identity of the relevant polymerases and their possible role in viral RNA replication.[95]

C. Identification of a Host-Encoded RNA-Dependent RNA Polymerase in Cowpea Leaves

The RNA-dependent polymerase activity released by washing the crude membrane fraction can be stripped of its endogenous template by DEAE Sepharose 2B chromatography and further purified by affinity chromatography on poly(U)-Sepharose 4B and glycerol gradient centrifugation.[95] When the last step is replaced by phosphocellulose chromatography, an improved recovery of enzyme activity and a higher degree of purification is obtained.[129] The purified enzyme consists of a single polypeptide of 130,000 daltons. Using the same purification procedure, a similar RNA-dependent RNA polymerase has been purified from the membrane fraction of uninfected leaves. Antibodies raised against such purified enzyme preparations have subsequently been used to develop an antibody-linked polymerase assay (ALPA) which allows the detection of RNA-dependent RNA polymerase activity on protein blots from SDS polyacrylamide gels.[130] By means of ALPA the 130K polypeptide could be definitely identified as a host-encoded RNA-dependent RNA polymerase. By comparing the amounts of proteins from the crude membrane fraction of uninfected leaves and CPMV-infected leaves the increase of 130K polypeptide 3 to 4 days after inoculation has been estimated to be at least 20-fold. Being provided with a sensitive detection method for the host-encoded enzyme from cowpea leaves, a possible role of this enzyme in viral RNA replication has been examined (see Section IV.E).

D. Purification and Analysis of the CPMV RNA Replication Complex

Whereas washing of the crude membrane fraction of CPMV-infected cowpea leaves with Mg^{2+}-deficient buffer releases the host-encoded RNA-dependent RNA polymerase from the membranes, the viral RNA replication complex remains firmly bound. Analysis of the RNA synthesized by such washed membranes in vitro shows that the replication complex has retained its main characteristics and is still capable of elongating nascent viral RNA strands already initiated in vivo. To solubilize the CPMV RNA replication complex the membranes should be treated with Triton® X-100. Solubilization does not lead to loss of activity, while virtually no activity remains associated with the detergent-insoluble membrane residue. The solubilized complex has been further purified by subsequent Sepharose 2B exclusion chromatography, and finally by glycerol gradient centrifugation.[96] While these purification steps remove the vast majority of the contaminating proteins, the capacity of the replication complex to elongate in vitro nascent viral RNA chains, is preserved. Using Western blot analysis the various fractions obtained upon purification of the replication complex have been examined for the occurrence of specific viral-encoded proteins, the host-encoded 130K RNA-dependent RNA polymerase and other host proteins. Such analyses showed that the 130K host enzyme is completely separated from the viral replication complex after Sepharose 2B chromatography, which indicates that this polymerase is not involved in viral RNA replication.[96] On the contrary, analysis of preparations further purified by glycerol gradient centrifugation revealed that they contain three major polypeptides of 110K, 68K, and 57K. Figure 8 gives an example of such analysis of viral RNA-replication complexes after purification by glycerol gradient centrifugation. Using antisera against various viral polypeptides it has been shown (Figure 8 and Reference 96) that of these polypeptides, the 110K polypeptide represents the B-RNA-encoded protein, derived from the C-terminal part of the 200K polyprotein (see Figure 2). The 68K and 57K polypeptides are probably host-encoded polypeptides. Together with the evidence presented in Section V, these analyses indicate that the viral-encoded 110K polypeptide represents the core polymerase in the viral replication complex.

E. The Puzzling Host-Encoded RNA-Dependent RNA Polymerase

The available evidence points out that the host-encoded RNA-dependent RNA polymerase in cowpea leaves has no function in CPMV-RNA replication. This enzyme is absent in purified RNA replication complexes, which are capable of synthesizing full-length viral RNAs.[129]

As far as the characteristics of the host-encoded RNA-dependent RNA polymerase have been studied, the enzyme occurs membrane bound and, upon isolation, the membrane-bound enzyme transcribes in vitro plant (and after infection also viral) RNAs into small (4 to 5S) RNA chains of negative polarity.[95] Partially purified enzyme stripped of endogenous template does not exhibit template specificity and transcribes for example exogenous CPMV RNA into small pieces of complementary RNA, which appear to have arisen from initiation at

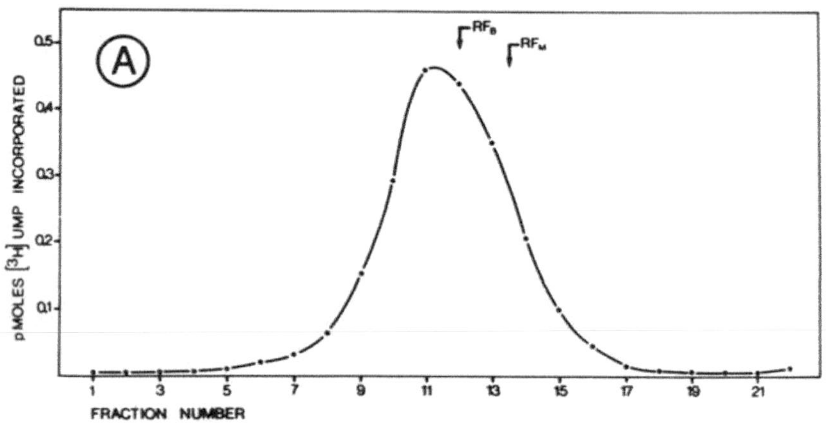

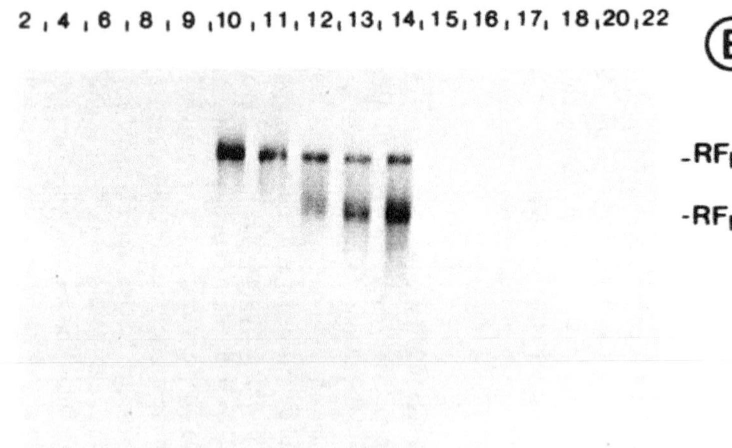

FIGURE 11. Nucleotide sequences (CaMV isolate Strasbourg) upstream of the 5′-ends (cap sites denoted + 1) of 35S RNA (nt7435) and mP62 (nt5764) with some similarities to eukaryotic promotors including a −30 TATA box and more remote upstream regions related to enhancer core sequences. See text for further discussion.

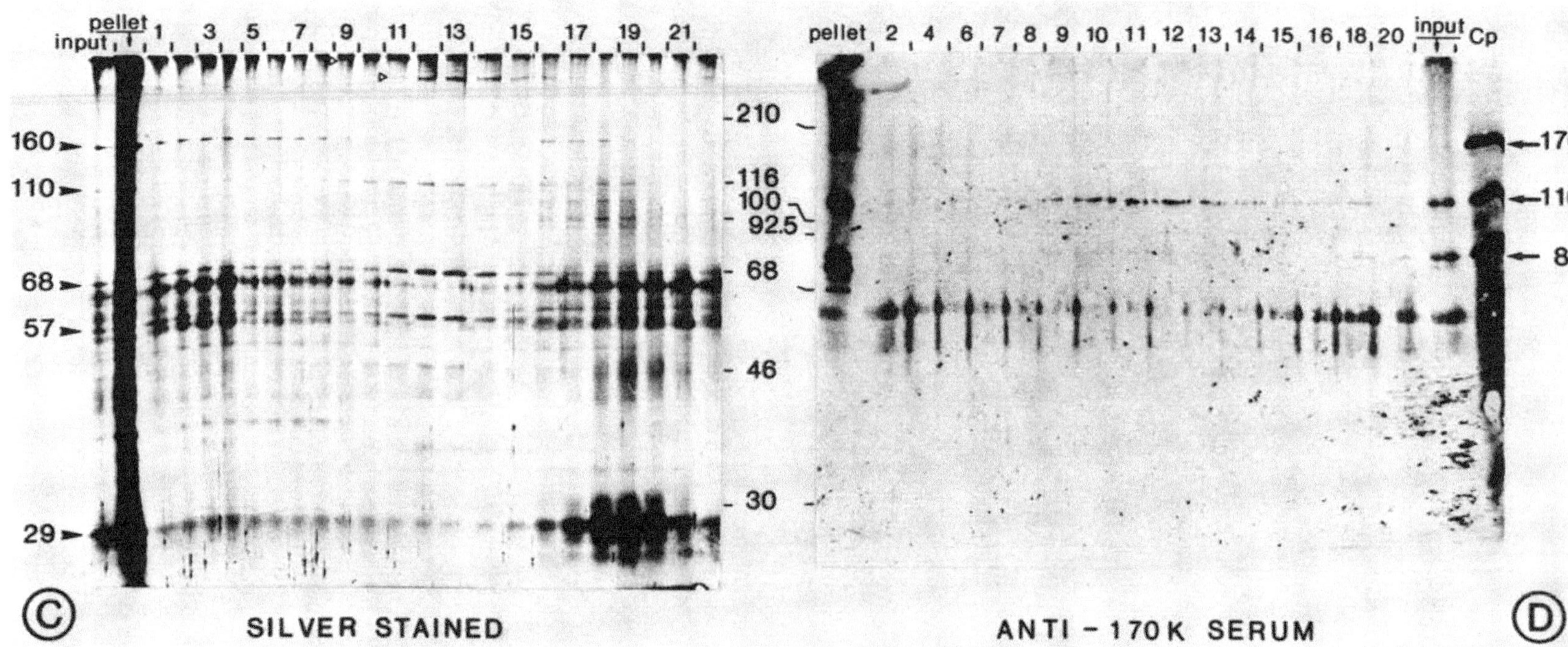

FIGURE 8. Analysis of the CPMV-RNA replication complex after purification by glycerol gradient centrifugation. (A) CPMV-RNA replication complexes solubilized from the membranes of infected leaf cells and purified by Sepharose 2B column chromatography were analyzed by glycerol gradient centrifugation. Purified CPMV double-stranded RNAs (RF_B and RF_M) were run on a parallel gradient as sedimentation markers. The replication complexes, measured as [^3H]-UMP-incorporating activity, sediment as a peak at 20-25S just ahead of the RF_B and RF_M markers. (B) RNA products synthesized *in vitro* by the glycerol gradient fractions of A were analyzed by agarose gel electrophoresis under nondenaturing conditions. RF_B and RF_M indicate the positions of purified double-stranded CPMV RNAs. The analysis of the products synthesized by different fractions show that replication complexes involved in replicating B-RNA sediment slightly faster than those involved in M-RNA replication. (C) Aliquots of the glycerol gradient fractions of A were analyzed by SDS-PAGE after heating at 65°C. Polypeptide bands (as well as both double-stranded RNAs) were visualized by silverstaining. The major bands in the gradient fractions containing the replication complexes represent polypeptides of approximately 110K, 68K, and 57K. (D) After electrophoresis the polypeptides were blotted into nitrocellulose and the blot was subsequently incubated with anti-170K serum and ^{125}I-labeled protein A to detect B-RNA-encoded polypeptides. The 110K polypeptide is the only viral-encoded polypeptide detectable in the replication complexes. Bands occurring in the 53K to 65K region of the gel are due to background staining. For experimental details see Reference 96. (Reprinted from Dorssers, L., Van der Krol, S., van der Meer, J., Van Kammen, A., and Zabel, P., *Proc. Nat. Acad. Sci. U.S.A.* With permission.)

multiple sites.[131] Since in such cases the added template RNA remains intact during RNA synthesis, the small size of the products is apparently not due to contaminating endonuclease activity. The production of small-sized complementary RNAs might therefore be an intrinsic property of the host-encoded RNA polymerase. Similarly, small heterogenous RNAs are produced by a host-encoded RNA-dependent RNA polymerase purified from cucumber mosaic virus-infected cucumber leaves.[132,133]

The activity of the host polymerase is greatly enhanced upon virus infection. As already mentioned, in infected cowpea leaves the activity increases 20-fold compared to uninfected leaves. This increase is restricted to CPMV-infected cells within the leaf tissue.[129] Very notably the amount of host polymerase does not increase in mesophyll protoplasts upon infection. Similar differential increases of enzyme activity in reponse to virus inoculation have been observed in connection with resistance and hypersensitivity of host plants to virus infection. Such host reactions appear tissue specific and are not exhibited in isolated leaf cell protoplasts.[134-138]

The function of the polymerase in plant cells and the importance of its increase after virus infection has remained obscure. Since it can also transcribe small parts of viral RNA upon virus infection, it may be suggested that the enzyme interferes in some way with virus multiplication. One can only speculate about the possible function of this interesting enzyme activity. It will certainly be of great help for elucidating the function of the host-encoded RNA-dependent RNA polymerase, if the nature of its endogeneous template in healthy plants is determined.

F. Evaluation of the Present Studies on CPMV-Replicase

The purification of the CPMV-RNA replication complex and the identification of a virus-encoded 110K polypeptide in this complex may be of general significance for the study of plant viral RNA replication.

Various findings have suggested that plant RNA viruses encode information essential to viral RNA replication. (1) Temperature sensitive mutants of TMV[139,140] and AlMV[141,142] have been described with defects in viral RNA synthesis. (2) In plant viruses with a bipartite or tripartite genome including TRV,[76] CPMV,[19] TBRV,[80] AlMV,[143] and BMV,[144] it has been demonstrated that replication of at least one of the genomic RNAs is completely dependent on the expression and replication of other genomic RNAs. (3) Replication of the satellite virus of TNV[145] and of satellite RNAs of TBRV, TRV, and CMV[146] is entirely dependent on the expression and replication of the helper virus.

In spite of these findings, the involvement of host-encoded RNA-dependent RNA polymerases in plant viral RNA replication has repeatedly been proposed, mainly because attempts to identify a virus-specific RNA polymerase run into a strong, infection-induced, enhancement of a host-encoded RNA-dependent RNA polymerase.[133,147-155]

Contradictory to this hypothesis, it has now been demonstrated that CPMV-RNA replication does not require the host-encoded RNA polymerase of cowpea, but uses a different polymerase activity specified by the virus itself. Compared to the host enzyme, the activity of which increased dramatically upon infection with CPMV, the activity of the virus-specific polymerase is low. The viral activity appears to represent only 5% of the total activity in the crude membrane fraction of CPMV-infected leaves 4 days after inoculation when the rate of virus multiplication in the leaves has reached its maximum. For this reason the virus-specific polymerase can easily be overlooked and only by employing an assay which keeps track of the synthesis of viral RNA, it will be recognized.[95] It seems conceivable that in some cases where the purification of a plant viral replicase has been attempted unsuccessfully, the virus-specific polymerase has been overlooked because of improper assay conditions and the very small amount of viral replicase relative to the host-encoded RNA-dependent RNA polymerase. It now seems almost imperative in purifying plant viral RNA replicase to use test methods which keep control on the ability of the enzyme to produce full-length viral RNAs. In particular, this should apply when the purification of a viral RNA replication complex is frustrated because of rapid loss of endogeneous template. Indeed, recently it has been shown that by using a proper assay, a BMV-replicase preparation from BMV-infected barley with considerable template specificity can be obtained.[156,157]

The purified CPMV-RNA replication complex is capable of elongating viral RNA strands of which the synthesis has been initiated in vitro, but is unable to start RNA synthesis. Neither has it been possible to demonstrate that the 110K protein or any of the other virus-specific proteins, not associated with the active replication complex, is capable of transcribing viral RNA template added in vitro. It appears possible therefore that the 110K protein alone is unable to start viral RNA synthesis but needs additional proteins for initiation. If initiation of CPMV-RNA synthesis is mediated by the generation of VPg from its precursor by analogy to what has been suggested for poliovirus RNA,[43] then it is conceivable that this step does not occur on added template. Similarly it is possible that specific initiation of viral RNA synthesis requires host proteins, in addition to the virus-encoded polymerase. Clearly the proper assay conditions for establishing initiation of viral RNA synthesis has still to be developed. In the next section (V) first recently found analogies between CPMV and the animal picornaviruses will be discussed, to return to the possible mechanism of viral RNA-replication in the final section.

V. ANALOGY BETWEEN COMOVIRUSES AND ANIMAL PICORNAVIRUSES

The plant comoviruses and animal picornaviruses exhibit striking analogy in various respects. True enough, the groups of viruses have quite different biological properties, but they share several features with regard to genome structure, replication, and translation strategy. The genomic RNAs of both comoviruses and picornaviruses have a VPg at the 5′ terminus and a 3′-terminal poly(A) tail. The RNAs of both groups of viruses are expressed by translation into large-sized polyproteins, which are subsequently processed to functional proteins by virus-encoded protease activities. Both como- and picornaviruses induce, in inoculated cells, the formation of a vesicular membrane structure, which represents the site of viral RNA synthesis. The capsids of comoviruses are built up of two proteins present in 60 copies each, which are probably processed from a common precursor protein. Such structure is rare with small icosahedral plant viruses but resembles the picornaviral capsids, which are made up of 60 copies of each of four proteins derived from a common precursor.[158,159]

With the recently obtained data on the nucleotide sequences, a more detailed comparison of the organization of the genes of comoviruses and picornaviruses has become possible. The complete nucleotide sequence of poliovirus has been determined[114,160,161] and the viral peptides have been precisely mapped within the single polyprotein derived from this RNA. As illustrated in Figure 9, the 246K polyprotein of poliovirus has been divided into three regions: P1 from which the four poliocapsid proteins are derived, P2 the central region with as yet unknown function, and P3, the region from which VPg, the polioviral protease (P3-7c) and RNA replicase (P3-4b) are derived.[114] For a second picornavirus, foot-and-mouth-disease virus (FMDV), the nucleotide sequence of two parts of the genome has been determined, corresponding to the coding region of the capsid proteins and the viral polymerase.[162,163] Computer analysis has revealed that the 200K polyprotein specified by CPMV B-RNA contains two regions with marked sequence homology to polypeptide P2-X from polio, and to the polymerases of both polio and FMDV, respectively[97] (Figure 9). On the other hand, the amino acid sequences of the overlapping polyproteins from M-RNA do not exhibit significant homology to any of the picornaviral proteins (Figure 9).

The sequence homologous to polypeptide P2-X of poliovirus could be mapped within the

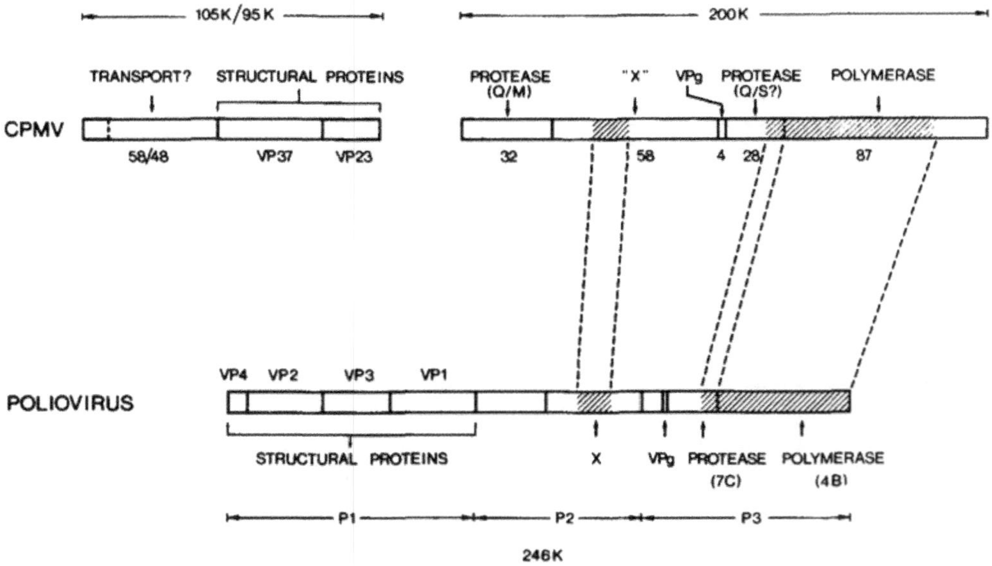

FIGURE 9. Comparison of the domain organization of the CPMV and polioviral polyproteins. M-RNA of CPMV specifies two overlapping polyproteins of 95K and 105K which are cleaved at a glutamine-methionine sequence by the 32K protease encoded by B-RNA (see also Figure 2), to give overlapping 58K and 48K polypeptides, possibly involved in viral RNA transport,[74] and a 60K precursor to both capsid proteins VP37 and VP23.[58] B-RNA of CPMV encodes a 200K polyprotein which contains two regions (shaded in the figure) homologous to sequences in the polioviral polyprotein.[97] The 200K B polyprotein is proposed to consist of five different domains, which are separated by proteolytic cleavages at glutamine-serine dipeptide sequences.[61,97,119] The responsible proteolytic activity probably resides in the 28K polypeptide (see also Section II.E). In view of the homology to polioviral polypeptides the 87K polypeptide must represent the core polymerase domain of the polyprotein and the 58K polypeptide might be a counterpart of polioviral protein P2-X. Note that the larger of the two regions of homology between the polio and CPMV polyproteins extends into the sequences of the polio protease (P3-7C) and the 28K polypeptide, respectively. (From Franssen, H., Levnissen, J., Goldbach, R., Lomonossoff, G., and Zimmern, D., *EMBO J.*, 3, 855, 1984. With permission.)

58K polypeptide encoded by B RNA. The homology to the polymerases of poliovirus and FMDV maps within the 87K polypeptide of CPMV (Figure 9). For the rest it should be noted that the region of homology between this part of the polioviral and CPMV B-polyprotein extends into the sequence of the polio protease P3-7c and the 28K polypeptide region, respectively. Actually the sequence of the 87K polypeptide exhibits 20.9% homology to both polio- and FMDV polymerases.[97] This may be compared with 30% homology between the sequences of the polioviral and the FMDV RNA polymerases.[163] The marked similarity of the 87K polypeptide to the RNA polymerases from two different picornaviruses presents strong evidence that this polypeptide constitutes the core polymerase domain of the CPMV replicase. This matches the finding of the 110K polypeptide, which contains the complete sequence of this 87K polypeptide, in purified CPMV replication complexes.[96]

The homology between the sequences of the polioviral protease P3-7c and the CPMV 28K polypeptide, which adjoin the polymerase sequences in both cases, supports the suggestions, based on results obtained by in vitro processing experiments,[61] that the 28K polypeptide represents a protease involved in the processing of the 200K polyprotein from B-RNA.

These recent findings indicate that the analogy between comoviruses and picornaviruses is not limited to structural features of the genomic RNAs and their mode of expression but includes the organization of the genes involved in viral RNA replication. Within the 200K polyprotein from B-RNA, the 58K, VPg, protease, and polymerase polypeptides are found one after the other, which corresponds to the P2-X, VPg, protease, and polymerase polypeptides within the polioviral polyprotein with both sets of polypeptides showing homology in amino acid sequences and probably representing analogous functions.

The functions of P2-X from poliovirus and the 58K polypeptide have not yet been biochemically defined, but as noted in Section II.E.3, the 58K polypeptide is always found firmly membrane bound in CPMV-infected cells, while there is also evidence that one or more proteins derived from the P2 region of the polioviral polyprotein are associated with the vesicular membranes induced in poliovirus-infected cells.[164] This might suggest corresponding functions for both proteins in the membrane structures, which are the sites of viral RNA synthesis.

The resemblance between CPMV and picornaviruses in the organization of RNA-replicating functions (including homology in amino acid sequence) suggests together with the similar translation and processing strategy used by these viruses that such combination of features is crucial to the replication of the type of RNA genome of both groups and surmise a common ancestor of comoviruses and picornaviruses.

Homology may be found among other plant and animal viruses, as is illustrated by the recent observation that a protein encoded by cauliflower mosaic virus shows sequence homology to retroviral reverse transcriptase.[165] Since the replication of the genome of cauliflower mosaic virus follows a mechanism similar to that of retroviruses, such observations suggest a common origin for both groups of viruses or convergent evolution. More recently it has been discovered that three other plant RNA viruses, AlMV, BMV, and TMV, encode proteins homologous to polypeptide ns72 from Sindbis virus[166] and this may provide a further example of evolutionary relationship among groups of plant and animal RNA viruses.

If the analogy between comoviruses and picornaviruses reflects a common evolutionary origin, the subsequent divergence of the viruses has resulted in different biological properties and differences in additional functions. Comoviruses have acquired (or maintained?) the interesting condition of a divided genome, which provides for separate regulation of the production of structural and nonstructural viral proteins. Apparently as a consequence of the divided CPMV genome, B-RNA encodes an additional function, the 32K protease, charged with the processing of the polyproteins from M-RNA. The completely different host ranges selected by the two groups of viruses will have contributed to the origin of some major differences between them. For example, a remarkable difference between comoviruses and picornaviruses is that picornaviruses are capable of inducing shut-off of host protein synthesis in infected cells, whereas this has not been observed for comoviruses. It is not known to which viral encoded function this property should be attributed.

From the point of view of the molecular evolution of viruses, it will be of considerable interest to examine to which extent comoviruses have diverged in sequences among each other in comparison to the divergency with respect to picornaviruses. Unfortunately very few data are available about this. As indicated in Section II.F there appears to be little or no homology in nucleotide sequence between the M-RNAs of CPMV and four other comoviruses. Furthermore antisera raised against various polypeptides from CPMV B-RNA did not react with B-specified proteins of CPSMV with the exception of antiserum against the 87K polypeptide, which is capable of precipitating the 85K polypeptide encoded by CPSMV B-RNA.[98] Since the 87K polypeptide of CPMV represents the RNA polymerase activity, the 85K polypeptide presumably is the analogous protein of CPSMV. Such cross-reaction may indicate that among comoviruses homology in amino acid sequence is maintained as far as the replicase function is concerned. On the other hand, other corresponding functions encoded by different comoviruses have diverged rapidly.

Another group of bipartite plant RNA viruses, the nepoviruses, have a genome structure and translation strategy similar to comoviruses. Nepoviruses differ from comoviruses, apart from a number of biological properties, by the fact that they have only a single species of capsid protein. But the two genomic RNAs of nepoviruses have both a VPg and a poly(A) tail and, as far as known, the distribution of functions between the two RNAs is similar to that of comoviruses. The RNAs of nepoviruses are also translated into polyproteins, and recently preliminary evidence for processing of these polyproteins has been published.[167] Since data on the nucleotide sequence of nepoviral RNAs are not available, a detailed comparison of nepoviruses with comoviruses and picornaviruses is not yet possible.

VI. CONCLUSION

Substantial progress has been made in elucidating the translation and processing strategy of CPMV and the identification of functions encoded by its bipartite genome, but our understanding of the viral RNA replication process is still far from complete.

It may be taken for granted that replication of the genomic RNAs will proceed through the synthesis of complementary (minus) strands, which in turn are used as template for the synthesis of progeny viral RNAs. So far, experimental data on the transcription of virion RNAs into complementary RNA are lacking. With poliovirus it has been shown that both minus strands and nascent RNA strands possess VPg molecules,[168] suggesting that initiation of viral RNA synthesis always involves VPg. Considering the analogy between CPMV and poliovirus it appears plausible to assume that (minus) strands of CPMV RNAs will also be supplied with VPg. This can be made certain by analyzing the 5′-terminal structures of (minus) strand RNA present in the replication complexes. Such analysis might furthermore reveal if (part of) the poly(A) tail is transcribed into (minus) strands, which would provide information on the initiation site of transcription on the virion RNAs. With poliovirus it has been demonstrated that the in vitro transcription of virion RNA into complementary copies requires the virus-specific polymerase P3-4b and a host factor with a molecular weight of 67K.[169] Antibodies raised against this 67K protein have been shown to precipitate a specific complex of this host factor and the polio polymerase.[170] One might presume that a host factor is similarly involved in CPMV-RNA replication. In this context it is noteworthy that the highly purified viral RNA replication complex contained, besides the viral 110K polymerase, 68K and 57K polypeptides which appear to be host proteins.[96] Whether the 68K and 57K polypeptides represent functional subunits required for virus specific polymerase activity or host proteins still contaminating the analyzed replication complex remains to be determined but they may offer a starting point for elucidating the involvement of one or more host factors in viral RNA replication.

Replication of CPMV RNAs in vivo is associated with vesicular membrane structures, the proliferation of which is induced upon virus infection. The induction of the proliferation of membranes in which the viral replication complex is integrated seems to be achieved by a B-RNA-encoded function which has not yet been identified. This leads to questions as how the replication complex is established and how closely the posttranslational processing of the B-RNA-encoded 200K polyprotein is linked to viral RNA replication. The processing scheme for this protein (see Figure 2) points out that nine cleavage products have been detected, of which five are final cleavage products, derived in the order: NH_2 — 32K — 58K — 4K(VPg) — 28K — 87K — COOH. The other (170K, 110K, 84K, and 60K) polypeptides are intermediates which occur in infected cells in remarkably large amounts (Figure 4). Therefore, these molecules may represent polypeptides with specific functions in virus multiplication. Indeed, the 110K polypeptide which contains the sequences of both the 28K and the 87K polypeptide has been found in purified viral replication complexes. Whether both the 110K polypeptide and the free 87K polypeptide are both active in RNA synthesis remains to be determined.

A description of a putative model for CPMV-RNA replication, taking into account the various observations and comments presented in this chapter, may help to illustrate a possible close linkage between protein processing and viral RNA replication, namely, infecting B-RNA directs the synthesis of a 200K polyprotein. The first cleavages of the polyprotein generating 32K and 170K polypeptides is already achieved at the nascent polypeptide chain by the internal proteolytic activity located in the 28K domain of the polyprotein. At this stage, proliferation of cellular membranes may be induced by a yet unknown stimulus. Further processing of the 170K polypeptide seems to occur in association with those membranes since upon cleavage of the 170K polypeptide into 60K and 110K polypeptides, the 60K polypeptide, which is the direct precursor to VPg, is deposited in the membranes. The viral RNA replication complex is then assembled using the 60K VPg-precursor, the 110K viral replicase, and one or more host proteins, and is integrated in the membranes. Synthesis of viral RNA is subsequently initiated when VPg is released from the 60K polypeptide and forms a covalent bond to the first nucleotide of the new chain. There upon, chain elongation can proceed by the 110K replicase. Possibly the 110K replicase is cleaved into 87K and 28K polypeptides either during chain elongation or when the synthesis of a new viral RNA strand has been terminated. This model would imply an important regulatory role of protein processing in viral RNA replication in which only a single RNA chain can be produced each time a 170K polypeptide is processed, because only one VPg molecule can be supplied at every turn. The model described above does not account for the generation of the 84K polypeptide which is found in infected cells in about equal amounts to the 110K and 60K processing intermediates. The 84K polypeptide containing the sequences of the 60K VPg precursor and the 28K polypeptide is produced together with the 87K polypeptide by an alternative first cleavage of the 170K polypeptide. If it is arbitrary whether the 170K polypeptide is first cleaved into 60K and 110K polypeptides, or into 84K and 87K polypeptides, (a viral RNA replication complex can equally well be assembled from both pairs of polypeptides), then this would imply that the 84K polypeptide also provides for VPg and the 87K polypeptide is also active as polymerase. Alternatively, the possibility of a second cleavage pathway for the 170K polypeptide may be significant in regulating viral RNA replication, in a way not yet understood.

Furthermore the model points out that initiation of viral RNA synthesis by the CPMV-replication complex may start as the result of a series of interactions between virus- and host-encoded proteins which undergo processing by proteolytic cleavages. The different proteolytic cleavages indicated in the model are all achieved by the activity located in the B-encoded 28K polypeptide sequence as part of a larger precursor sequence. So far it is not known if any proteolytic cleavage activity is conducted by the 28K polypeptide in its free form. Perhaps the 28K polypeptide is only active as a domain of larger polypeptide sequences to guide the assembly of the replication complex in a defined direction making this process virtually irreversible.

With respect to the processing of the 105K and 95K polyproteins translated from M-RNA, it is very noticeable that it has not been possible so far to detect in vivo the 60K precursor of both capsid proteins VP37 and VP23, nor the 58K and 48K cleavage products from the M-RNA polyproteins. Apparently, the processing of the M-RNA translation products by the 32K protease is very rapidly achieved.[58] It has not yet been possible to identify the protease involved in the further cleavage of the 60K precursor in VP37 and VP23 capsid proteins. It is tempting to assume, in accordance with a similar hypothesis for poliovirus, that the further cleavage of the 60K capsid protein precursor occurs upon assembly and provides for stabilization of the viral capsids. Since no 60K precursors accumulate, assembly of virus capsids apparently proceeds very fast. If rapid assembly also implies rapid encapsidation of newly synthesized viral RNA, such a mechanism might explain why in virus-infected cells, the number of progeny (plus)strand viral RNAs is much greater than the number of (minus)strands. Alternatively, minus- and plus-strand synthesis may be separately regulated by a mechanism which is not yet understood. Further analysis of viral RNA replication in protoplasts inoculated with only B components, where no encapsidation occurs, might prove which hypothesis is correct.

It will be clear from the different hypotheses discussed so far that further biochemical identification of the activities of the various virus-encoded polypeptides and host proteins involved is required to elucidate their role in virus multiplication. For unraveling the in vivo functions encoded on the viral RNAs, it appears equally important to have the disposal of a set of mutants with each just a single mutation at well-defined positions in the genomic RNAs. With respect to the latter approach, it appears worthwhile to investigate the possibility of producing such mutants by means of site-directed mutagenesis of cloned DNA copies of the viral RNAs. Recently, full-size DNA copies of M-RNA have been cloned in our laboratory and it has been shown that such copies can be transcribed in vitro into RNA molecules, which direct in rabbit reticulocyte lysates 105K and 95K polypeptides similar to the polyproteins produced by natural M-RNA. The identity of these polypeptides was substantiated by processing with extracts from B-inoculated protoplasts (see Section II.A) which produced 48K, 58K, and 60K polypeptides, as might be expected for proper M-RNA-specified polyproteins. Such faithful transcription of the cloned viral M-DNA enables the investigation of the effect of defined mutations introduced in M-DNA, on the expression of the subsequently transcribed M-RNA sequences upon in vitro translation. Furthermore, it serves the purpose of producing mutated M-RNAs which can be used for studies on virus multiplication in vivo. The construction of full-size DNA copies of B-RNA is under way, and will enable similar studies for B-RNA. An important goal of such studies will be the production of B-RNA and M-RNA from cloned B-DNA and M-DNA copies, which will be infectious. But even before this aim is reached, cloned DNA copies of defined segments of B-RNA may prove useful for producing different B-encoded polypeptides in larger amounts than can be obtained from infected cells. This may greatly increase the possibilities for the study of specific properties and functions of these polypeptides. Ultimately it may even help to achieve the assembly of a viral RNA replication complex capable of initiating and completing viral RNA synthesis in vitro. In addition to the experimental approaches described in this chapter, studies of the type mentioned above may contribute to the further elucidation of the molecular mechanism of plant virus replication.

ACKNOWLEDGMENT

We thank Piet Madern for photography and Marie-José van Neerven and Bea Verhaar for their help in preparing the manuscript. The virus research by our group was supported by the Netherlands Foundation for Chemical Research (S.O.N.) with financial aid from the Netherlands Organization for the Advancement of Pure Research (Z.W.O.).

REFERENCES

1. **Bruening, G.,** Comovirus group, *CMI/AAB Description of Plant Viruses,* No. 199, Harrison, B. D. and Murant, A. F., Eds., 1978.
2. **Van Kammen, A.,** Plant viruses with a divided genome, *Annu. Rev. Phytopathol.,* 10, 125, 1972.
3. **Jaspars, E. M. J.,** Plant viruses with a multipartite genome, *Adv. Virus Res.,* 19, 37, 1974.
4. **Bruening, G.,** Plant covirus systems: two component systems, in *Comprehensive Virology,* Vol. 11, Fraenkel-Conrat, H. and Wagner, R. R., Eds., Plenum Press, New York, 1977, 55.
5. **Stace-Smith, R.,** Comoviruses, in *Handbook of Plant Virus Infections and Comparative Diagnosis,* Kurstak, E., Ed., Elsevier/North-Holland Biomedical Press, 1981, 171.

5a. **Davies, J. W.,** personal communication, 1982.
6. **Harrison, B. D. and Murant, A. F., Eds.,** Description of Plant Viruses No. 1 to 275, Commonwealth Agricultural Bureau, Slough, England.
7. **Van Kammen, A. and de Jager, C. P.,** Cowpea mosaic virus, *CMI/AAB Descriptions of Plant Viruses, No. 197,* Harrison, B. D. and Murant, A. F., Eds., 1978.
8. **Swaans, H. and van Kammen, A.,** Reconsideration of the distinction between the severe and yellow strains of cowpea mosaic virus, *Neth. J. Plant Pathol.,* 79, 257, 1973.
9. **De Jager, C. P.,** Cowpea severe mosaic virus, *CMI/AAB Descriptions of Plant Viruses, No. 209,* Harrison, B. D. and Murant, A. F., Eds., 1979.
10. **Beier, H., Siler, D. J., Russell, M. L., and Bruening, G.,** Survey of susceptibility to cowpea mosaic virus among protoplasts and intact plants from *Vigna sinensis* lines, *Phytopathology,* 67, 917, 1977.
11. **Beier, H., Issinger, D. G., Deuschle, M., and Mundry, K. W.,** Translation of the RNA of cowpea severe mosaic virus in vitro and in cowpea protoplasts, *J. Gen. Virol.,* 54, 379, 1981.
12. **De Jager, C. P.,** personal communication, 1982.
13. **Fulton, J. P. and Scott, H. A.,** A serogrouping concept for legume comoviruses, *Phytopathology,* 69, 305, 1979.
14. **Goldbach, R. and Rezelman, G.,** unpublished results, 1982.
15. **Zabel, P., Weenen-Swaans, H., and van Kammen, A.,** In vitro replication of cowpea mosaic virus RNA. I. Isolation and properties of the membrane-bound replicase, *J. Virol.,* 14, 1049, 1974.
16. **Rottier, P. J. M., Rezelman, G., and van Kammen, A.,** Protein synthesis in cowpea mosaic virus-infected cowpea protoplasts: detection of virus-related proteins, *J. Gen. Virol.,* 51, 359, 1980.
17. **Van Kammen, A.,** The relationship between the components of cowpea mosaic virus. I. Two ribonucleoprotein particles necessary for the infectivity of CPMV, *Virology,* 34, 312, 1968.
18. **De Jager, C. P.,** Genetic analysis of cowpea mosaic virus mutants by supplementation and reassortment tests, *Virology,* 70, 151, 1976.
19. **Goldbach, R., Rezelman, G., and van Kammen, A.,** Independent replication and expression of B-component RNA of cowpea mosaic virus, *Nature (London),* 286, 297, 1980.
20. **Wu, G-J. and Bruening, G.,** Two proteins from cowpea mosaic virus, *Virology,* 46, 596, 1971.
21. **Geelen, J. L. M. C.,** Structuur en Eigenschappen van Cowpea-Mosaïekvirus, Ph.D. thesis, Agricultural University, Wageningen, 1974.
22. **Rottier, P. J. M., Rezelman, G., and van Kammen, A.,** The inhibition of cowpea mosaic virus replication by actinomycin D, *Virology,* 92, 299, 1979.
23. **Partridge, J. E., Shannon, L. M., Gumph, D. J., and Colbaugh, P.,** Glycoprotein in the capsid of plant viruses as a possible determinant of seed transmissibility, *Nature (London),* 247, 391, 1974.
24. **El Manna, M. M. and Bruening, G.,** Polyadenylate sequences in the ribonucleic acids of cowpea mosaic virus, *Virology,* 56, 198, 1973.
25. **Steele, K. P. and Frist, R. H.,** Characterization of the 3′-termini of the RNAs of cowpea mosaic virus, *J. Virol.,* 26, 243, 1978.
26. **Ahlquist, P. and Kaesberg, P.,** Determination of the length distribution of poly A at the 3′-terminus of the virion RNAs of EMC virus, poliovirus, rhinovirus, RAV-61 and CPMV and of mouse globin mRNA, *Nucl. Acids Res.,* 7, 1195, 1979.
27. **Daubert, S. D., Bruening, G., and Najarian, R. C.,** Protein bound to the genome RNAs of cowpea mosaic virus, *Eur. J. Biochem.,* 92, 45, 1978.
28. **Stanley, J., Rottier, P., Davies, J. W., Zabel, P., and van Kammen, A.,** A protein linked to the 5′ termini of both RNA components of the cowpea mosaic virus genome, *Nucl. Acids Res.,* 5, 4505, 1978.
29. **Daubert, S. D. and Bruening, G.,** Genome-associated proteins of comoviruses, *Virology,* 98, 246, 1979.
30. **Harrison, B. D. and Barker, H.,** Protease-sensitive structure needed for infectivity of nepovirus RNA, *J. Gen. Virol.,* 40, 711, 1978.
31. **Mayo, M. A. Barker, H., and Harrison, B. D.,** Evidence for a protein covalently linked to tobacco ringspot virus RNA, *J. Gen. Virol.,* 43, 735, 1979.
32. **Mayo, M. A., Barker, H., and Harrison, B. D.,** Polyadenylate in the RNA of five nepoviruses, *J. Gen. Virol.,* 43, 603, 1979.
33. **Koenig, I. and Fritsch, C.,** A protein linked at the 5′ end of satellite and genomic tomato black ring virus RNAs: study of in vitro translation after protease treatment, *J. Gen. Virol.,* 60, 343, 1982.
34. **Mayo, M. A., Barker, H., and Harrison, B. D.,** Specificity and properties of the genome-linked proteins of nepoviruses, *J. Gen. Virol.,* 59, 149, 1982.
35. **Reisman, D. and De Zoeten, G. A.,** A covalently linked protein at the 5′ ends of the genomic RNAs of pea enation mosaic virus, *J. Gen. Virol.,* 62, 187, 1982.
36. **Hari, V., Siegel, A., Rozek, C., and Timberlake, W. E.,** The RNA of tobacco etch virus contains poly(A), *Virology,* 92, 568, 1979.
37. **Hari, V.,** The RNA of tobacco etch virus: further characterization and detection of protein linked to RNA, *Virology,* 112, 391, 1981.

38. **Lee, Y. -F., Nomoto, A., Detjen, B. M., and Wimmer, E.,** A protein covalently linked to poliovirus genome RNA, *Proc. Natl. Acad. Sci. U.S.A.*, 74, 59, 1977.
39. **Flanegan, J. B., Pettersson, R. F., Ambros, V., Hewlett, M. J., and Baltimore, D.,** Covalent linkage of a protein to a defined nucleotide sequence at the 5′-terminus of the virion and replicative intermediate RNAs of poliovirus, *Proc. Natl. Acad. Sci. U.S.A.*, 74, 961, 1977.
40. **Sangar, D. V., Rowlands, D. J., Harris, T. J. R., and Brown, F.,** Protein covalently linked to foot-and-mouth disease virus RNA, *Nature (London)*, 268, 648, 1977.
41. **Golini, F., Nomoto, A., and Wimmer, E.,** The genome-linked protein of picornaviruses. IV. Difference in the VPg's of encephalomyocarditis virus and poliovirus as evidence that the genome-linked proteins are virus-coded, *Virology*, 89, 112, 1978.
42. **Hruby, D. E. and Roberts, W. K.,** Encephalomyocarditis virus RNA. III. Presence of a genome associated protein, *J. Virol.*, 25, 413, 1978.
43. **Wimmer, E.,** Genome-linked proteins of viruses, *Cell*, 28, 199, 1982.
44. **Burroughs, J. N. and Brown, F.,** Presence of a covalently linked protein on calicivirus RNA, *J. Gen. Virol.*, 41, 443, 1978.
45. **Veerisetty, V. and Seghal, O. P.,** Proteinase K-sensitive factor essential for the infectivity of southern bean mosaic virus ribonucleic acid, *Phytophathology*, 70, 282, 1980.
46. **Ghosh, A. R., Dasgupta, T., Salerno-Rife, T., Rutgers, T., and Kaesberg, P.,** Southern bean mosaic virus has a 5′-linked protein but lacks a 3′-terminal poly (A), *Nucl. Acids Res.*, 7, 2139, 1979.
47. **Hull, R. and Morris-Krsinich, B. A. M.,** Protein Covalently Linked to the RNA of Turnip Rosette Virus, *Annual Report*, John Innes Institute, Norwich, U.K., 1979, 98.
48. **Mayo, M. A., Barker, H., Robinson, D. J., Tamada, T., and Harrison, B. D.,** Evidence that potato leafroll virus RNA is positive-stranded, is linked to a small protein and does not contain polyadenylate, *J. Gen. Virol.*, 59, 163, 1982.
49. **Reijnders, L., Aalbers, A. M. J., van Kammen, A., and Thuring, R. W. J.,** Molecular weights of plant viral RNAs determined by gel electrophoresis under denaturing conditions, *Virology*, 60, 515, 1974.
50. **Murant, A. F., Taylor, M., Duncan, G. H., and Raschké, J. H.,** Improved estimates of molecular weight of plant virus RNA by agarose gel electrophoresis and electron microscopy after denaturation with glyoxal, *J. Gen. Virol.*, 53, 321, 1981.
51. **Van Wezenbeek, P., Verver, J., Harmsen, J., Vos, P., and van Kammen, A.,** Primary structure and gene organization of the middle component RNA of cowpea mosaic virus, *EMBO J.*, 2, 941, 1983.
52. **Lomonossoff, G. P. and Shanks, M.,** The nucleotide sequence of cowpea mosaic virus B RNA, *EMBO J.*, 2, 2153, 1983.
53. **Pelham, H. R. B. and Stuik, E. J.,** Translation of cowpea mosaic virus RNA in a messenger dependent cell-free system from rabbit reticulocytes, in Proc. Coll. Nucl. Acids and Protein Synthesis in Plants, Centre National de la Recherche Scientifique, Paris, 1977, 691.
54. **Davies, J. W., Aalbers, A. M. H., Stuik, E. J., and van Kammen, A.,** Translation of cowpea mosaic virus RNA in a cell-free extract from wheat germ, *FEBS Lett.*, 77, 265, 1977.
55. **Pelham, H. R. B.,** Synthesis and proteolytic processing of cowpea mosaic virus proteins in reticulocyte lysates, *Virology*, 96, 463, 1979.
56. **Goldbach, R. W., Schilthuis, J. G., and Rezelman, G.,** Comparison of *in vivo* and *in vitro* translation of cowpea mosaic virus RNAs, *Biochem. Biophys. Res. Commun.*, 99, 89, 1981.
57. **Rezelman, G., Goldbach, R., and van Kammen, A.,** Expression of bottom component RNA of cowpea mosaic virus in cowpea protoplasts, *J. Virol.*, 36, 366, 1980.
58. **Franssen, H., Goldbach, R., Broekhuijsen, M., Moerman, M., and van Kammen, A.,** Expression of middle-component RNA of cowpea mosaic virus: in vitro generation of a precursor to both capsid proteins by a bottom-component RNA-encoded protease from infected cells, *J. Virol.*, 41, 8, 1982.
59. **Kalnins, A., Otto, K., Rüther, H., and Müller-Hill, B.,** Sequence of the *lac* Z gene of *Escherichia coli*, *EMBO J.*, 2, 593, 1983.
60. **Stuik, E.,** Protein Synthesis Directed by Cowpea Mosaic Virus RNAs, Ph.D. thesis, Agricultural University, Wageningen, 1979.
61. **Franssen, H., Goldbach, R., and van Kammen, A.,** Translation of bottom component RNA of cowpea mosaic virus in reticulocyte lysate: faithful proteolytic processing of the primary translation product, *Virus Res.*, 1, 39, 1984.
62. **Peng, X. X. and Shih, D. S.,** Proteolytic processing of the proteins translated from the bottom component RNA of cowpea mosaic virus. The primary and secondary cleavage reactions, *J. Biol. Chem.*, 1984, in press.
63. **Broekhuijsen, M. and Goldbach, R.,** unpublished results, 1981.
64. **Franssen, H., Moerman, M., Rezelman, G., and Goldbach, R.,** Evidence that the 32,000-dalton protein encoded by the bottom-component of cowpea mosaic virus is a proteolytic processing enzyme, *J. Virol.*, 50, 183, 1984.

65. **Gopo, J. M. and Frist, R. H.,** Location of the gene specifiying the smaller protein of the cowpea mosaic virus capsid, *Virology*, 79, 259, 1977.
66. **Mühlbach, H. -P.,** Plant cell cultures and protoplasts in plant virus research, in *Current Topics in Microbiology and Immunology*, Vol. 99, Cooper, M., Henle, W., Hofschneider, P. H., Kopowski, H., Melchers, F., Rott, R., Schweiger, H. G., Vogt, P. K., and Zinkernagel, R., Eds., Springer-Verlag, Berlin, 1982, 81.
67. **Harrison, B. D. and Mayo, M. A.,** The use of protoplasts in plant virus research, in *Use of Tissue Culture and Protoplasts in Plant Pathology*, Helgeson, J. P. and Deverall, B. J., Eds., Academic Press, Sydney, Australia, 1983, 69.
68. **Hibi, T., Rezelman, G., and van Kammen, A.,** Infection of cowpea mesophyll protoplasts with cowpea mosaic virus, *Virology*, 64, 308, 1975.
69. **Beier, H. and Bruening, G.,** The use of an abrasive in the isolation of cowpea leaf cell protoplasts, which support the multiplication of cowpea mosaic virus, *Virology*, 64, 272, 1975.
70. **Rottier, P. J. M., Rezelman, G., and van Kammen, A.,** Protein synthesis in cowpea mosaic virus-infected cowpea protoplasts: further characterization of virus-related protein synthesis, *J. Gen. Virol.*, 51, 1980, 373.
71. **Goldbach, R., Rezelman, G., Zabel, P., and van Kammen, A.,** Expression of the bottom component RNA of cowpea mosaic virus: evidence that the 60 kilodalton VPg-precursor is cleaved into single VPg and a 58 kilodalton polypeptide, *J. Virol.*, 42, 630, 1982.
72. **Dorssers, L.,** RNA-Dependent RNA Polymerases from Cowpea Mosaic Virus-Infected Cowpea Leaves, Ph.D. thesis, Agricultural University, Wageningen, 1983.
73. **Zabel, P., Moerman, M., Van Straaten, F., Goldbach, R., and van Kammen, A.,** Antibodies against the genome-linked protein VPg of cowpea mosaic virus recognize a 60,000 dalton percursor polypeptide, *J. Virol.*, 41, 1083, 1982.
74. **Rezelman, G., Franssen, H. J., Goldbach, R. W., Ie, T. S., and van Kammen, A.,** Limits to the independence of bottom component RNA of cowpea mosaic virus, *J. Gen. Virol.*, 60, 335, 1982.
75. **Assink, A. M., Swaans, H., and van Kammen, A.,** The localisation of virus-specific double-stranded RNA of cowpea mosaic virus in subcellular fractions of infected *Vigna* leaves, *Virology*, 53, 384, 1973.
76. **Lister, R. M.,** Functional relationship between virus-specific products of infection by viruses of the tobacco rattle type, *J. Gen. Virol.*, 2, 43, 1968.
77. **Lister, R. M.,** Tobacco rattle NETU virus in relation to functional heterogeneity of plant viruses, *Fed. Proc.*, 28, 1875, 1969.
78. **Cadman, C. H.,** Evidence for association of tobacco rattle virus nucleic acid with a cell component, *Nature (London)*, 193, 49, 1962.
79. **Sänger, H. L. and Brandenburg, E.,** Über die Gewinnung von Infectiosem aus 'Wintertyp'-Pflanzen des Tabak-Rattle-Virus durch Phenolextraktion, *Naturwissenschaften*, 48, 391, 1961.
80. **Robinson, D. J., Barker, H., Harrison, B. D., and Mayo, M. A.,** Replication of RNA-1 of tomato black ring virus independently of RNA-2, *J. Gen. Virol.*, 51, 317, 1980.
81. **Sänger, H. L.,** Characteristics of tobacco rattle virus. I. Evidence that its two particles are functionally defective and mutually complementing, *Mol. Gen. Genet.*, 101, 346, 1968.
82. **Randles, J. W., Harrison, B. D., Murant, A. F., and Mayo, M. A.,** Packaging and biological activity of the two essential RNA species of tomato black ring virus, *J. Gen. Virol.*, 36, 187, 1977.
83. **Fritsch, C., Mayo, M. A., and Murant, A. F.,** Translation products of genome and satellite RNAs of tomato black ring virus, *J. Gen. Virol.*, 46, 381, 1980.
84. **Ghabrial, S. A. and Lister, R. M.,** Coat protein and symptom specification in tobacco rattle virus, *Virology*, 52, 1, 1973.
85. **Mayo, M. A., Fritsch, C., and Hirth, L.,** Translation of tobacco rattle virus RNA *in vitro* using wheat germ extracts, *Virology*, 67, 408, 1976.
86. **Fritsch, C., Mayo, M. A., and Hirth, L.,** Further studies on the translation products of tobacco rattle virus RNA *in vitro*, *Virology*, 77, 722, 1977.
87. **Stanley, J., Goldbach, R., and van Kammen, A.,** The genome-linked protein of cowpea mosaic virus is coded by RNA from the bottom component, *Virology*, 106, 180, 1980.
88. **Goldbach, R. and Rezelman, G.,** Orientation of the cleavage map of the 200-kilodalton polypeptide encoded by the bottom-component RNA of cowpea mosaic virus, *J. Virol.*, 46, 614, 1983.
89. **Goldbach, R. and Rezelman, G.,** unpublished results, 1983.
90. **Thongmeearkom, P. and Goodman, R. M.,** Complementation and pseudorecombination between ribonucleic acids from two natural isolates of cowpea mosaic virus (severe sub-group), *Virology*, 85, 75, 1978.
91. **Hiebert, E. and Purcifull, D. E.,** Mapping of the two coat protein genes on the middle RNA component of squash mosaic virus (comovirus group), *Virology*, 113, 1981, 630.
92. **Leonard, D. A. and Zaitlin, M.,** A temperature-sensitive strain of tobacco mosaic virus defective in cell-to-cell movement generates an altered viral-coded protein, *Virology*, 117, 416, 1982.

93. **Zimmern, D. and Hunter, T.,** Point mutation in the 30-K open reading frame of TMV implicated in temperature-sensitive assembly and local lesion spreading of mutant Ni 2519, *EMBO J.*, 2, 1893, 1983.
94. **De Zoeten, G. A., Assink, A. M., and van Kammen, A.,** Association of cowpea mosaic virus-induced double-stranded RNA with a cytopathological structure in infected cells, *Virology*, 59, 341, 1974.
95. **Dorssers, L., Van der Meer, J., van Kammen, A., and Zabel, P.,** The cowpea mosaic virus RNA replication complex and the host-encoded RNA-dependent RNA polymerase-template complex are functionally different, *Virology*, 125, 155, 1983.
96. **Dorssers, L., Van der Krol, S., Van der Meer, J., van Kammen, A., and Zabel, P.,** The cowpea mosaic virus B-RNA-encoded 110K polypeptide is involved in viral RNA replication, *Proc. Natl. Acad. Sci. U.S.A.*, 31, 1951, 1984.
97. **Franssen, H., Leunissen, J., Goldbach, R., Lomonossoff, G., and Zimmern, D.,** Homologous sequences in non-structural proteins from cowpea mosaic virus and picornaviruses, *EMBO J.*, 3, 855, 1984.
98. **Van der Krol, S., Rezelman, G., and Goldbach, R.,** unpublished results, 1983.
99. **Goldbach, R. and Krijt, J.,** Cowpea mosaic virus-encoded protease does not recognize primary translation products of M RNAs from other comoviruses, *J. Virol.*, 43, 1151, 1982.
100. **Gabriel, C. J., Derrick, K. S., and Shih, D. S.,** The synthesis and processing of the proteins of bean pod mottle virus in rabbit reticulocyte lysates, *Virology*, 122, 476, 1982.
101. **Goelet, P., Lomonossoff, G. P., Butler, P. J. G., Akam, M. E., Gait, M. J., and Karn, J.,** Nucleotide sequence of tobacco mosaic virus RNA, *Proc. Natl. Acad. Sci. U.S.A.*, 79, 5818, 1982.
102. **Brederode, F. Th., Koper-Zwarthoff, E. C., and Bol., J. F.,** Complete nucleotide sequence of alfalfa mosaic virus RNA 4, *Nucl. Acids Res.*, 8, 2213, 1980.
103. **Cornelissen, B. J. C., Brederode, F. Th., Moormann, R. J. M., and Bol, J. F.,** Complete nucleotide sequence of alfalfa mosaic virus RNA 1, *Nucl. Acids Res.*, 11, 1253, 1983.
104. **Cornelissen, B. J. C., Brederode, F. Th., Veeneman, G. H., Van Boom, J. H., and Bol, J. F.,** Complete nucleotide sequence of alfalfa mosaic virus RNA 2, *Nucl. Acids Res.*, 10, 3019, 1983.
105. **Barker, R. F., Jarvis, N. P., Thompson, D. V., Loesch-Fries, L. S., and Hall, T. C.,** Complete nucleotide sequence of alfalfa mosaic virus RNA 3, *Nucl. Acids Res.*, 11, 2881, 1983.
106. **Davies, J. W., Stanley, J., and van Kammen, A.,** Sequence homology adjacent to the 3′-terminal poly (A) of cowpea mosaic virus RNAs, *Nucl. Acids Res.*, 7, 493, 1979.
107. **Najarian, R. C. and Bruening, G.,** Similar sequences from the 5′ end of cowpea mosaic virus RNAs, *Virology*, 106, 301, 1980.
108. **Stanley, J. and Van Kammen, A.,** Nucleotide sequences adjacent to the proteins covalently linked to the cowpea mosaic virus genome, *Eur. J. Biochem.*, 101, 45, 1979.
109. **Lomonossoff, G., Shanks, M., Matthes, H. D., Singh, M., and Gait, M. J.,** The location of the first AUG codons in cowpea mosaic virus RNAs, *Nucl. Acids Res.*, 10, 4861, 1982.
110. **Messing, J. and Vieira, J.,** A new pair of M13 vectors for selecting either DNA strand of double-digest restriction fragments, *Gene*, 19, 269, 1982.
111. **Sanger, F., Nicklen, S., and Coulson, A. R.,** DNA sequencing with chain-terminating inhibitors, *Proc. Natl. Acad. Sci. U.S.A.*, 74, 5463, 1977.
112. **Hunter, T. R., Hunt, T., Knowland, J., and Zimmern, D.,** Messenger RNA for the coat protein of tobacco mosaic virus, *Nature (London)*, 260, 759, 1976.
113. **Pelham, H. R. B.,** Leaky UAG termination codon in tobacco mosaic virus RNA, *Nature (London)*, 272, 469, 1978.
114. **Kitamura, N., Semler, B. L., Rothberg, P. G., Larsen, G. R., Adler, C. J., Dorner, A. J., Emini, E. A., Hanecak, R., Lee, J. L., Van der Werf, S., Anderson, C. W., and Wimmer, E.,** Primary structure, gene organisation and polypeptide expression of poliovirus RNA, *Nature (London)*, 291, 547, 1981.
115. **Rueckert, R. R., Matthews, T. J., Kew, O. M., Pallansch, M., McLean, C., and Omilianowski, D.,** Synthesis and processing of picornaviral polyprotein, in *The Molecular Biology of Picornaviruses*, Pérez-Bercoff, R., Ed., Plenum Press, New York, 1979, 113.
116. **Kozak, M.,** Possible role of flanking nucleotides in recognition of the AUG initiator codon by enkaryotic ribosomes, *Nucl. Acids Res.*, 9, 5233, 1981.
117. **Bruening, G.,** Biochemistry of plant viruses, in *The Biochemistry of Plants. A Comprehensive Treatise*, Vol. 6, Stumpf, P. K. and Conn, E. E., Eds., Marcus, A., Ed., Academic Press, 1981, 571.
118. **Franssen, H. and Goldbach, R.,** unpublished results, 1983.
119. **Zabel, P., Moerman, M., Lomonossoff, G., Shanks, M., and Beyreuther, K.,** Cowpea mosaic virus VPg: sequencing of radiochemically modified protein allows mapping of the gene on B-RNA, *EMBO J.*, 3, 1629, 1984.
120. **Lockhart, B. E. L. and Semancik, J. S.,** Inhibition of a plant virus by Actinomycin D, *Virology*, 36, 504, 1968.
121. **Lockhart, B. E. L. and Semancik, J. S.,** Differential effect of Actinomycin D on plant virus multiplication, *Virology*, 39, 362, 1969.

122. **Otsuki, Y., Takebe, I., Honda, Y., Kajita, S., and Matsui, C.,** Infection of tobacco mesophyll protoplasts by potato virus X, *J. Gen. Virol.*, 22, 375, 1974.
123. **Alblas, F. and Bol, J. F.,** Factors influencing the infection of cowpea mesophyll protoplasts by alfalfa mosaic virus, *J. Gen. Virol.*, 36, 175, 1977.
124. **Dawson, W. D. and Schlegel, D. E.,** The sequence of inhibition of tobacco mosaic virus synthesis by actinomycin D, 2 thiouracil and cycloheximide in a synchronous infection, *Phytopathology*, 66, 177, 1976.
125. **Dawson, W. D.,** Time-course of actinomycin D inhibition of tobacco mosaic virus multiplication relative to the rate of spread of infection, *Intervirology*, 9, 304, 1978.
125a. **Renaudin, J. and Bové, J. M.,** Effet de l'actinomycin D sur la production de virions par les protoplasts de Chou de Chine infectés *in vitro* par le virus de la mosaique jaune du navet, *C. R. Acad. Sci. Paris*, 284, 783, 1977.
125b. **Mouches, C., Candresse, T., and Bové, J. M.,** Turnip yellow mosaic virus RNA-replicase contains host and virus-encoded subunits, *Virology*, 134, 78, 1984.
126. **Morris-Krsinich, B. A. M., Hull, R., and Russo, M.,** Infection of turnip leaf protoplasts with turnip rosette virus, *J. Gen. Virol.*, 43, 339, 1979.
127. **Morris-Krsinich, B. A. M. and Hull, R.,** Replication of turnip rosette virus RNA in inoculated protoplasts, *J. Gen. Virol.*, 64, 2661, 1983.
128. **Zabel, P., Jongen-Neven, I., and van Kammen, A.,** In vitro replication of cowpea mosaic virus RNA II. Solubilization of membrane-bound replicase and the partial purification of the solubilized enzyme, *J. Virol.*, 17, 679, 1976.
129. **Van der Meer, J., Dorssers, L., van Kammen, A., and Zabel, P.,** The RNA-dependent RNA polymerase of cowpea is not involved in cowpea mosaic virus RNA replication: immunological evidence, *Virology*, 132, 413, 1984.
130. **Van der Meer, J., Dorssers, L., and Zabel, P.,** Antibody-linked polymerase assay on protein blots: a novel method for identifying polymerase following SDS-polyacrylamide gel electrophoresis, *EMBO J.*, 2, 233, 1983.
131. **Dorssers, L.,** RNA-Dependent RNA Polymerases from Cowpea Mosaic Virus-infected Cowpea Leaves, Ph.D. thesis, Agricultural University, Wageningen, The Netherlands, 1983.
132. **Kumarasamy, R. and Symons, R. H.,** Extensive purification of the cucumber mosaic virus-induced RNA replicase, *Virology*, 96, 622, 1979.
133. **Symons, R. H., Gill, D. S., Gordon, K. H. J., and Gould, A. R.,** *Manipulation and Expression of Genes in Eukaryotes*, Nagley, P., Linnane, A. W., Peacock, W. J., and Patman, J. A., Eds., Academic Press, New York, 1983, 373.
134. **Matthews, R. E. F.,** *Plant Virology*, 2nd ed., Academic Press, New York, 1981.
135. **Otsuki, Y., Shimomura, T., and Takebe, I.,** Tobacco mosaic virus multiplication and expression of the N gene in necrotic responding tobacco varieties, *Virology*, 50, 45, 1972.
136. **Motoyoshi, F. and Oshima, N.,** Infection with tobacco mosaic virus of leaf mesophyll protoplasts from susceptible and resistant lines of tomato, *J. Gen. Virol.*, 29, 81, 1975.
137. **Motoyoshi, F. and Oshima, N.,** Expression of genetically controlled resistance to tobacco mosaic virus infection in isolated tomato leaf mesophyll protoplasts, *J. Gen. Virol.*, 34, 499, 1977.
138. **Beier, H., Bruening, G., Russel, M. L., and Tucker, C. L.,** Replication of cowpea mosaic virus in protoplasts isolated from immune lines of cowpea, *Virology*, 95, 165, 1979.
139. **Dawson, W. O. and White, J. L.,** Characterization of a temperature-sensitive mutant of tobacco mosaic virus deficient in synthesis of all RNA species, *Virology*, 90, 209, 1978.
140. **Dawson, W. O. and White, J. L.,** A temperature-sensitive mutant of tobacco mosaic virus deficient in synthesis of single-stranded RNA, *Virology*, 93, 104, 1979.
141. **Sarachu, A. N., Nassuth, A., Roosien, J., Van Vloten-Doting, L., and Bol, J. F.,** Replication of temperature-sensitive mutants of alfalfa mosaic virus in protoplasts, *Virology*, 125, 64, 1983.
142. **Roosien, J. and Van Vloten-Doting, L.,** Complementation and interference of UV Mts mutants of alfalfa mosaic virus, *J. Gen. Virol.*, 63, 189, 1982.
143. **Nassuth, A., Alblas, F., and Bol, J. F.,** Localization of genetic information involved in the replication of alfalfa mosaic virus, *J. Gen. Virol.*, 53, 207, 1981.
144. **Kiberstis, P. A., Loesch-Fries, L. S., and Hall, T. C.,** Viral protein synthesis in barley inoculated with native and fractionated brome mosaic virus RNA, *Virology*, 112, 804, 1981.
145. **Kassanis, B.,** Properties and behavior of a virus depending for its multiplication on another, *J. Gen. Microbiol.*, 27, 477, 1962.
146. **Murant, A. F. and Mayo, M. A.,** Satellites of plant viruses, *Annu. Rev. Phytopathol.*, 20, 49, 1982.
147. **Duda, C. T., Zaitlin, M., and Siegel, A.,** *In vitro* synthesis of double-stranded RNA by an enzyme system isolated from tobacco leaves, *Biochim. Biophys. Acta*, 319, 62, 1973.
148. **Fraenkel-Conrat, H.,** RNA polymerase from tobacco necrosis virus infected and uninfected tobacco: purification of the membrane-associated enzyme, *Virology*, 72, 23, 1976.

149. **Romaine, C. P. and Zaitlin, M.,** RNA-dependent RNA polymerase in uninfected and tobacco mosaic virus-infected tobacco leaves: viral-induced stimulation of a host polymerase activity, *Virology*, 86, 241, 1978.
150. **Clerx, C. M. and Bol, J. F.,** Properties of solubilized RNA-dependent RNA polymerase from alfalfa mosaic virus-infected and healthy tobacco plants, *Virology*, 91, 453, 1978.
151. **Ikegami, M. and Fraenkel-Conrat, H.,** RNA dependent RNA polymerase of tobacco plants, *Proc. Natl. Acad. Sci. U.S.A.*, 75, 2122, 1978.
152. **Ikegami, M. and Fraenkel-Conrat, H.,** The RNA-dependent RNA polymerase of cowpea, *FEBS Lett.*, 96, 197, 1978.
153. **Ikegami, H. and Fraenkel-Conrat, H.,** Characterization of the RNA-dependent RNA polymerase of tobacco leaves, *J. Biol. Chem.*, 254, 149, 1979.
154. **Hall, T. C., Miller, W. A., and Bujarski, J. J.,** Enzymes involved in the replication of plant viral RNAs, in *Advances in Plant Pathology, Vol. 1*, Ingram, D. and Williams, P., Eds., Academic Press, New York, 1982, 179.
155. **Fraenkel-Conrat, H.,** RNA-dependent RNA polymerases of plants, *Proc. Natl. Acad. Sci. U.S.A.*, 80, 422, 1983.
156. **Bujarski, J. J., Hardy, S. F., Miller, W. A., and Hall, T. C.,** Use of dodecyl-β-D-maltoside in the purification and stabilization of RNA polymerase from Brome mosaic virus-infected barley, *Virology*, 119, 465, 1982.
157. **Miller, W. A. and Hall, T. C.,** Use of micrococcal nuclease in the purification of highly template dependent RNA-dependent RNA polymerase from Brome mosaic virus-infected barley, *Virology*, 125, 236, 1983.
158. **Putnak, J. R. and Phillips, B. A.,** Picornaviral structure and assembly, *Microbiol. Rev.*, 45, 287, 1981.
159. **Strauss, E. G. and Strauss, J. H.,** Replication strategies of the single-stranded RNA viruses of eukaryotes, *Curr. Top. Microbiol. Immunol.*, 105, 1, 1983.
160. **Nomoto, A., Omata, T., Toyoda, H., Kuge, S., Horie, H., Kataoka, Y, Genba, Y., Nakano, Y., and Imura, N.,** Complete nucleotide sequence of the attenuated poliovirus Sabin 1 strain genome, *Proc. Natl. Acad. Sci. U.S.A.*, 79, 5793, 1981.
161. **Racaniello, V. R. and Baltimore, D.,** Molecular cloning of poliovirus cDNA and determination of the complete nucleotide sequence of the viral genome, *Proc. Natl. Acad. Sci. U.S.A.*, 78, 5887, 1981.
162. **Boothroyd, J. C., Highfield, P. E., Cross, A. M., Rowlands, D. J., Lowe, D. A., Brown, F., and Harris, T. J. R.,** Molecular cloning of foot and mouth disease virus genome and nucleotide sequences in the structural protein genes, *Nature (London)*, 290, 800, 1981.
163. **Robertson, B. H., Morgan, D. O., Moore, D. M., Grubman, M. J., Card, J., Fisher, T., Weddell, G., Dowbenko, D., and Yansura, D.,** Identification of amino acid and nucleotide sequence of the foot-and-mouth disease virus RNA polymerase, *Virology*, 126, 614, 1983.
164. **Bienz, K., Egger, D., Rasser, Y., and Bossart, W.,** Intracellular distribution of poliovirus proteins and the induction of virus-specific cytoplasmic structures, *Virology*, 131, 39, 1983.
165. **Toh, H., Hayashida, H., and Miyata, T.,** Sequence homology between retroviral reverse transcriptase and putative polymerases of hepatitis B virus and cauliflower mosaic virus, *Nature (London)*, 305, 827, 1983.
166. **Haseloff, J., Goelet, P., Zimmern, D., Ahlquist, P., Dasgupta, R., and Kaesberg, P.,** Striking similarities in amino acid sequence among nonstructural proteins encoded by RNA viruses that have dissimilar genomic organisation, *Proc. Natl. Acad. Sci. U.S.A.*, 81, 4358, 1984.
167. **Morris-Krsinich, B. A. M., Forster, R. L. S., and Mossop, D. W.,** The synthesis and processing of the nepovirus grapevine fanleaf virus proteins in rabbit reticulocyte lysate, *Virology*, 130, 523, 1983.
168. **Petterson, R. F., Ambros, V., and Baltimore, D.,** Identification of a protein linked to nascent poliovirus RNA and to the polyuridylic acid of negative-strand RNA, *J. Virol.*, 27, 357, 1978.
169. **Dasgupta, A., Zabel, P., and Baltimore, D.,** Dependence of the activity of the poliovirus replicase on a host cell protein, *Cell*, 19, 423, 1980.
170. **Dasgupta, A.,** Antibody to host factor precipitates poliovirus RNA polymerase from poliovirus-infected HeLa cells, *Virology*, 128, 252, 1983.
171. **Brunt, A. A., Philips, S., Barton, R. J., and Jones, R. A. C.,** Ullucus virus C, *CMI/AAB Descriptions of Plant Viruses*, No. 277, 1984.

Chapter 4

ORGANIZATION AND EXPRESSION OF THE CAULIFLOWER MOSAIC VIRUS GENOME

S. N. Covey

TABLE OF CONTENTS

I. INTRODUCTION

The caulimoviruses are the only group of plant viruses known to encapsidate a genome of double-stranded DNA. They have attracted considerable interest as possible vectors for the genetic manipulation of plant cells and also as systems for studying plant gene expression and replication. As a result, the type member of the caulimovirus group, cauliflower mosaic virus (CaMV), is one of the best characterized plant viruses at the molecular level.

Recently, a number of observations have revealed similarities between CaMV and certain animal viruses, in particular the retroviruses, which replicate by reverse transcription. Some of the similarities are so striking that it is now accepted by many that CaMV (and probably all caulimoviruses) replicates by reverse transcription of an RNA intermediate and so represents the first example of this phenomenon in plant cells. This feature could have significant implications, not only for plant molecular biology, especially in the light of the fact that CaMV is also structurally related to a group of transposable elements (see section VI) widespread in eukaryotic organisms.

The emphasis in this chapter will be to consider the CaMV genome as a discrete group of specialized plant genes and also to draw wider comparisons with other eukaryotic genetic elements. This is not intended as a definitive review of CaMV research, but is rather a distillation of recent findings relevant to the organization of the CaMV genome, the structure, function, and regulation of its genes and gene products. Some of the discussion will be of a speculative nature and this is intended to provoke argument and stimulate experimental work in particular areas. A consideration of the pertinent features of caulimoviruses other than CaMV is included where appropriate.

II. BIOLOGICAL FEATURES OF CAULIMOVIRUSES

A brief description of the biological features of caulimoviruses in general and CaMV in particular will follow; for more detailed coverage, the reader is referred to comprehensive reviews,[1-4] and to Chapter 5 in this volume.

The caulimovirus group has six certain members and a further six likely or possible members.[4] At least 33 variants or isolates of CaMV have been described and it is likely that there are natural varients of the other caulimoviruses too. Caulimovirus host-range is fairly narrow with susceptible hosts representing 17 plant families. CaMV, for instance, is limited to the cruciferae in the natural environment.

Following the inoculation of leaves of susceptible plants with virus or viral DNA, the virus subsequently spreads systemically through the plant and, for CaMV at least, particles can be found in most cells which emerge from the plant apex after the initial infection event. Systemic symptoms are variable and with CaMV depend upon the particular isolate but in general a mosaic and mottling of the leaves with vein-clearing is observed. In severe infections, the smaller leaves can become stunted and crinkled.

Of those caulimoviruses investigated, transmission is normally by aphids in a semipersistent manner. Some isolates of CaMV cannot be transmitted by aphids and it has been shown that a virus-specified factor, not produced by the aphid nontransmissible isolates, is required in order that aphids can acquire and transmit the virus from plant to plant[6,7] (see Section IV.B.3).

It has been estimated that there are of the order of 10^5 CaMV particles in each systemically infected turnip leaf cell[8] and the virus is found almost exclusively within, or in close association with, cytoplasmic inclusion bodies (sometimes called viroplasms). Other caulimoviruses also induce the formation of inclusion bodies in which virus particles accumulate. Exceptions are carnation etched-ring virus (CERV), dahlia mosaic virus (DaMV), and petunia vein-clearing virus (PVCV), where particles occur free in the cytoplasm[9] in addition to those found in inclusion bodies. Strong denaturing agents and detergents (e.g., urea and Triton® X-100) are required to disrupt inclusion bodies and liberate virus particles for further purification.[10]

Purified caulimovirus particles are isometric nucleoprotein structures about 40 to 50 nm in diameter.[11] Virions of CaMV contain DNA (about 17% by weight[11]) and neutron diffraction studies of CaMV suggest that a shell of DNA is sandwiched between two shells of protein with an apparently empty interior to the particle.[12] Caulimovirus DNA, extracted from virions, consists largely of double-stranded molecules of 7.8 to 8.0 kilobase (kb) pairs.[4] Both the virus particles and the DNA purified from isolated virions, for those caulimoviruses tested, are infectious when used to inoculate plants mechanically.

III. STRUCTURE OF VIRION DNA

A. Molecular forms

1. Isolation

Purified particles of CaMV are highly resistant to disruption by many chemical reagents. Extraction of DNA from virions is best effected by partial enzymatic digestion of capsid protein using pronase or a similar enzyme.[10,13] Nonvirion DNA which contaminates a virus preparation can be removed by deoxyribonuclease I (DNAse I) treatment prior to virion disruption; DNA within the virion is not accessible to DNAse I and is therefore not affected by this treatment. DNA liberated from virions can be further purified by phenol/chloroform extraction to remove residual protein, isopycnic banding in cesium chloride gradients, and then concentrated by precipitation with ethanol. DNA thus purified is infectious when rubbed on the leaves of a susceptible host plant. It is also a suitable substrate for restriction endonuclease digestion and other enzymatic modifications and can be cloned into bacterial plasmids. A similar purification procedure is effective for other caulimoviruses and *E. coli* recombinant plasmid clones containing the genomes of DaMV (8.0 kbp), thistle mottle virus (ThMV, 7.8kbp), mirabilis mosaic virus (MMV, 7.8kbp), carnation-etched ring virus (CERV, 8.0kbp), and figwort mosaic virus (FMV, 7.8kbp) have been constructed.[16,19] In addition to cloned DNA of CaMV,[14,15] that of MMV, CERV, and FMV is infectious to plants.[16]

2. Twists, Circles, and Gaps

DNA isolated from CaMV virions exhibits a variety of structural conformations with unusual topology. Electron microscope observations have revealed various degrees of twistedness in the DNA. Because CaMV DNA contains single-strand discontinuities, the twisted forms could not have arisen as a result of supercoiling. Molecules appear to be twisted about a single locus but restriction enzyme analysis has suggested that this locus is positioned randomly with respect to known restriction enzyme sites in the viral DNA.[17] Preliminary evidence for a protein[17] stabilizing the twists has not been confirmed and this needs further investigation. Hohn et al.[3] have proposed that the twisted topology results from base-pair interactions close to the single-strand discontinuities, possibly stabilized by RNA. Whatever effects the stabilization of twists in the DNA following its liberation from virions, it is likely to be a remnant of packaging constraints within the virus particle.

Twisted molecules form multiple bands upon electrophoresis in agarose gels. These bands have also been observed in the virion DNAs of CERV, FMV, MMV, ThMV, and DaMV,[16,18-19] and so it is a characteristic feature of caulimoviruses. In addition to the twisted forms, open-circular molecules are observed and also linear forms thought to be produced by breakage of ciruclar molecules at one of the single-strand discontinuities. The discontinuities in CaMV virion DNA were first observed by Volovitch et al.[20,21] These are regions of the DNA susceptible to digestion by the single-strand-specific nuclease S_1.[22] When circular DNA molecules of most CaMV isolates are electrophoresed in agarose gels under denaturing conditions, three major single-stranded DNA species are observed. The largest is 8kb long and the sum of the other two species is also 8kb. Thus, the DNA of most CaMV isolates consists of three single-stranded components maintained in a circular double-stranded conformation because the termini of each strand are located at different overlapping positions in the molecule (see Figure 1). The discontinuities have been mapped with restriction endonucleases; they are at fixed sites with one in one strand and two in the other.[23] In one isolate of CaMV, CM4-184, there is just one discontinuity in each strand because the region containing the third discontinuity is deleted.[22] Other caulimovirus DNAs also have discontinuities, always one in one strand and two in the other strand of CERV and ThMV[18] and three in the second DNA strand of FMV, DaMV, and MMV.[16,19] At least one isolate of DaMV appears to have only two second strand discontinutities.[18a] Thus, caulimovirus DNAs have only one discontinuity in one strand (designated G1 or Δ1) and one to three discontinuities in the other (G2, G3, G4 or Δ2, Δ3, Δ4) (Figure 1). In CaMV, the strand always with one discontinuity is transcribed into RNA[23,24] and is now called the minus-strand. Asymmetric transcription probably occurs in the other caulimoviruses too, but this has not yet been investigated. G1 in the minus-strand (of CaMV DNA) is now accepted as the zero position in DNA maps of caulimoviruses. The DNA strand with a variable number of discontinuities in different isolates or caulimoviruses is known as the plus-strand. In the current theory of CaMV replication[87] it is supposed that the gaps are close to priming sites for DNA synthesis (see Section VI and Chapter 5 of this volume).

3. Virion RNA and Small DNAs

Experiments with purified CaMV virion DNA have demonstrated the presence of ribonucleotides which are released following treatment of the DNA with alkali.[17,25] It has been estimated that 0.6% of virion associated nucleotides are ribonucleotides. The circular DNA molecules of chloroplasts[26] and mitochondria[27] have alkali-labile regions due to the presence of ribonucleotides and these are though to be remnants of RNA primers of DNA synthesis. Renaturation studies have suggested that CaMV DNA-associated RNA is more complex (i.e., 120 nucleotides) than would be expected for a primer of DNA synthesis. Guilley et al.[28] have shown that the 5′ extremities of each of the three discontinuities in virion DNA contain one or a few covalently linked ribonucleotides. They have also detected a small virion-associated single-stranded DNA fragment of minus-strand polarity with a single 5′-terminal ribonucleotide. This would not account for the RNA complexity reported by Hull[17] and the virion-associated RNA could be adventitiously encapsidated cellular RNA. However,

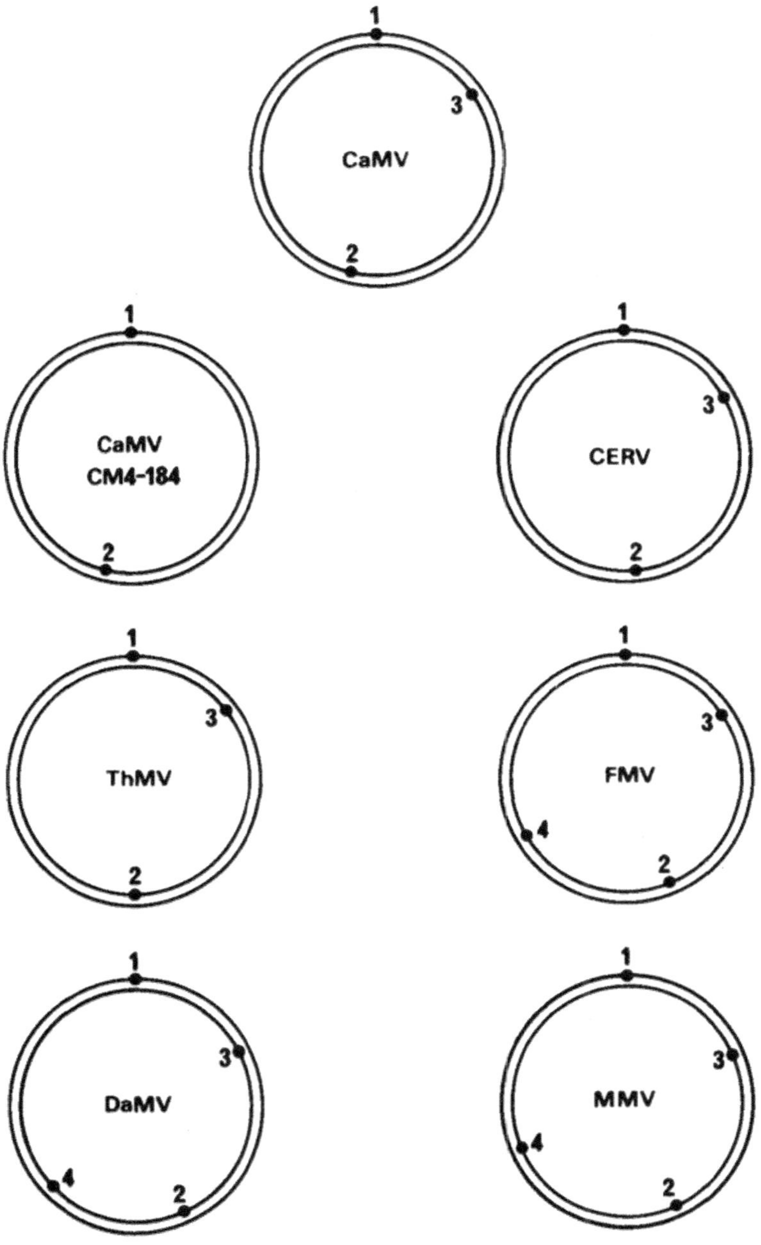

FIGURE 1. Structure of caulimovirus virion DNAs orientated relative to the unique single-strand discontinuity G1 (Δ1) and showing the locations of the second strand discontinuities G2-4 (Δ2 to 4) in CaMV (cauliflower mosaic virus), the mutant CaMV isolate CM4-184 which has a 421bp deletion encompassing G3, CERV (carnation etched-ring virus), ThMV (thistle mottle virus), FMV (figwort mosaic virus), DaMV (dahlia mosaic virus), and MMV (mirabilis mosaic virus).

Covey et al.[29] have recently characterized a host cell tRNA covalently linked to a small CaMV-DNA fragment which originates from virion DNA in the region of G1 and is found encapsidated in CaMV particles.[30] These DNA-associated RNA fragments are probably related to CaMV DNA replication (see Chapter 5).

4. Other Forms of Virion DNA

A variety of CaMV-DNA forms, which appear not to be encapsidated in virions,[31-32a] have also been detected in virus-infected cells. They exist either as free molecules or are in a loose and as yet uncharacterized association with protein. These DNA forms are largely truncated molecules delineated in various combinations by each of the three discontinuities. Some molecules have single-strand extensions of the minus-strand of variable length. In addition, a spectrum of small DNA molecules orginating from the G1 region of virion DNA has been observed.[29,30] The free DNA forms could represent replication intermediates, defective molecules, or in some cases be produced by nuclease action in vivo. Unencapsidated viral DNA forms also occur in plants infected by other caulimoviruses.[18a]

Covalently closed circular CaMV DNA with a supercoiled conformation[31,33] has been recovered from fractions of infected turnip cells.[31-33] A minichromosome containing such supercoiled CaMV DNA is transcriptionally active and synthesizes virus-specific RNA in isolated nuclei.[34,35] Less than genome-length CaMV supercoiled DNAs have also been detected.[125-127]

B. Fine Structure of Virion DNA

1. Restriction Enzyme Maps

Derivation of restriction endonuclease maps of viral DNAs provides a relatively rapid comparison of different but related molecules. However, a knowledge of restriction enzyme sites is perhaps most useful in recombinant DNA manipulations and the mapping of functional domains. Detailed restriction maps of at least 33 different isolates of CaMV are available[5] and work on mapping 5 other caulimoviruses has begun.[16,18,19] In Figure 2 the restriction maps of five CaMV isolates[36-39] and four other caulimoviruses are shown. The cutting sites of only four enzymes are included, but these are representative of typical variations that occur. Only limited information is available for ThMV[16] with all known sites shown in Figure 2.

Several of the enzyme sites are common to each CaMV isolate but there is some variation. The Bari isolate is the least typical. It has been observed[5] that only one restriction enzyme site is common to all CaMV isolates; this is an *Eco*R1 site at nt6107 (according to the numbering system of Franck et al.[41]). However, a single base-pair change can create or destroy a restriction enzyme recognition site, so it is not surprising to find variation between isolates. Hull[5] and Hohn et al.[36] have suggested that certain DNA regions of different isolates can be classified as "constant" or "variable" with respect to restriction enzyme sites. In contrast, Volovitch et al.[40] concluded that variations were randomly distributed and could be accounted for by point mutations. This has largely been borne out from nucleotide sequencing which, in the case of three CaMV isolates for which sequence data is available,[38,41,42] shows that differences are due, in the main, to single base changes, small deletions, and insertions with no major rearrangements. An exception is isolate CM4-184 which has a 421 bp deletion between nt1382 and nt1803[37] (Figure 2).

When the restriction maps of the six caulimoviruses, for which there are data, are compared as a whole, it appears that no similarities exist in the position of enzyme sites relative to the minus-strand discontinuity (G1). A single *Eco*R1 site in CERV is in a similar relative position to that *Eco*R1 site conserved between CaMV isolates. This is probably a coincidence, however, because DNA/DNA hybridization studies between caulimoviruses have revealed virtually no homology.[16] Each caulimovirus is thus distinct and it remains to be seen how the genome organization of the less-well-characterized members of the group compares with that of CaMV.

2. The Nucleotide Sequence

Information specifying the mechanisms of virus propagation inside infected cells (host

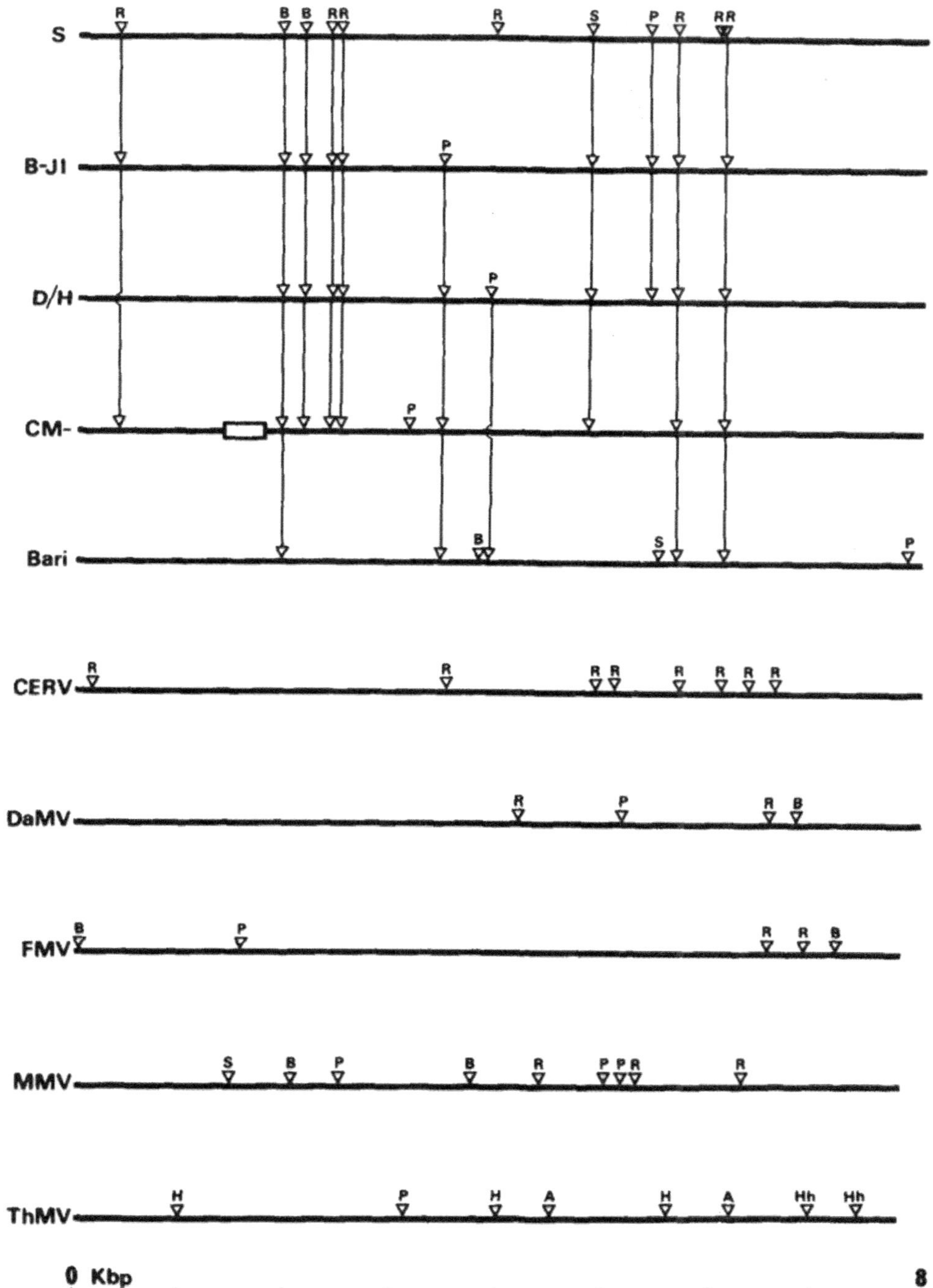

FIGURE 2. Restriction enzyme maps of caulimovirus DNAs drawn in a linear configuration relative to G1 (see Figure 1) which is at the terminus of each line. The cutting sites of four restriction enzymes (R, *Eco*R1; B, *Bam*H1; S, *Sal*1; P, *Pst*1) are shown (open triangles) except for ThMV in which the complete known map is presented (H, *Hind*III; A, *Ava*1; Hh, *Hha*1). Vertical lines in CaMV isolates S (Strasbourg), B-J1 (Cabb B-J1), D/H (D/Hungary), CM- (CM4-184) and Bari indicate enzyme sites in common. The box in CaMV isolate CM4-184 shows the position of a 421bp deletion. For key to caulimovirus names see Figure 1.

genes apart) resides primarily in the arrangement of nucleotides that constitute the virus genome. The availability of vast tracts of sequence data, made possible by modern techniques of rapid sequencing, is a tantalizing prospect for the investigator who has the task of assigning specific functions to particular nucleotide sequences.

The complete nucleotide sequences of three CaMV isolates have already been determined and no doubt several more will appear in due course. These three isolates produce infections in plants which exhibit distinct biological characteristics. CaMV isolate Strasbourg (S) was the first to be sequenced[41] and the biological features of its infection of plants can be considered as "typical". CM1841, the second isolate sequenced,[42] is deficient in the aphid acquisition/transmissibility function. Isolate D/Hungary (D/H)[38] induces attenuated symptoms in plants and produces a typical cytoplasmic inclusion bodies. These isolates vary only slightly in the size of their DNA: S, 8024bp; CM1841, 8031bp; D/H, 8016bp.

Computer analysis of the DNA sequences has been used to locate protein synthesis termination codons (TGA, TAG, TAA) by scanning each of the three possible reading frames in both DNA strands. Long regions lacking these termination codons (here termed "open regions" to distinguish them from the three possible open reading frames of a nucleic acid sequence) are present in each of the three reading frames but in only one of the two strands. CaMV DNA has eight open regions long enough to code for proteins of 10K molecular weight or larger. Many polypeptides with a molecular weight less than 10K possess biological function in other systems, and CaMV DNA has many open regions which could code for small proteins. The 10K threshold is therefore somewhat arbitrary. However, the DNA strand of CaMV with only short open regions probably does not code for protein since transcription is limited to the DNA strand containing long open regions (the minus-strand) in vivo.[23,24] The eight[3] long open regions (Figure 3) are conserved in each of the three sequenced isolates and this might imply that they are all functional and specify polypeptides. In-frame protein synthesis initiation codons occur close to the start of each open region and the predicted sizes of polypeptides that each could encode are shown in Table 1. One feature of the open regions is that they are clustered; the end of each open region closely abuts or even overlaps the beginning of the next. Two exceptions are (1) the 90-nucleotide sequence between open regions V and VI (the small intergenic region) and (2) a large intergenic region of 700 nucleotides between open regions VI and VII (Figure 3).

Balàzs et al.[38] have compared the three sequenced CaMV isolates with respect to base substitutions, deletions, and insertions. They found that the nucleotide sequences differ by only 4 to 5% from one another. The changes are due mostly to base substitutions but with some small deletions and insertions. The longest deletion is one of 21 nucleotides in isolate D/H compared with isolates S and CM1841 located in open region V. Insertions in D/H are no longer than eight consecutive nucleotides. The base substitutions are fairly uniformly distributed except in the C-terminal half of open region V and the two intergenic regions which are relatively highly conserved. The intergenic regions contain important regulatory sequences.

Base substitutions in the putative protein-coding open regions are mostly in the third (silent) base and so amino acid assignments remain unchanged. Table 2 shows amino acid changes, predicted from the nucleotide sequences, in CaMV isolates S, CM1841, and D/H[38] considered pairwise. Out of a total of about 2500 amino acid-coding triplets, there are 102 amino acid changes in S relative to D/H, 133 in CM1841 relative to D/H, and 105 in S relative to CM1841. Of these, 70 amino acid changes are common to both S and CM1841. The less highly conserved regions are probably located in protein domains in which some variation does not adversely change the protein function. The highly conserved C-terminal half of open region V implies that this protein domain has a precise function which presumably cannot accommodate much variation. The greatest variation between CaMV isolates is in open regions VII and VIII, perhaps casting doubt on their protein-coding

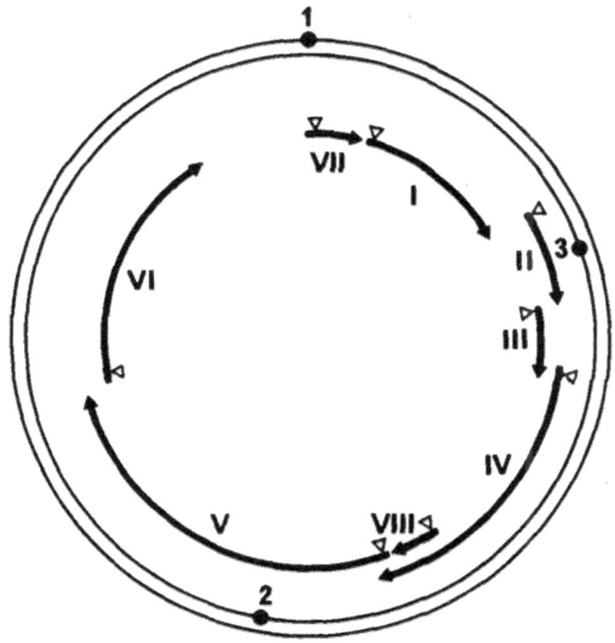

FIGURE 3. Open regions (arrows labeled with Roman numerals) in CaMV DNA located by computer search of the nucleotide sequence, for protein synthesis termination codons TGA, TAA, TAG, in each of the three reading frames. The first in-phase ATG (AUG) initiation codon in each open region is indicated (open triangle). The long open regions shown are found exclusively on the messenger-sense DNA strand (plus-strand) which in virion DNA contains the discontinuities G2 and G3. The minus-strand in a supercoiled form of CaMV DNA is transcribed in vivo into messenger-sense RNA transcripts.

Table 1
OPEN REGIONS AND POLYPEPTIDE SIZES PREDICTED FROM THE DNA NUCLEOTIDE SEQUENCE OF CaMV ISOLATE S

	Nucleotide			
Open region	Last stop codon	ATG	Stop codon	Polypeptide (mol wt)
I	331	364	1347	38K
II	1328	1349	1828	18K
III	1812	1830	2219	15K
IV	2168	2201	3670	57K
V	3591	3633	5672	79K
VI	5713	5776	7338	61K
VII	7888	13	302	11K
VIII	3216	3266	3590	12K

capacity. Variation in open region VI of D/H may be related to the unusual appearance of D/H-induced inclusion bodies which contain a gene product expressed from this open region (see Section IV.B).

Table 2
VARIATIONS IN AMINO ACID CODINGS DEDUCED FROM THE NUCLEOTIDE SEQUENCES OF CaMV ISOLATES STRASBOURG (S), D/HUNGARY (D/H), AND CM1841 (CM)

Open region	Number of amino acids			Substitutions			Additions			Deletions			Total changes (%)			
	S	CM	DH	S/M	S/DH	DH/CM	S/CM	S/DH	DH/CM	S/CM	S/DH	DH/CM	S/CM	S/DH	DH/CM	Mean
I	327	327	327	7	7	8	—	—	—	—	—	—	2.1	2.1	2.4	2.2
II	159	159	159	6	4	8	—	—	—	—	—	—	3.8	2.5	5.0	3.8
III	129	129	129	5	3	6	—	—	—	—	—	—	3.9	2.3	4.6	3.6
IV	489	488	490	16	13	24	1	1	1	2	2	3	3.9	3.3	5.7	4.3
V	680	680	675	9	14	19	—	7	7	—	2	2	1.3	3.4	4.1	3.6
VI	520	520	522	28	33	27	—	—	—	—	2	2	5.4	6.7	5.6	5.9
VII	96	96	96	11	3	11	—	—	—	—	—	—	11.5	3.1	11.5	8.7
VIII	108	107	107	19	11	15	—	1	—	1	—	—	17.8	12.1	14.0	14.6
Total	2508	2506	2505	101	88	118	1	8	8	3	6	7	4.2	4.1	5.3	4.2

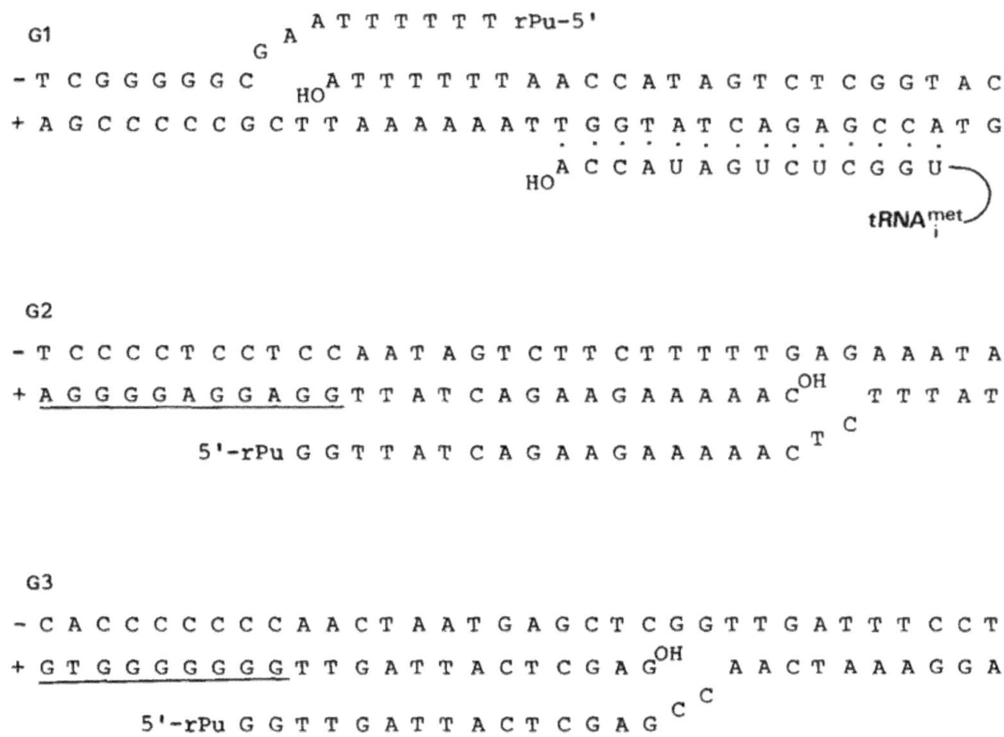

FIGURE 4. Nucleotide sequence around the DNA single-strand discontinuities of CaMV isolate Strasbourg.[41,44] The 5′-end of each discontinuity is at a fixed site with terminal purine ribonucleotides[28] (rPu) covalently linked to each 5′-extremity. The 3′-ends are of variable length in the population of virion DNA molecules. The sequences underlined are thought to be important in priming synthesis of the DNA plus-strand (+) at G2 and G3.[45,46,87] The sequence homologous with plant tRNA$_i^{met}$ adjacent to G1[30] is considered to be the origin of minus-strand (−) synthesis.

3. Gap Structures

Early experiments by Volovitch et al.[20] and Hull and Howell[22] showed that three sites in most isolates of CaMV DNA were susceptible to digestion by the single-strand-specific nuclease S1. This was interpreted by suggesting that a nick or short gap was present in one of the DNA strands at each S1-sensitive site. The conclusion reached from "fill-in" experiments using DNA polymerase I was that the discontinuities were gaps in one of the two strands of 8 to 10 nucleotides.[23,43] Subsequently, sequencing of CaMV virion DNA revealed an unexpected structure for the discontinuities. They were shown to be short regions of strand overlap potentially producing a triple-stranded structure. Further work by Richards et al.[44] revealed the size variation of the overlaps at each discontinuity. The G1 (Δ1) overlap was predominantly 8 to 10 nucleotides long, G2 (Δ2) 19 nucleotides long, and G3 (Δ3) predominantly 15 nucleotides long but, in a small proportion of molecules, up to 35 nucleotides long. Thus, variation exists at each discontinuity presumably due to heterogeneity within the population of virion DNA molecules used in the study. Guilley et al.[28] demonstrated that in 30% of molecules the terminal nucleotide at G1 is ribo A. In 5% of molecules at G2 there is an oligo ribopurine tract eight to ten nucleotides long and in 5% of molecules a single A residue. G3 also has a terminal ribonucleotide in a proportion of molecules. The current view of the gap structures in isolate S[28,44] is shown in Figure 4. In the model of CaMV replication[45,46] G1 is located close to the origin of DNA minus-strand synthesis, and G2 and G3 are close to the origins of plus-strand DNA synthesis. The ribonucleotides are thought to be remnants of RNA primers of DNA replication. The overlaps at each gap are probably produced by short-lived displacement synthesis towards the end of the replication cycle. It is likely, although yet to be determined, that the gap structures of all other CaMV isolates and those of other caulimoviruses are similar to those of CaMV isolate S.

IV. FUNCTIONS OF CaMV GENES

The nucleotide sequence of CaMV DNA predicts that as many as eight polypeptides could be encoded by the virus genome. There are several possible functions that virus-specific polypetides might fulfill including enzymes required in replication or transcription, structural proteins and others involved in virus transmission within and between plants, and host-range specificity. Two approaches to identifying CaMV gene functions and products are (1) to characterize virus-specific polypeptides and RNA transcripts isolated from infected plants and (2) predictions made from the DNA sequence together with in vitro manipulation of cloned DNA (e.g., site-specific mutagenesis, in vitro recombination, and use of expression vectors). Both general approaches have been adopted.

A. Virus-Specific Proteins

The spectrum of proteins present in a plant cell changes as a consequence of virus infection due to the synthesis of viral genome-encoded polypeptides. In addition, specific host genes could be activated or repressed in response to the virus infection. Thus, in analyzing virus-specific proteins found in infected cells, a distinction must be drawn between virus- and host-coded gene products.

1. Coat Protein

There has been some confusion in the literature regarding the polypeptide composition of purified CaMV virions. It has been claimed[47-54] that from 2 to 11 polypeptides with molecular weights of between 27K and 100K comprise virion structural polypeptides as determined by polyacrylamide gel electrophoresis. Al Ani et al.[51] concluded, however, that multiple polypeptides observed by others were artifacts produced by degradation and partial polymerization of the authentic coat polypeptide; they claim[51] that the CaMV coat protein consists of a single polypeptide species of 42K. This work was extended by Hahn and Shepherd,[52,53] who detected two virion polypeptides, one of 58K and a second of 44K (probably the same as the 42K species reported by Al Ani et al.[51]) both of which were found to be phosphorylated in vivo. Pulse-labeling experiments with ^{32}P-orthophosphate showed a precursor-product relationship between the 58K and 44K CaMV virion polypeptides. It was concluded that the primary translation product of the coat protein gene is a 58K polypeptide which is phosphorylated and assembled into virions where it undergoes gradual conversion in vivo to produce the 44K species. It is not known whether the 58K polypeptide is the mature coat protein which is degraded to the 44K species, or if a true precursor/product relationship exists between the two. Subsequent degradation of the 44K polypeptide either in vivo or fragmentation and polymerization during virus purification would produce the multiplicity of polypeptide size variants observed in polyacrylamide gels.

Phosphorylation appears not to be the only modification to the coat protein as evidence for glycosylation has been reported.[50,54] How these modifications function in the assembly of CaMV virions is not at all clear.

2. Inclusion Body Proteins

A characteristic feature of cells infected by all known caulimoviruses is the accumulation of cytoplasmic inclusion bodies. Most of the virus particles observed in infected cells are associated with inclusion bodies in which they appear to be embedded. Inclusion bodies

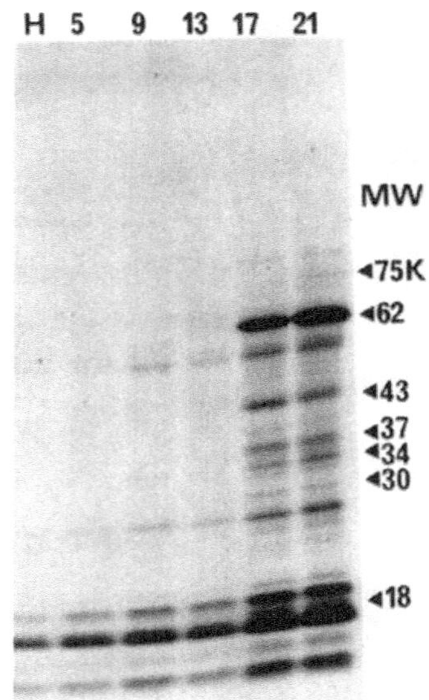

FIGURE 5. SDS-polyacrylamide gel electrophoresis of CaMV-specific polypeptides, stained with Coomassie Blue, in inclusion body preparations from systemically infected turnip leaves 5, 9, 13, 17, and 21 days after inoculation of healthy (H) plants. The sizes of polypeptides novel to infection are indicated.

contain most of the novel viral-specific polypeptides found in infected plants. They can be prepared from leaf extracts by cycles of low-speed centrifugation and further purified by centrifugation in sucrose or cesium chloride gradients. The most abundant component of inclusion bodies is a polypeptide the molecular weight of which has been variously described as being 55K,[55] 61K,[56] 62K[57,58] and 66K.[59,60] These differences could result from variations in the gel electrophoresis methods used to determine their size. They may also relate to virus isolate differences and in most cases probably refer to the same polypeptide. However, Shepherd et al.[55] claim that two different types of inclusion bodies induced by isolate CM1841 have distinct polypeptide components. The major polypeptides of vacuolated virion-containing inclusion bodies are 64K, 55K, and 43K molecular weight. Only one major polypeptide, of 55K, was found in granular nonvirion-containing inclusion bodies. Xiong et al.[60] also observed two types of inclusion body in cells infected with CaMV isolates S and D/H. A major 66K polypeptide was present in S inclusion bodies but in lesser amounts in those of D/H. They correlate this latter observation with an unusual cytological appearance of D/H inclusion bodies which were apparently deficient in matrix material. Polypeptides of 55K, 43K, and 37K were found in both types. Limited protein sequencing data suggests that the 55K polypeptide found in granular CM1841 inclusion bodies is not virus coded and not related to chloroplast fraction 1 protein large subunit[61] of similar size. Exactly what function the 55K polypeptide serves remains uncertain, but presumably it is produced by the host in response to virus infection. It is now generally accepted that the major nonvirion protein of virion-containing inclusion bodies is the 61K to 66K species which constitutes the bulk of the inclusion body matrix. The 42K to 44K and 34K to 37K polypeptide species associated with inclusion body preparations (see Figure 5) are probably related to the virus coat protein.

An additional polypeptide associated with inclusion bodies, of 18K (P18), has been reported by Woolston et al.[62] This polypeptide was detectable in plants infected by an aphid-

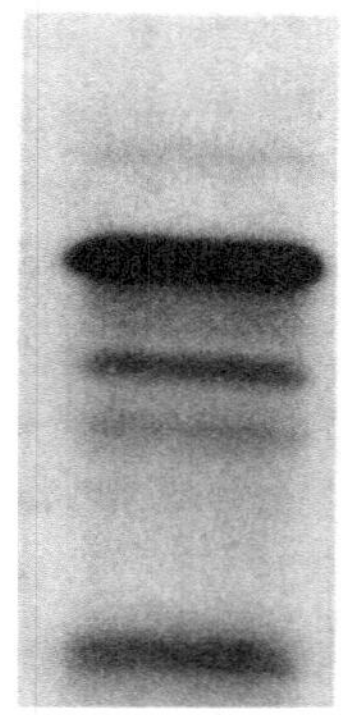

FIGURE 6. Hybrid select translation (HST) of CaMV-specific poly(A)+RNA. RNA was isolated from infected turnip leaves and hybridized to CaMV DNA immobilized on DBM paper. Selected RNA was melted off the DNA and translated in vitro using a messenger-dependent rabbit reticulocyte lysate. Polypeptides synthesized in vitro, labeled with ^{35}S-methionine were separated on an SDS-polyacrylamide gel and detected by fluorography. (Modified from Covey, S. N. and Hull, R., *Virology*, 111, 463, 1981. With permission.)

transmissible CaMV isolate, Cabb. B-J1, but was not found in inclusion body preparations from plants infected with aphid nontransmissible isolates Campbell and CM4-184. Polypeptide P18 was also not detectable in plants infected with Cabb. B-J1 rendered aphid-nontransmissible by in vitro deletion mutagenesis.

It should be pointed out that some of the additional polypeptides recovered from low-speed pellet fractions containing virus inclusion bodies, but not present in low-speed pellet fractions from noninfected plants, could be host proteins that artifactually adhere to inclusion bodies during isolation.

3. In Vitro Translation Studies

In vitro translation, in heterologous cell-free systems, of cellular total RNA or poly(A)+RNA from CaMV-infected leaves has produced a major additional 61K to 66K M_r polypeptide (P61 to P66) not found in translation products of RNA from healthy plants.[57-59] This translation product comigrated in polyacrylamide gels with the 61K to 66K matrix protein of virion-containing inclusion bodies. Comigration on gels is not sufficient evidence to identify a protein; in addition, these results do not tell us whether P61 to P66 is virus or host encoded. However, hybrid arrest translation (HART) and hybrid select translation (HST) experiments have resolved the latter question. In HART, translation of a specific messenger RNA is arrested when the RNA is hybridized to the DNA sequence from which it is transcribed. This is monitored by nonappearance of a specific translation product. Al Ani et al.[57] and Odell and Howell[59] showed by HART that the P61 to 66 mRNA was virus coded, and *Eco*R1 fragment b of CaMV isolate B-JI contained the coding sequence for P61 to 66. Covey and Hull[58] came to similar conclusions using HST. Virus-specific poly(A)+RNA was purified from host-cell RNA by hybridization to CaMV (Cabb. B-J1) DNA immobilized on diazobenzyloxymethyl (DBM) paper. The major translation product from hybrid selected RNA was P61 to 66 together with minor virus-specific polypeptides of 75K, 52K, 44K, and 34K (see Figure 6).

The identity of in vitro synthesized P61 to 66 was demonstrated by Covey and Hull,[58] who showed that it had a similar pattern of peptide fragments, produced by partial digestion with V_8 protease, to those from the major matrix protein (P61 to 66) of inclusion bodies labeled in vivo. Xiong et al.[63] subsequently confirmed that in vitro synthesized P61 to 66 was serologically related to that protein of similar size isolated from inclusion bodies.

Little information is available at present upon the identity of other polypeptides synthesized by in vitro translation of CaMV-specific RNA. The translation of one CaMV-specific polypeptide of 75K (P75), observed by HST[58] (Figure 6), has been hybrid arrested by a CaMV DNA *Eco*R1 fragment containing open region V[128] which is sufficiently long to encode a 79K polypeptide. Future in vitro translation experiments should provide further information about the less-well-characterized CaMV polypeptides and their mRNAs.

B. Identification of CaMV Genes

1. The Coat Protein Gene

Compelling evidence[41] first suggesting that open region IV of the CaMV genome encoded the virus coat protein came from an amino acid composition analysis of virion structural protein which has a high lysine content. The amino acid composition of proteins potentially coded by each of the CaMV DNA open regions was compared with that of the coat protein. Only open region IV showed a very close match with the coat protein amino acid composition. Further evidence came from experiments of Daubert et al;[64] who cloned a CaMV DNA *Eco*R1-*Hin*dIII fragment, containing open region IV, into the *E. coli* expression vector pBR325. The open reading frame of the CaMV insert was fused with the plasmid chloramphenicol acetyl transferase gene. Antigen synthesized in *E. coli* containing pBR325 with the inserted CaMV DNA fragment cross-reacted with antiserum raised against purified coat protein. Only CaMV DNA fragments containing open region IV in the correct orientation produced coat protein antigen. Open region IV is long enough to encode a 57K polypeptide corresponding closely to the size of 58K which, as claimed by Hahn and Shepherd,[52,53] is the likely primary translation product of the coat protein gene.

2. The Inclusion Body Matrix Protein Gene

The DNA sequence encoding the P61 to 66 matrix protein of inclusion bodies was located by hybridization experiments with its polyadenylated mRNA. Odell and Howell[59] showed that the mRNA for P61 to 66 sedimented as a 19S species in sucrose gradients and its translation was arrested by CaMV (B-J1) DNA fragment *Eco*R1-b. Covey and Hull[58] identified, by in vivo ^{32}P-labeling, a 19S RNA species which comigrated with P61 to 66 translation activity in agarose slab gels. Cross-contact hybridization studies revealed that the P61 to 66 mRNA hybridized to the contiguous CaMV (B-J1) DNA *Eco*R1 fragments −d and −b. Open region VI is the only open region common to both restriction enzyme fragments and it was concluded that this is the P61 to 66 inclusion body matrix protein gene.[58]

This was confirmed by Xiong et al.,[63] who sequenced the N-terminal amino acids of in vitro synthesized P61 to 66 and concluded that the first in-phase ATG codon (AUG in the RNA sense) of open region VI specified the initiating amino acid of P61 to 66. Open region VI is long enough to encode a 61K mol wt polypeptide (in CaMV isolate S — see Table 1).

3. The Aphid Transmission Factor

The transmission of CaMV from plant to plant by aphids is dependent upon the production of a virus-specified factor, the aphid transmission factor (ATF). This was the suggestion of Lung and Pirone,[6] who found that nontransmissible CaMV isolates CM1841 and Campbell were transmitted by aphids from plants co-infected with transmissible isolates or by aphids which had previously probed plants containing transmissible isolates. Also, aphids only

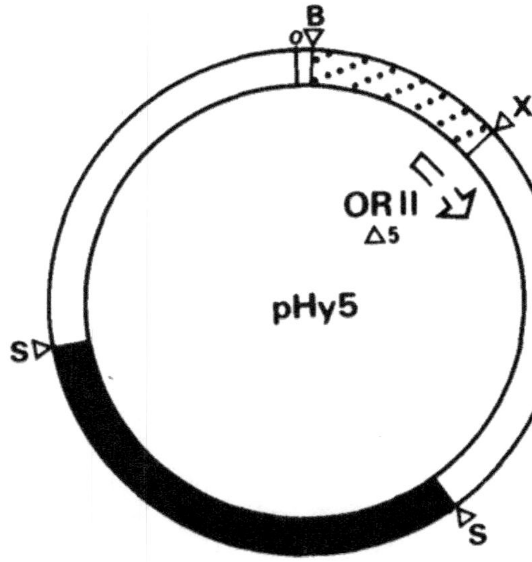

FIGURE 7. A hybrid CaMV genome (pHy5) of an aphid nontransmissible isolate Campbell (unshaded area) rendered transmissible when the *Bst* EII (B) to *Xho*I (X) fragment (halftone) is replaced by that fragment from a transmissible isolate Cabb B-JI. The dark shaded area is pAT153 DNA. An in vitro constructed deletion (Δ5) within open region (OR) II also renders Cabb B-JI aphid nontransmissible.

transmit purified virus (either of transmissible or nontransmissible isolates) if they previously probe plants infected with a transmissible isolate.[7] A factor mediating virus transmission by aphids is therefore specified only by transmissible isolates of CaMV and is produced in infected plants.

Woolston et al.[62] extended these studies by constructing, in vitro, hybrid genomes between a transmissible CaMV isolate (B-J1) and a nontransmissible isolate (Campbell). The transmissible B-J1 was rendered nontransmissible when recombined with the larger *Bst*EII-*Sal*I fragment of Campbell. The reciprocal recombinant construct conferred transmissibility on Campbell. The location of the aphid transmissibility function was narrowed down in similar hybrid genome constructions to a region between the *Xho*I site and the *Bst*EII site which contains open region I and parts of open regions II and VII (see Figure 7). Concomitant with the aphid transmissibility function was the presence of an 18K polypeptide (P18) associated with inclusion bodies isolated from infected leaves. In all the hybrid constructs, P18 was present only in plants infected with CaMV DNA containing the *Xho*I-*Bst*EII fragment of the transmissible isolate B-JI. Open region II, within the *Xho*I-*Bst*EII fragment, is sufficiently long to encode an 18K polypeptide and is therefore a suitable candidate for the helper factor gene.

This view is supported by the finding that another nontransmissible CaMV isolate, CM4-184, has a 421bp deletion in open region II and plants infected with this isolate do not produce P18 either. Woolston et al.[62] have shown that deletion of 126bp in open region II of the normally transmissible (P18-producing) isolate B-J1 renders it nontransmissible and non-P18 producing. However, yet another aphid nontransmissible isolate, CM1841, appears to have no significant insertions or deletions in open region II or between the *Xho*I-*Bst*EII sites.[42] Moreover, sequencing of the nontransmissible Campbell isolate transmission-related *Xho*I-*Bst*EII fragment has revealed no major insertions or deletions which might account for the virus phenotype[62a] although a premature termination codon was found in the middle of open region VII of Campbell, but not B-JI. However, open region VII in the nontransmissible isolate CM1841 is intact suggesting either that open region VII is not involved in aphid transmission, or that a third region of the genome (presumably containing a lesion in isolate CM1841) is required to complete the function. Armour et al.[65] also showed, by in vitro deletion mutagenesis, that open region II is involved in aphid transmission of isolate NY8153. Gardner[69] has performed hybrid recombination experiments between the transmissible isolate Cabb B-D and the nontransmissible isolate CM1841 obtaining results similar to those of Woolston et al.[62] In conclusion, there is compelling evidence implicating open region II in aphid transmissibility of CaMV probably via production of an 18K polypeptide. It still remains unresolved, however, whether some other part (or parts) of the genome is (are) required in the helper function.

4. The Reverse Transcriptase Gene?

The striking similarities between CaMV and animal viruses which replicate by reverse transcription[45,46] have stimulated the search for a CaMV-specific polymerase which uses an RNA template to synthesize DNA (see Chapter 5). Of the unassigned open regions which could code for this enzyme, only I and V are obvious possibilities and as suggested by Hull and Covey,[45] open region V is perhaps the best candidate since it would produce a 79K polypeptide, one of similar size to some retrovirus reverse transcriptase polypeptides and in particular to that of murine sarcoma virus reverse transcriptase monomers.[66] A protein of 79K, with reverse transcriptase activity, has not yet been detected in CaMV-infected plants. However, a minor product of about 75K mol wt is synthesized in vitro by translation of virus-specific poly(A) + RNA[58,128] (and Figure 6). Low levels of a novel 75K mol wt polypeptide have also been detected in inclusion body preparations (Figure 5).

Toh et al.[67] have provided further evidence for the proposition that open region V encodes a reverse transcriptase. They compared the amino acid sequences of retrovirus reverse transcriptases from Moloney murine leukemia virus (Mo-MuLV) and Rous sarcoma virus (RSV) with that predicted from the putative polymerase gene of hepatitis B virus and CaMV open region V. Some similarities in amino acid sequence were found to be very close, with ten amino acids in identical relative positions and other domains exhibiting a high degree of homology. Toh et al.[67] argue that similarities in amino acid sequence would be expected even in distantly related eukaryotic viruses in proteins which conserve a highly specific function. Volovitch et al.[67a] also arrived at similar conclusions. It is interesting to note that Balazs et al.[38] comment that part of open region V is the most highly conserved CaMV coding sequence between CaMV isolates.

5. Other Viral Genes

Open regions I, III, VII, and VIII remain, as yet, unassigned with respect to protein-encoding function. In vitro mutagenesis[68,69] experiments have shown that DNA in open regions I and III is essential for virus propagation in plants. More recently, Daubert et al.[118] and Dixon et al.[119] have extended these studies. Insertions of small DNA fragments into open regions I, III, IV, V, and the 5′ portion of VI were found to be lethal. However, mutagenesis of the 3′ regions of open regions IV and VI resulted in no loss of infectivity, but the latter modification reduced the severity of symptoms in plants,[118] while the former retarded symptom development.[119] Open region VII appears to be nonessential for virus propagation within plants since a small insertion, and a large deletion of 105bp within it resulted in no loss of DNA infectivity.[119] The tentative evidence linking the dispensable open region VII with the aphid transmissibility function[62a] was discussed in Section III. There is a relatively high degree of sequence variation within open regions VII and VIII, so it is possible that they do not encode protein at all.

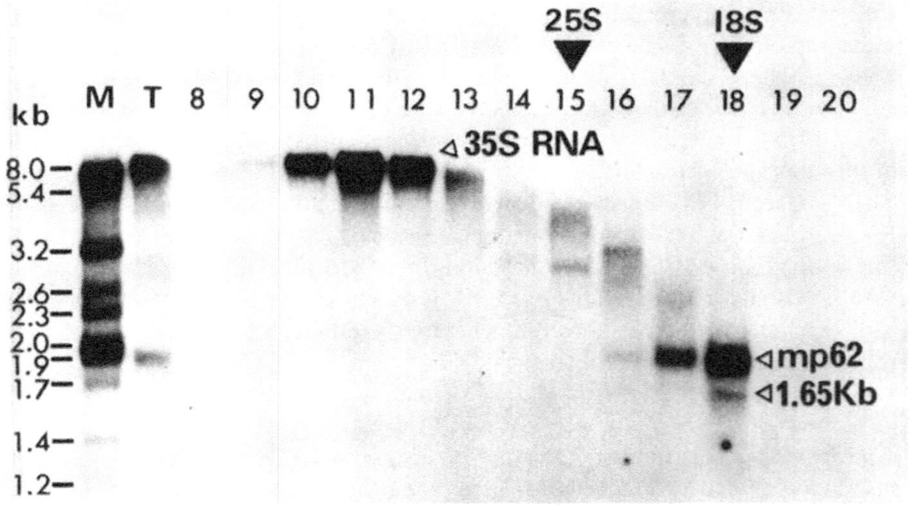

FIGURE 8. Northern blot-hybridization of CaMV-specific polyadenylated transcripts (T) sedimented through a sucrose gradient (fractions 8 to 20). A major 35S RNA comigrates during denaturing agarose gel electrophoresis with an 8kb size marker (M). The major 1.9kb RNA, recovered from fraction 18 and cosedimenting with 18S ribosomal RNA is the messenger RNA (mP62) for the P61 to 66 inclusion body matrix protein. The minor 1.65kb transcript (open triangle) is, as yet, uncharacterized. (Modified from Covey, S. N., Lomonossoff, G. P., and Hull, R., *Nucl. Acids Res.*, 9, 6735, 1981. with permission.)

Woolston et al.[62] noted, as a result of hybrid construction experiments, that symptom expression in plants was specified by the small *Xho*1-*Sal*1 fragment which contains all of open regions III, IV, and VIII and parts of II and V. Whether this is related to expression of the coat protein gene or one of the other open regions in the restriction fragment is, as yet, unknown.

V. EXPRESSION OF CaMV DNA

A. RNA Transcripts

CaMV-specific RNA was first detected in infected turnip protoplasts.[24] A single polyadenylated RNA species was observed which migrated in polyacrylamide gels more slowly than ribosomal RNA and hybridized exclusively to the DNA minus-strand. Virus-specific RNA in the poly(A)+RNA fraction, isolated from infected whole leaves, has also been shown to be complementary only to the minus-strand.[23] The conclusion that the minus-strand is the transcribed strand is supported by the CaMV DNA sequence which shows that the potential protein coding capacity is located only in the strand complementary to the minus-strand[38,41,42] (i.e., the plus-strand). CaMV transcripts have been identified in RNA preparations by blot-hybridization methods following denaturing agarose gel electrophoresis. Only two abundant CaMV transcripts have been detected and characterized[58,70-72] (Figure 8). Transcripts of low abundance have been reported,[58,72,123,124] but their functions remain obscure. It has been estimated[58] that about 5% of the poly(A)+RNA fraction of infected cells is virus-specific RNA and so represents a sizeable proportion of the cell messenger RNA complement.

The major CaMV transcripts are (1) the mRNA for the 61K to 66K inclusion protein, here designated for simplicity mP62 and variously described as 19S RNA, mVIBP, 2.3kb RNA, and 1.9kb RNA; (2) 35S RNA, a genome length molecule otherwise called the large transcript, 8kb RNA or 8.2kb RNA.

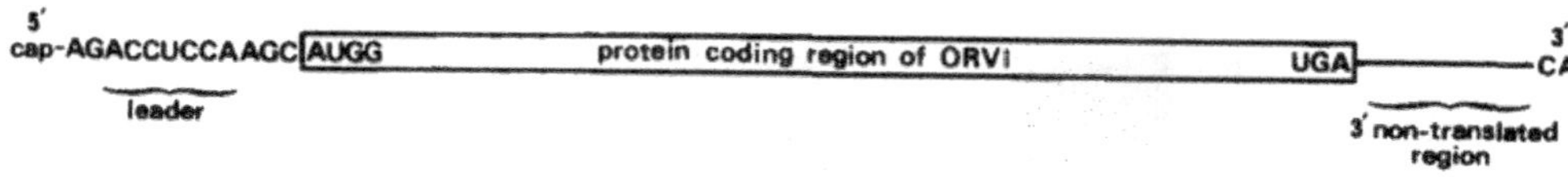

FIGURE 9. Structural and functional domains of the messenger RNA (mP62) for the P61 to 66 inclusion body matrix protein.

1. mP62

In an earlier section (IV.B.2) I discussed the evidence demonstrating that mP62 is the polyadenylated messenger for the major inclusion body protein (P61 to 66) encoded by open region VI. The RNA sediments in sucrose gradients as a 19S species[59] and its size, determined by blot-hybridization following fully denaturing agarose gel electrophoresis, is 1.9kb.[71] The fine structure of mP62 and the DNA sequence from which it is transcribed has been determined by S1 nuclease mapping and sequencing techniques.[71-73] The RNA 5′ terminus is transcribed from within CaMV DNA (B-J1) EcoR1 fragment-d. This fragment contains the start of open region VI and the small intergenic region. The 3′ terminus was mapped to the large intergenic region some 400 nucleotides upstream of the virion DNA minus-strand discontinuity G1. The S1 mapping experiments showed that mP62 is transcribed from a contiguous DNA sequence and is therefore not spliced during maturation. This is in agreement with the structure of open region VI and also precludes donation of a leader sequence transcribed elsewhere in the genome.

High-resolution S1 mapping located the 3′-end of mP62 at nt7615 (in isolate S[72]). The 5′ end was mapped by primer extension and dideoxy sequencing of a cDNA copy of mP62 and was shown to be located at nt5764[72] or nt5765.[71] The length of mP62 is 1850 ± 3 nucleotides plus the poly(A) segment and its structure is shown in Figure 9. A 5′ cap[59] precedes a short leader sequence of 12 nucleotides before the first in-phase AUG initiation codon. The protein coding sequence is 1560 nucleotides long equivalent to 520 amino acid residues. A TGA stop codon precedes a 276 nucleotide 3′ nontranslated region adjacent to the poly(A) tail. Messenger P62 is judged to be polyadenylated on the basis of its affinity for oligo (dT)-cellulose[57-59] and is probably also capped.[59]

2. 35S RNA

35S RNA is transcribed from the complete CaMV DNA minus-strand and denaturing gel electrophoresis has shown that this polyadenylated RNA is about the same size as the minus-strand.[70-72] It was originally considered most likely that 35S RNA transcription was initiated just downstream of the virion DNA minus-strand discontinuity G1 and terminated, after one circuit, just upstream of G1. One conclusion of UV-mapping experiments by Howell[74] was that the 35S RNA 5′ terminus was not adjacent to G1. This would require a traversing of the discontinuity during transcription or some splicing mechanism to produce the mature 35S transcript. Subsequently, it was demonstrated that the 35S RNA 3′ end mapped close to that of mP62 and the 5′ end was located close to its own 3′-end.[71-72] Various explanations were suggested to account for likely difficulties a polymerase might encounter to negotiate the discontinuity in the minus-strand. However, this was obviated when a transcriptionally active supercoiled form of CaMV DNA was isolated from nuclei of infected cells[35] and this molecule is now thought to be the template for transcription of 35S RNA.

A surprising conclusion reached from the work of Covey et al.[71] was that the 5′ end of 35S RNA mapped about 200 nucleotides upstream of its own 3′ end. This finding was confirmed by Guilley et al.[72] The 35S RNA 3′ end is now known to be coterminal with that of mP62 at nt7615. The 5′ end maps at nt7435, 180nt upstream of the 3′ end. This ''overlap'' produces a direct terminal repeated sequence in 35S RNA (Figure 10) (disregarding the poly

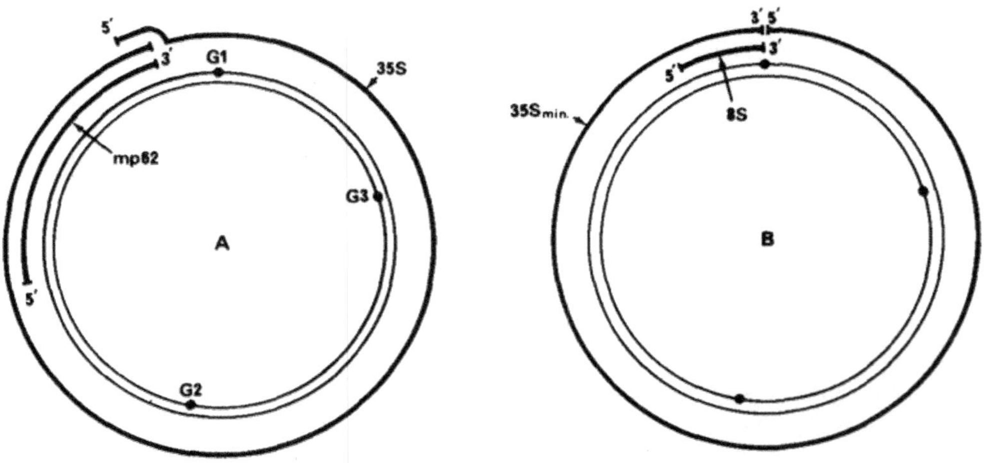

FIGURE 10. Major (A) and minor (B) CaMV transcripts (thick lines) aligned with CaMV virion DNA (thin lines). The 5′-end of 35S RNA is located about 600nts upstream of the DNA minus-strand discontinuity G1. The 35S RNA 3′-end is 180nts downstream of its own 5′-end, producing a direct terminal repeat, and is 400nts from G1. 35S RNA is 3′ co-terminal with mP62. The 5′-end of the minor 8S transcript is co-terminal with that of 35S RNA and its 3′-end is at G1 where the termini of 35S RNA_{min} are located. See text for references.

(A) tail). 35S RNA is, therefore, a copy of the complete protein coding functions residing in CaMV DNA (if, as it is believed, they are all located on one DNA strand). Strong evidence implicating 35S RNA in the replication of CaMV DNA is detailed elsewhere. Does 35S RNA also function in the translation of CaMV genes? Two possibilities are (1) that 35S RNA is a polycistronic messenger or (2) that it is a precursor to smaller discrete mRNAs for each of the open regions. There is little published experimental evidence for or against either proposition (see Section V.B.4).

3. Minor Transcripts

Two minor transcripts, apparently polyadenylated, have been described by Guilley et al.[72] One, 35S RNA^{min}, has termini mapping at G1. The other, 8S RNA, has a 5′ end coterminal with 35S RNA and a 3′ end at G1 (Figure 10). No function has been assigned, with any certainty, to either RNA. 35S RNA^{min} could be a polycistronic template for translation or a precursor to smaller RNAs. 8S RNA does not contain an open region and therefore is an unlikely messenger. The involvement of both RNAs in CaMV replication is an open question. It is possible that neither molecule functions in infected cells but could have arisen from fortuitous transcription of virion DNA.

Other minor CaMV transcripts have been reported by Condit et al.[123,124] some of which appear to be of subgenomic size while others are 35S molecules. The latter population consist of genome-length RNAs with 5′ and 3′ termini mapping at several locations on the viral genome. Whether these function in protein synthesis, are artifacts of S1-mapping, or are contaminating DNA forms is difficult to assess at present. A subgenomic polyadenylated RNA of 1.8kb was found which apparently originates from a region of the viral genome containing open region I.[124]

None of the minor CaMV transcripts detected by northern blot-hybridization or S1 mapping experiments have yet been shown to be mRNAs by in vitro translation. Moreover, other transcripts could be nonpolyadenylated and detection of these is more difficult on gels because ribosomal RNAs interfere with transfer of minor RNA species to nitrocellulose. Preliminary S1 mapping experiments[122] suggest that the population of virus-specific poly(A)-minus RNAs is different from that of poly(A)-plus RNA. Similar results have been reported by Guilfoyle et al.[120] Minor RNAs in the poly(A) + RNA fraction might be obscured by fragments of 35S RNA which produce a high background during blot-hybridization. Investigation of polyribosome-associated RNA should help to resolve some of these questions.

B. DNA Sequences Controlling Expression

1. The Transcription Complex

Nuclei isolated from CaMV-infected leaves synthesize virus-specific RNA in vitro.[34] This RNA was complementary to the minus-strand, known to be transcribed in vivo. In vitro transcription was sensitive to α-amanitin strongly implicating involvement of host cell RNA polymerase II. Olszewski et al.[35] have isolated a transcriptionally active minichromosome containing covalently closed circular CaMV-DNA molecules from turnip nuclei. The minichromosome,[33] treated with micrococcal nuclease,[35] exhibited a typical nucleosome structure and contained turnip RNA polymerase II actively transcribing RNA from the CaMV minus strand. This nuclear minichromosome is probably the major CaMV transcriptional complex producing (at least) 35S RNA; it is not known whether virion DNA is also transcribed. The involvement of RNA polymerase II in CaMV transcription implies that nucleotide sequences modulating this process are related to host cell sequences of similar function.

2. Promotors

The characterization of nucleotide sequences controlling gene expression is an important step towards understanding functional chromosome domains. Many eukaryotic genes, particularly in animal and animal virus systems, have been sequenced. The nucleotide sequences upstream of the RNA transcription initiation site (the cap site, designated + 1) which constitute promotors have several features in common. The general picture that emerges is that eukaryotic promotors invariably possess an A-T rich sequence located about 30 nucleotides upstream of the cap site. This sequence motif, known as the TATA box[75] (consensus TATAT/AAT/A), is thought to fix the precise positioning of the cap site. At − 50nt to − 100nt a sequence of consenus GGC/TCAATCT (the CAT box[76]) appears to influence the quantitative level of transcription. At more remote locations from the cap site, and not necessarily upstream of it, additional sequences affecting the level of transcription, known as enhancers, have been detected and characterized.[77]

The upstream sequences of relatively few genes encoding plant proteins have been determined. Those plant genes and upstream sequences that are known share similar overall features of the better characterized animal genes.[78] Plant genes have a − 30 TATA box (consensus TC/GTATAT/AA_{1-3}C/TA) and the cap site nucleotide is invariably an A residue. The CAT box consensus sequence of animal genes is not always found in plant genes but a − 70 to − 100 consensus plant sequence, the AGGA box (C/TA_{2-5}G/$TNGA_{2-4}$C/TC/T), proposed by Messing et al.,[78] may function in a similar manner to the CAT box of animal cells.

A search of the CaMV-DNA sequence for TATA box consensus sequences has revealed several possible promotors.[3] Many of these are probably not functional components of promotors because of their location relative to known RNAs and open regions. Exceptions are the TATA boxes just upstream of the cap sites of both mP62 and 35S RNA (Figure 11). The cap site of each RNA is at an A residue with a TATA box at positions − 31 and − 33, respectively. A CAT box-like sequence (ACAAT) is at − 64 from the 35S RNA cap site but one is not upstream of the mP62 cap site. However, sequences resembling the plant AGGA box are located at − 70 and − 90 upstream of mP62 and 35S RNA cap sites, respectively.

There are additional conserved sequence motifs upstream of both major CaMV RNA cap sites. One, at − 60 to − 75, is the sequence TCCCAC and a second, TGGAGA, is located

							nt7435
	−261	−166	−144	−64	−35	−9	+1
35S RNA	GTGGAAAAGGAAGGTGG..	..ACCCCCACCCAC..	..GTGGAAAAAG..	..ACAATCCCAC..	..CCTCTATATAAG..	..TTGGAGA..	..A
					T		
mP62	..TTGGAAGAAA..	..ACAACCAC..		..AGATTCCCAC..	..C CTATATAAA..	..CTGGAGA..	. A
	−216	−143		−75	−35	−18	+1
							nt5764

FIGURE 11. Nucleotide sequences (CaMV isolate Strasbourg) upstream of the 5′ ends (cap sites denoted + 1) of 35S RNA (nt7435) and mP62 (nt5764) with some similarities to eukaryotic promotors including a −30 TATA box and more remote upstream regions related to enhancer core sequences. See text for further discussion.

between the TATA box and cap site (Figure 11). The functional significance of these heptameric sequences is not known but it is interesting to note that the plant genes encoding zein (maize),[79] the small subunit polypeptide of ribulose bis(phosphate) carboxylase (wheat,[80] soybean[80a]), actin (soybean[81]), and five leghemoglobin genes (soybean[82]) have a similarly located sequence related to the CaMV TGGAGA motif.

The two CaMV "promotor" sequences are very highly conserved in the three isolates for which data are available. A single T substitute for an A residue in the mP62 TATA box of Strasbourg (S) isolate does not appear to adversely affect transcription of open region VI, at least, as far as it is known.

Upstream sequences of both CaMV RNAs are recognized as promotors in a cell-free transcription system derived from HeLa cells which accurately initiates transcription of these RNA at their respective cap sites.[72] Thus, Howell's[74] contention that the two major CaMV transcripts are transcribed from separate promotors is largely borne out by other experimental evidence. However, the recognition of CaMV promotors by an animal in vitro transcription system presumably reflects only those features common to animal and CaMV promotors. Sequence and chromosomal domains controlling subtleties of transcription initiation might well be dissimilar and thus not recognized in CaMV by the heterologous polymerase in HeLa cells. Such questions about CaMV will be better answered by use of a homologous plant in vitro transcription system or by plant cell transfection experiments.

To my knowledge, there is no direct evidence for the existence of functional enhancer DNA sequences in plant chromosomes or in CaMV; experiments designed to detect these domains are yet to be reported. An enhancer "core" sequence has been proposed from studies of several animal and animal virus genes.[77] This sequence approximates to the consensus GGTGTGGAAAG about which variation has been observed. Inspection of the CaMV DNA sequence upstream of the 35S RNA cap site reveals two possible enhancer sequences (at nt7290 and nt7175) which closely match the enhancer core sequence. One, TCGTGGAAAAAG, is at −140, and a second, AGTGGAAAAGG, is at −260 from the 35S RNA cap site. Similarly, an enchancer-like sequence TTGGAAGAAA is found at nt5550, 210nt upstream of the mP62 cap site. An AC-rich sequence is also present in the −140 to −170 region of both cap sites (Figure 11).

3. *Terminators*

A conserved sequence, AATAAA, located about 20 nucleotides from the 3′ terminus of mRNAs in eukaryotes, is thought to specify postranscriptional poly(A) addition.[111] It is not known whether this sequence is directly involved with transcription termination per se, but can be considered, because of its proximity to transcript 3′ ends, as one of several possible components of the transcription termination process. This sequence is also found in the 3′ segment of plant mRNAs for which limited data is available.[78] However, the location of the poly(A) signal sequence in plants is at variance compared with other eukaryotes. Several plant genes appear to possess two signal sequences and in these cases AATAAA is located closer to the protein synthesis termination codon and the second sequence, with a consensus of G/AATAAN, is found 15 to 40 nucleotides from the RNA 3′ end.

Both CaMV major transcripts coterminate downstream of a single classical poly(A) signal sequence, AATAAA positioned 12 nucleotides from the RNA 3′ ends. However, transcription termination in CaMV is likely to exhibit unusual features, since, for 35S RNA synthesis at least, termination apparently occurs only on the second circuit of transcription in order to produce a molecule with a terminal direct repeat. There are several possible mechanisms by which this might be achieved. One is that transcription, once initiated, does not terminate and that mature RNA is processed back to the poly(A) signal sequence from larger molecules. A situation somewhat analogous to this exists for polyoma virus.[83] DNA sequences or RNA secondary structure upstream of the 35S RNA 5′ end might be required for termination.

```
nt7591                             ▼
5'..TTCTATCAATAAAATTTCTAATTCCTAAAACCAAAATCCA...3'

                      nt7634:   CTAAAA
                        7636:           AAAATCC
                        7649:  CCTAAA
                        7682:      AAACCA
                          29:         ACCAAAA
                          46:    TAAAACC
                          55:         ACCAAAAT
                          81:   CTAAAA
                         225:   CTAAAA
                         252:           AAAATC
```

FIGURE 12. Nucleotide sequence in the region of the 3′-ends (mapping at closed triangle) of the two major CaMV transcripts. The poly(A) signal sequence is boxed as well as the symmetrical sequence together with fragmentary downstream repeats of it.

This could explain how mP62 correctly terminates at the same site and would infer that significant upstream termination sequences reside within the mP62 coding sequence. If this were not the case, then molecules longer than mP62, but with the same 5′ end, would be expected to accumulate. Such molecules have not been observed. This would make an alternative possibility, that of "leaky" termination signals permitting read-through, less likely as well. Whatever the mechanism of transcription termination in CaMV, it is interesting to note that an unusual nucleotide sequence is located in the termination region. The two major RNA 3′ ends map within the central part of a 70 nucleotide DNA sequence virtually devoid of G residues. In addition, repeating elements flank the poly(A)-signal sequence. The mP62 and 35S RNA 3′ ends map within a 17-nucleotide symmetrical sequence, CCTAAAACCAAAATCC (see Figure 12). A nine-nucleotide direct repeat containing 7 nucleotides of the distal portion of the sequence is located 4 nucleotides downstream of it. The proximal 6 nucleotides of the symmetrical sequence constitute part of a second direct repeat of 7 nucleotides located 18 nucleotides downstream of it. Parts of the sequence 6 to 8 nucleotides long are repeated in at least 8 other sites more remote from it but downstream and within the large intergenic region. Whether this has any functional significance for transcription termination is difficult to assess and these questions will probably be best answered by DNA transfection of cells following site-directed mutagenesis. The termination-proximal sequences are very highly conserved between CaMV isolates.

The generation of RNAs with direct repeats, produced by read-through (or whatever mechanism) of a first, and recognition of a second termination signal, occurs during transcription of retrovirus proviral DNA and some eukaryotic transposable elements such as yeast Ty elements and *Drosophila copia*-like elements. It is interesting to note that sequences closely related to the CaMV symmetrical sequence downstream of the poly(A)-signal sequence are similarly placed in several of the *Drosophila copia*-like elements (see Section VI).

4. *Sequences Controlling Translation*

The current hypothesis concerning the initiation of protein synthesis on eukaryotic messenger RNAs propounded by Kozak[84] can be summarized as follows. It is thought that the 40S ribosome subunit binds close to the 5′ end of the RNA, probably in a region lacking extensive secondary structure, and scans downstream until it reaches an AUG initiation codon located in a favorable sequence environment. This is often the first AUG it encounters

```
                              G
CONSENSUS:                    A N N [A U G] G
```

REGION Ia	U U G A G | A U G | G		Ib	A U A A U | A U G | C	
IIa	A U A A U | A U G | A		IIb	*	
IIIa	C U A A A | A U G | G		IIIb	A A U C C | A U G | A	
IVa	G C A G G | A U G | G		IVb	A U C U U | A U G | A	
Va	A A G U G | A U G | G		Vb	A G G U G | A U G | A	
VIa	C A A G C | A U G | G		VIb	*	
VIIa	G A G C C | A U G | A		VIIb	*	
VIIIa	A G A A U | A U G | G		VIIIb	*	

FIGURE 13. Nucleotide sequences (RNA sense) proximal to the first (a) and second (b) in-phase AUG initiation codons of CaMV open regions I-VIII aligned below the consensus eukaryotic sequence.

but several examples are known where an upstream AUG does not have the appropriate flanking sequences to permit initiation which subsequently occurs at an AUG codon in a favorable configuration. The favored consensus sequence is G/ANN*AUG*G. The only identified CaMV RNA known to have a messenger function, mP62, conforms to this consensus. Two important questions remain concerning expression of the other CaMV genes. One is the identification of the first functional initiating codon in each of the open regions. The second concerns the structure of the RNA(s) involved in the translation of these open regions. Can anything be inferred from the DNA sequence which may shed light on these problems? The first AUG codon in each CaMV open region conforms to preferred eukaryotic initiation proximal sequences (Figure 13). The nucleotide following AUG in open regions II and VII is an A rather than a G residue but these open regions have only one appropriate in-frame initiation codon. If the first initiating codon in each open region is recognized, where, in the messenger RNA expressing them, does the small ribosome subunit bind? Two possibilities are (1) that subgenomic messengers are generated from separate promotors (as is mP62) or by processing from 35S RNA (or some other precursor molecule) and (2) they are expressed from (a) polycistronic messengers(s). The latter case suggests that ribosomes initiate on internal AUG codons. Although there is little published evidence that this actually happens in CaMV, Sieg and Gronenborn[128a] have proposed from mutagenesis studies a "relay race" model for the translation of open regions I to III in which ribosomes move directly from the end of one reading frame to the begining of the next.

Hull[129] has proposed a mechanism for translation of 35S RNA by which ribosomes might initiate at internal AUG codons. The leader sequence of 35S RNA, from its 5′ end to open region I, is about 1kb in length. Potential intramolecular base pairing within the leader of 35S RNA (determined by computer analysis of the nucleotide sequence[130]) can bring the RNA 5′ end, where 40S ribosome subunits are thought to bind, to an RNA domain immediately adjacent to the initiating codon of open region I. A similar mechanism has been suggested for the translation of retrovirus genomic RNA.[131] Furthermore, Hull has also observed[129] that there is a short sequence in the leader of 35S RNA, complements of which are found upstream of each CaMV open region. Thus, a ribosome subunit would be brought

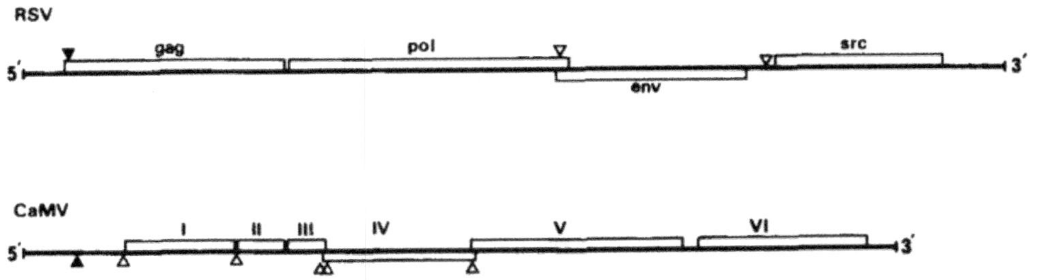

FIGURE 14. (A) Genetic organization of RSV genomic RNA showing location of open regions and the position of the splice donor sequence (closed triangle) in the RNA leader and splice acceptor (open triangles) sequences at the 5′ ends of the *env* and *src* cistrons. (B) CaMV 35S RNA showing similarities to RSV and location of possible splice donor (closed triangle) and acceptor (open triangle) sequences.

to the start of each open region by appropriate folding of the RNA. However, there have been no reports of the successful translation of 35S RNA in vitro although additional specific proteins, found only in infected cells might be required to stabilize 35S RNA secondary structure.

Yet another means of expression of the CaMV genes is suggested by comparison with that exhibited by the animal retrovirus Rous sarcoma virus (RSV). The close similarities in genome organization and replication between CaMV and retroviruses will be detailed later (see Section VI and Chapter 5) but the significance of this might also extend to gene expression. Genomic RNA of RSV contains four long open regions the *gag, pol, env,* and *src* genes (see References 86, 104, and Figure 14). The *gag* and *pol* genes are expressed by translation of an RNA molecule similar in structure to genomic RNA. The primary translation product is a *gag-pol* polyprotein which is subsequently cleaved to mature polypeptides. However, the *pol* cistron is not contiguous with that of *gag* (in fact it is in a different reading frame) and it is thought that translation is effected either by some frameshift mechanism or by a mini splice between the *gag* and *pol* reading frames which fuses the two genes. Subgenomic messenger RNAs containing the *env* and *src* cistrons are generated from the genomic RNA by splicing. A leader sequence at the 5′ end of RSV genomic RNA is spliced onto the *env* and *src* coding sequences, respectively. The leader splice donor sequence in RSV is located about 400 nucleotides from the RNA 5′ end. The *env* splice acceptor sequence is about 20 nucleotides inside its own open region; the *src* splice acceptor sequence is in an intergenic region between *env* and *src* (Figure 14). Within the RSV leader sequence there are several AUG codons upstream of the functional initiating AUG recognized in translation of viral mRNAs.

CaMV open regions I, II, IV, and V might be expressed by generation of subgenomic mRNAs similar to those of RSV. In the 5′ "leader" of CaMV 35S RNA, upstream of open region I, there is a unique octameric sequence (AAGGTAAG) at nt7881, some 445nts downstream of the RNA 5′ end, which is the only sequence within the leader to conform very closely with the consensus eukaryotic splice donor sequence CC/AAGGTA/GAGT.[121] Much variation about the consensus splice acceptor sequence exists but there is generally an oligo pyrimidine-rich region, lacking the dinucleotide AG, preceding an obiligatory dinucleotide AG at the acceptor site. Inspection of the CaMV sequence upstream of the first ATG triplet of open regions I, II, and IV and the second ATG of open regions IV and V, shows the presence of possible splice acceptor sequences (Figure 15). The motif AGCAGG is common to those sequences immediately preceding the ATG triplets of open regions IVa and Vb and other repeated elements precede AUG IVb and Vb (Figure 15b) although this is not known to be significant. As with the leader sequence of RSV, that of CaMV 35S

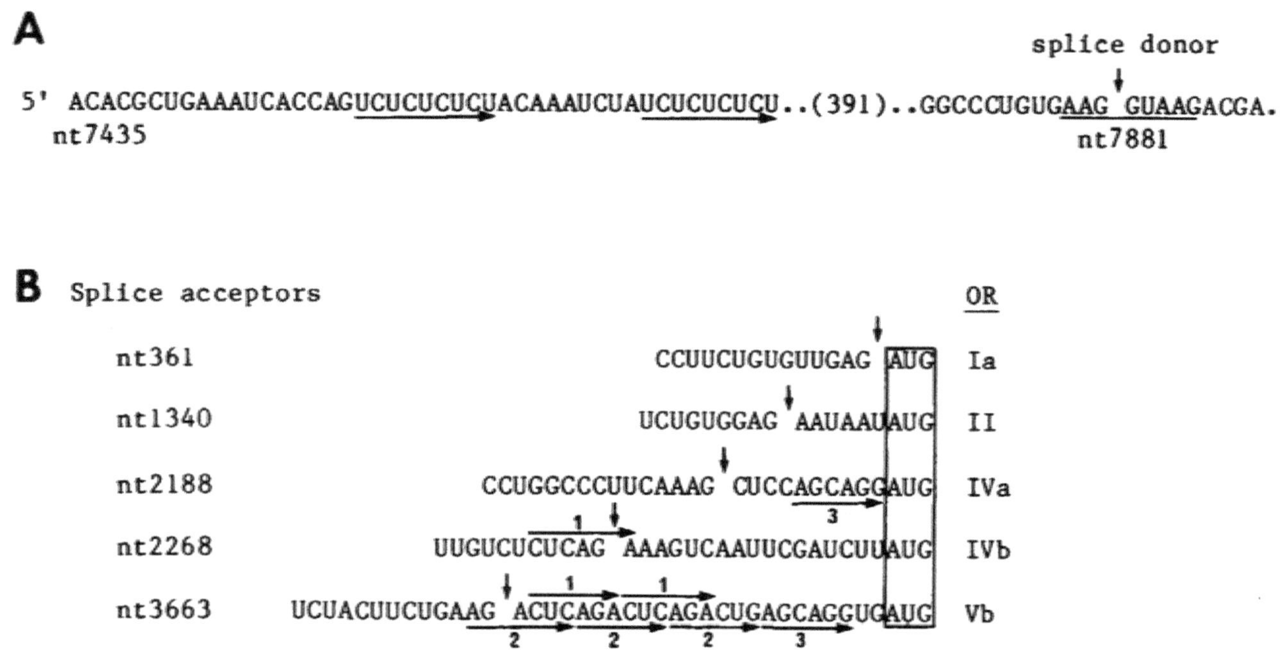

FIGURE 15. (A) A possible splice donor sequence 450 nucleotides from the 5′ end of CaMV 35S RNA. (B) Sequence similarities and possible splice acceptors upstream of initiation codons of the CaMV open regions (OR). Horizontal arrows denote repeated elements.

RNA (up to the putative splice donor sequence) has several ATG triplets before the supposed functional initation codon.

There is no obvious splice acceptor sequence upstream of open region III, but by analogy with RSV, frameshift read-through of CaMV region II into III might produce a small polyprotein which, as with the *gag-pol* precursor, is cleaved to give the mature proteins. There is some evidence of frameshift read-through between regions II and III: Brisson et al.[133] have replaced the latter half of region II with the *E. coli* dihydrofolate reductase (DHFR) gene which is expressed after infection of plants suggesting that the end of region II is not required for translation initiation of region III. This supports the "relay race" translation model of Sieg and Gronenborn and is evidence against part of the 35S RNA leader folding scheme suggested by Hull,[129] at least for the expression of region III. In addition, splicing may not be required in the expression of region I if it is recognized in a manner similar to the *gag* gene of RSV.

In summary, frameshift read-through could be the mechanism of expression of CaMV open regions I to III from a 35S RNA species, perhaps following translation initiation at the start of open region VII or I by folding of the 35S RNA leader sequence,[129] with possible subgenomic RNAs generated (by splicing?) for the translation of regions IV and V. Subgenomic CaMV RNAs have been detected[124] in infected plants but little is yet known about their structure or function. A further intriguing possibility is that distinct subgenomic CaMV minichromosomes[125-126] are templates for transcription of individual open regions.

These models are at present somewhat speculative but are presented in order to stimulate the search for messenger RNA(s) expressing open regions I to V. It is important that ideas are generated and can be tested experimentally as this is a crucial area of CaMV research in which several questions remain, as yet, unanswered.

VI. SIMILARITIES TO RETROVIRUSES AND EUKARYOTIC RETROID ELEMENTS

The genome organization of CaMV shares many features in common with integrating and movable genetic elements found in eukaryotic cells.[132] These include the DNA phase of animal retroviruses and structurally related "retroid" elements of yeast (Ty elements), *Drosophila* (*copia*-like elements) and of vertebrates which are thought to transpose to multiple chromosomal locations. Hepatitis B viruses which probably replicate by reverse transcription of an RNA intermediate also exhibit striking similarities to CaMV. In this section the discussion is limited to structural comparisons with these elements. The replication of CaMV is covered in Chapter 5.

A. Retroviruses

The encapsidated genome of animal retroviruses is a polyadenylated RNA (usually two copies) with distinguishing structural components (see reviews by Varmus[85,86]). These are a terminal repeated sequence (r); a unique sequence, U5, adjacent to the 5′ r and another, U3, adjacent to the 3′ r sequence. At the 3′ border of U5 there is a 15 to 20 nucleotide sequence, the primer binding site (pbs), complementary to the 3′ end of a host cell tRNA species which primes the first stage of reverse transcription. The long central domain (5kb to 8kb) of retrovirus genomic RNA is of plus sense and contains genes coding for virus-specific proteins. At the left-hand border of the RNA U3 sequence is a polypurine tract (ppt) delimiting the origin of plus-strand DNA synthesis. The DNA copy of retrovirus genomic RNA is integrated into host chromosomes as the double-stranded DNA provirus which is transcribed into genomic RNA and subgenomic messengers. The nucleotide sequences in retrovirus RNA are reorganized in proviral DNA. This integrated DNA has long terminal repeated elements (LTRs) flanking a central domain equivalent to the RNA central domain. These sequences are organized: left LTR: U3-r-U5; pbs; central domain; ppt; right LTR: U3-r-U5 (see Figure 16). The internal extremities of each LTR have perfect or near perfect inverted repeats of 10 to 20 nucleotides thought to function in the integration process.

The organization of CaMV genomic sequences share several features with those of retroviruses.[45,46,87] Protein coding sequences (open regions) are located on one DNA strand (the plus-strand) and are contained in 35S RNA following transcription of the DNA minus-strand. The polyadenylated 35S RNA has a terminal directly repeated sequence (r) of 180 nucleotides. A 14-nucleotide sequence of perfect complementarity with the 3′ end of plant $tRNA_i^{met}$ is located 600 nucleotides from the 5′ end of 35S RNA and is also adjacent to G1 in virion DNA. Located on the 5′ sides of each of the two plus-strand virion DNA discontinuities G2 and G3 are polypurine tracts reminicent of retrovirus plus-strand primer regions (ppts). CaMV differs from retroviruses in that its ppts are located within the protein coding domain and not flanking it (Figure 16).

The structure of CaMV 35S RNA is similar to that of retrovirus genomic RNA. The sequence analogous to retrovirus U5 in CaMV 35S RNA is located between the 3′ extremity of the 5′ r unit and the G1 position adjacent to pbs. However, the location of the 5′ extremity of the equivalent CaMV U3 sequence is not readily apparent because the ppts are located within, and not adjacnet to, the protein coding domain.

In retroviruses, inverted repeats at the borders of provirus DNA LTRs are adjacent to the pbs and ppt sequences. It is interesting to note that in CaMV a sequence adjacent to pbs (near G1) has long imperfect inverted repeats within (and therefore not adjacent to) the two ppt sequences (near G2 and G3). Both of these can form hairpin loop structures (Figure 17) which, in certain integrating DNAs (e.g., transposable elements and retroviruses), are thought to be intermediates in the integration process. There is no evidence for integration of CaMV DNA into host cell chromosomal DNA and so the significance of these inverted repeats remains questionable. They may simply be remnants of a previously functional structure.

Transcription of retrovirus provirus DNA to produce genomic RNA appears to be under the control of upstream sequences typical of eukaryotic promotors and these are located in the provirus LTRs. Like CaMV, read-through of the left LTR termination signals must precede correct termination of RNA synthesis specified in the right LTR.

B. Eukaryotic Retroid Elements

The structure of retroviral provirus DNA is shared by other distinct eukaryotic chromosomal elements found at multiple locations and which are thought to be mobile, like prokaryotic transposable elements. These include *copia* and *copia*-like elements of *Drosophila* (*copia*,[88] 17.6,[89] 297,[90] mdg1,[91] mdg3,[92] 412,[93] B104[94]) Yeast Ty elements,[95] and vertebrate retrovirus provirus-like structures.[96-97b] Because of their similarity to integrated retrovirus DNA, it has been proposed[98] that these "retroid" elements are evolutionarily related to retroviruses. Yeast Ty and *Drosophila copia* elements and the genes of mammalian intracisternal A particles (IAPs) are actively transcribed into full-length poly(A)+RNAs with terminal repeats analogous to retrovirus genomic RNA and CaMV 35S RNA. *Copia* RNA has been detected in retrovirus-like particles in *Drosophila*[99] and it has been mooted[100] that reverse transcription is involved in *copia* transposition. Such transposable elements have LTRs with regions reminiscent of retrovirus sequences U3-r-U5 containing transcription signals, pbs-like sequences, and polypurine tracts.

1. Polypurine Tracts

The polypurine tracts (ppts) which delimit the origins of plus-strand DNA synthesis of retroviruses and those of eukaryotic retroid elements are located outside of the 5′ extremity of the right LTR and in CaMV on the 5′ sides of the virion DNA discontinuities G2 and G3. In general, the ppts have an A-rich region preceding a G-rich region. A comparison of

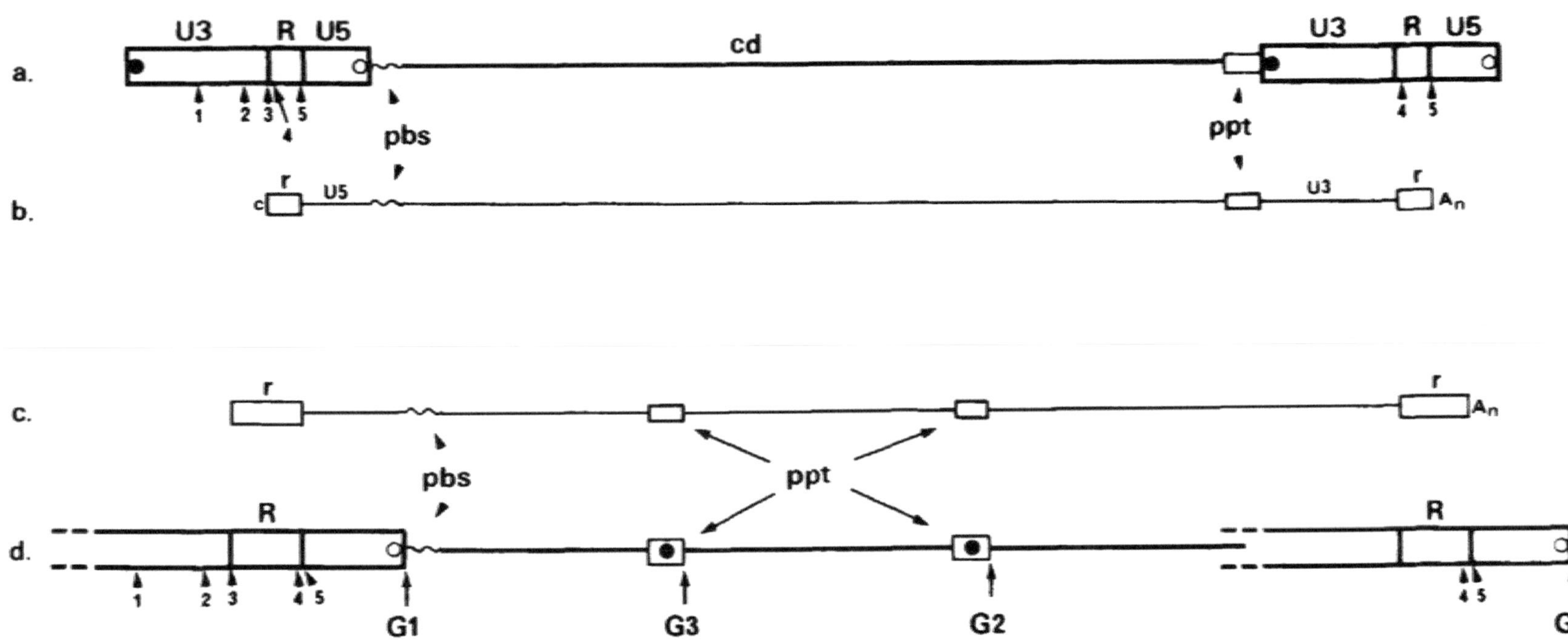

FIGURE 16. Organization of integrated retrovirus provirus DNA (a) and retrovirus genomic RNA (b) compared with CaMV 35S RNA (c) and CaMV virion DNA (d) drawn in a linear configuration such that the sequences specifying the termini of 35S RNA are represented twice. The long terminal repeats (LTRs) of retrovirus provirus DNA contain sequence elements U3, R, U5 which are rearranged in genomic RNA. Adjacent to U5 of the left LTR is a sequence (pbs) complementary to the 3′ end of a host cell tRNA which primes minus-strand synthesis. To the left of U3 in the right LTR a polypurine tract (ppt) delimits plus-strand synthesis. At the left border of each LTR is a short sequence (closed circle) an inverted repeat of which is located at the LTR right border (open circle). Sequences regulating transcription are 1, CAT box; 2, TATA box; 3, cap site; 4, poly(A) signal sequence; 5, the sequence delimiting the RNA 3′-end. The central domain (cd) encloses viral polypeptides on one strand only. Retrovirus RNA (b) has a 5′-terminal cap (c) and a 3′-terminal poly(A) segment (A_n) each adjacent to a terminal directly repeated-sequence (r). Symbols are the same for CaMV 35S RNA (c) and DNA (d). The pbs sequence in CaMV has homology with the 3′-end of plant $tRNA^{met}$ initiator.

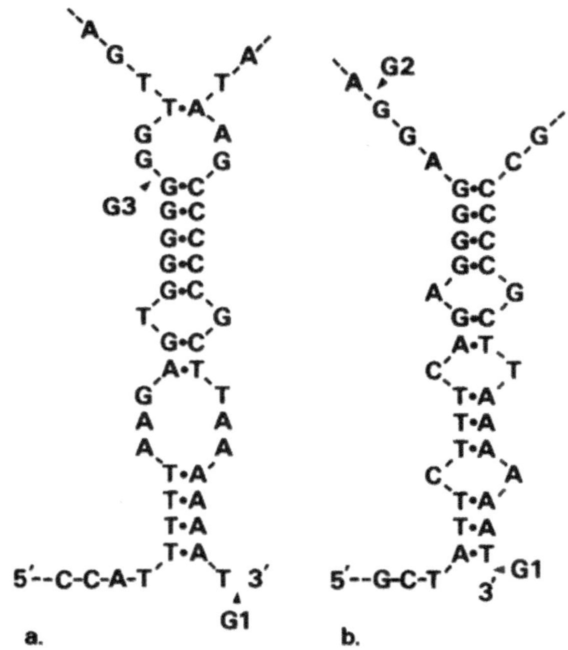

FIGURE 17. Hairpin loop structures (loop not shown) which could be formed from imperfect inverted repeat sequences located ajacent to CaMV virion DNA discontinuities (a) G3 and G1 and (b) G2 and G1.

these sequences extends the similarities to a T-rich segment to the left of the A-rich region (Figure 18). The general structure of the ppts (i.e., Ts-As-Gs) is conserved in retroviruses, retroid elements, and CaMV. In addition, a similar region is found close to the 5′ end of the plus-strand of hepatitis B virus (HBV)[108] and woodchuck hepatitis virus (WHV)[109] which are also thought to replicate by reverse transcription.[110] The ppt sequences in CaMV are highly conserved between isolates.

2. Primer Binding Sites and Inverted Repeats

The pbs sequences differ within and between groups of similar elements (except for the more highly conserved TGG complementary with the tRNA terminal nucleotides 3′ ACC) but are either perfectly homologous with or related to specific tRNAs present in the particular organism in which the elements are found. This is the case for retroviruses, retroid elements, and CaMV (Figure 19).

The terminal dinucleotides in the inverted repeats in the LTRs of retroviruses and retroid elements are invariably TG, CA. The imperfect inverted repeats adjacent to the discontinuities of CaMV virion DNA, discussed above, do not conform to this rule. This may be related to the apparent absence of CaMV integration into host chromosomes.

3. Transcription Terminator Sequences

The control of transcription termination in retrovirus, CaMV, and transcribed retroid elements is an important and distinct feature of these systems since read-through of termination signals is required to produce an RNA species with terminal repeats. The poly(A) signal sequence, AATAAA, is appropriately placed in the LTRs (and CaMV DNA) and is close to the RNA 3′ end. The unusual symmetrical sequence found in CaMV DNA near to the transcription termination site (Section V.B.3) exhibits varying degrees of homology with

	CTTCTTTT	AAAAAGAAAAAA	GGGGG	AAATG	MMTV
Retroviruses	TTTCC	AAGAAAAA	GGGGGG	AATG	MoLV/MSV
	CTTTT	ATA	GGGAGGGGG	AAATG	RSV/ev-1
	CTTTC	ATA	GGGAGGGGG	AGATG	FSV
		ACAAGA	GTGGGG	AATG	SNV
	TTTTTC	AAAA	GGAGGG	AATG	mdg 1/412
Drosophila	CTCTTT	AA	GGGAAGGG	AAGTG	17.6/297
			GAGGGGG		copia
	TTGTC	AGGA	GGCCGAG	TTG	mdg 3
Yeast	TCTATT		GGGTGG	TATG	Ty
Hepatitis B viruses	TTTGTTT	AAAGA	GGGGGAGGAG	ATTA	HBV
	TTATT	AAA	GGGAGGAGGG	ATTG	WHV
CaMV	TTCTTTC	AGA	GGGGAGGAGG	TTA	CaMV G2
	TTTTT	AAGA	GTGGGGGGG	TTG	CaMV G3

FIGURE 18. Sequence comparison of polypurine tracts, which delimit priming sites for DNA plus-strand synthesis in retroviruses, found adjacent to the right LTR of retroviruses (mouse mammary tumor virus MMTV;[101] Moloney leukemia virus, MoLV;[102] murine sarcoma virus, MSV;[103] Rous sarcoma virus, RSV;[104] avian endogenous provirus, ev-1;[105] Fujinami sarcoma virus, FSV;[106] spleen necrosis virus, SNV;[107] *Drosophila copia*-like elements;[88-94], Ty elements of yeast,[95] and upstream of the fixed 5′-termini of the DNA plus-strand of hepatitis B virus (HBV),[108] woodchuck hepatitis virus (WHV),[109] and CaMV DNA upstream of the discontinuities G2 and G3.

	T G G T G A C C C C G A C G T G A T	RSV/ev-1
Retroviruses	T G G G G G C T C G T C C G G G A T	MoLV/MSV/SNV
	T G G C G C C C G A A C A G G G A C	MMTV
	T G G T G T G C A C C T G G G T A G	FSV
	T G G C G C A G T C G A T G T G A T	17.6
	T G G C G C A G T C G A T A G G A T	297
Drosophila	T G G C G A C C G T G A C A A A G G A T	mdg 1
	T G G C G A C C G T G A C A G T	412
	G G T T A T G G G C C C A G T G C A	copia
Yeast	T G G T A G C G C C T G T G C T T C	Ty
CaMV	T G G T A T C A G A G C C A T	CaMV G1
Plant	A C C A U A G U C U C G G U	$tRNA_i^{met}$

FIGURE 19. Sequences homologous to the 3′ end of tRNAs which, in retroviruses, delimit the origin of DNA minus-strand synthesis at the beginning of genomic RNA reverse transcription, found on the right border of the left LTR in retroviruses, *Drosophila copia*-like elements and yeast Ty elements and in CaMV the sequence complementary to the 3′-end of plant $tRNA_i^{met}$[107a,b] is located immediately to the right of G1 in CaMV DNA.

a similar region of *copia*-like retroid elements (Figure 20). This homology extends only weakly to retroviruses so its significance for transcription termination is questionable.

4. Duplications

It has been observed that short sequences in retroid LTRs are repeated, often several times, which has been described as a characteristic feature of these elements.[89-91] It is

```
CaMV    AATAAA        ATTTCT       AATT        CCTAAAA C CAAAATCC
297     AATAAA (11)  ATTTCT        ATT         CCTAAAAACT     AT
17.6    AATAAA (19)  ATTT    (2)  ATT (13)   C AAAA CTC   AAT
B104    AATAAA (4)   ATTTCT  (2)              C AAAA CTC   AA C
copia   AATAAC        TT      (8) AAT          CATAAAA CTCAAAA C
MIF-I   AATAAC (3)   ATT     (4) AAT  (14)      AA    CAAA
RSV     AATAAA (4)   ATTT    (3)  ATT          C      AC CA     C
```

FIGURE 20. Nucleotide sequence in the region of transcription termination downstream of the poly(A) signal sequence AATAAA in CaMV DNA with regions of homology in similarly located sequences of eukaryotic retroid elements.

nt	sequence	nt	sequence
7357	TAAGGG AT	7374	T CCC A
7505	TAAGGGAAT	7393	ACCCA
7514	T AGGG T	7498	T CCC A
7525	T AGGG T	7558	ACCCT
7758	T AGGGAA	7655	T CCCT
7935	TAAGGGAAT	7953	TACCCTA
		7962	TACCCTA
7914	TACTATAC		
7925	TACTATAC		

FIGURE 21. Some short repeated sequences, a characteristic of retrovirus and retroid element LTRs, found in the large intergenic region of CaMV DNA. Nucleotide numbering is according to Franck et al.[41]

noteworthy that in the large intergenic region of CaMV (analogous to parts of an LTR), several oligonucleotide motifs are repeated, often in fragmentary form (Figure 21).

C. Hepatitis B Viruses

Hepatitis B virus (HBV) encapsidates a 3.2kb double-stranded DNA of ciruclar conformation. Like CaMV, the DNA of HBV[112] and related viruses from woodchucks (WHV),[113] ducks (DHV),[114] and ground squirrels (GSHV)[115] does not consist of covalently closed strands. One strand, the minus-strand, has a nick at a fixed position. The plus-strand (probably containing all the protein coding functions) has a fixed 5′ end but with a 3′ end of variable length (see Figure 22). It is thought that hepatitis B viruses replicate by reverse transcription[110] as has been proposed for CaMV. Again the similarities to CaMV are striking considering the separation of these two groups of viruses in evolutionary terms. For both virus groups, the minus-strand has a single discontinuity located in the major intergenic region; coding function has asymmetric polarity; a polypurine region is located adjacent to the 5′ ends of plus-sense DNA strands (Figure 18). HBV and WHV do not appear to have a pbs sequence; minus-strand synthesis is thought to be primed by a protein[116] rather than a tRNA molecule.

The conclusion of this section is that CaMV exhibits structural and organizational similarities to unusual groups of eukaryotic genetic elements The unifying theme is that reverse transcription is implicated in each system, although this has only been shown unambiguously to occur in retroviruses. A second feature is integration/transposition, neither of which has been shown to occur with CaMV. It is possible that CaMV evolved from an integrating/transposing element but became autonomous from chromosomal DNA. CaMV may be related to current transposable elements of plants. As yet only limited sequence data is available

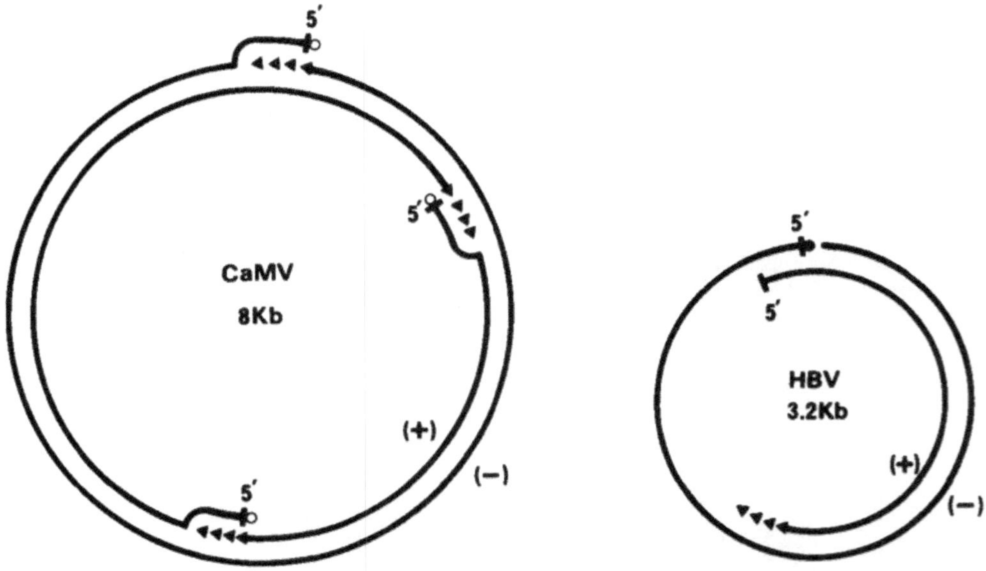

FIGURE 22. Comparison of the DNA genome of CaMV (left) with that of hepatitis B virus (HBV). Closed triangles represent variable length at the 3′ end of the plus-strands. Ribonucleotides are linked to the fixed 5′ ends of CaMV DNA strands (open circles) and a protein is covalently linked to HBV DNA (closed circle). In both virus DNAs, neither strand is covalently closed, but the molecules have circular conformations.

for such plant movable sequences, and extensive comparisons are not yet possible. The junction sequences of the *Antirrhinum* Tam1 element and the *Zea* cin element[117] follow the TG,CA rule at the integration site, a feature seemingly not shared by CaMV.

VII. OUTLOOK

The reader should now be aware that although much is known of the organization of the CaMV genome many questions concerning the expression of its genes remain unanswered. A major step towards providing some answers will be the development of cellular DNA transfection systems in order to probe for specific functions by site directed mutagenesis of genome subfragments. The identification and structural analysis of messenger RNA expressing most of the CaMV genes is still awaited. These are important steps in our understanding of the virus itself and plant genes in general. CaMV is also part of a unifying theme currently generating much excitement in eukaryotic molecular biology as a whole, and its relationship to systems which undergo reverse transcription and transposition will be clarified only after further intensive investigation.

ACKNOWLEDGMENTS

I am grateful to Drs. J. W. Davies, R. Hull, A. J. Maule, J. R. Penswick, R. Townsend, and C. J. Woolston for valuable discussion.

REFERENCES

1. **Shepherd, R. J.,** DNA plant viruses, *Annu. Rev. Plant Physiol.*, 30, 405, 1979.
2. **Shepherd, R. J.and Lawson, R. H.,** Caulimoviruses, in *Handbook of Plant Virus Infections and Comparative Diagnosis*, Kurstak, E., Ed., Elsevier/North-Holland Biomedical Press, Amsterdam, 1981, 847.

2a. **Hull, R.,** Caulimovirus group, *CMI/AAB Descriptions of Plant Viruses*, No. 295, 1984.

3. **Hohn, T., Richards, K., and Lebeurier, G.,** *Curr. Top. Microbiol. Immunol.*, 96, 193, 1982.
4. **Hull, R. and Davies, J. W.,** Genetic engineering with plant viruses, and their potential as vectors, *Adv. Virus Res.*, 28, 1, 1983.
5. **Hull, R.,** Structure of the cauliflower mosaic virus genome. III. Restriction endonuclease mapping of 33 isolates, *Virology*, 100, 76, 1980.
6. **Lung, M. C. Y. and Pirone, T. P.,** Studies on the reason for differential transmissibility of cauliflower mosaic virus isolates by aphids, *Phytopathology*, 63, 910, 1973.
7. **Lung, M. C. Y. and Pirone, T. P.,** Acquisition factor required for aphid transmission of purified cauliflower mosaic virus, *Virology*, 60, 260, 1974.
8. **Maule, A. J., Hull, R., and Donson, J.,** The application of spot hybridisation to the detection of DNA and RNA viruses in plant tissues, *J. Virol. Methods*, 6, 215, 1983.
9. **Lesemann, D. and Casper, R.,** Electron microscopy of petunia vein-clearing virus, an isometric plant virus associated with specific inclusions in petunia cells, *Phytopathology*, 63, 1118, 1973.
10. **Hull, R., Shepherd, R. J., and Harvey, J. D.,** Cauliflower mosaic virus: an improved purification procedure and some properties of the virus particles, *J. Gen. Virol.*, 31, 93, 1976.
11. **Pirone, T. P., Pound, G. S., and Shepherd, R. J.,** Properties and serology of purified cauliflower mosaic virus, *Phytopathology*, 51, 541, 1961.
12. **Chauvin, C., Jacrot, B., Lebeurier, G., and Hirth, L.,** The structure of cauliflower mosaic virus. A neutron diffraction study, *Virology*, 96, 640, 1979.
13. **Gardner, R. C. and Shepherd, R. J.,** A procedure for rapid isolation and analysis of cauliflower mosaic virus DNA, *Virology*, 106, 159, 1980.
14. **Howell, S. H., Walker, L. L., and Dudley, R. K.,** Cloned cauliflower mosaic virus DNA infects turnips *Brassica rapa*, *Science*, 208, 1265, 1980.
15. **Lebeurier, G., Hirth, L., Hohn, T., and Hohn, B.,** Infectivities of native and cloned DNA of cauliflower mosaic virus, *Gene*, 12, 139, 1980.
16. **Donson, J. and Hull, R.,** Physical mapping and molecular cloning of caulimovirus DNA, *J. Gen. Virol.*, 64, 2281, 1983.
17. **Hull, R.,** The DNA of Plant DNA viruses, in *Nucleic Acids in Plants*, Vol. 2, Hall, T. C. and Davies, J. W., Eds., CRC Press, Boca Raton, Fla., 1979, 3.
18. **Hull, R. and Donson, J.,** Physical mapping of the DNAs of carnation etched-ring and figwort mosaic viruses, *J. Gen. Virol.*, 60, 125, 1982.

18a. **Donson, J.,** Comparative Studies on Caulimoviruses, Ph.D thesis, University of East Anglia, U.K., 1983.

19. **Richins, R. D. and Shepherd, R. J.,** Physical maps of the genomes of dahlia mosaic virus and mirabilis mosaic virus — two members of the caulimovirus group, *Virology*, 124, 208, 1983.
20. **Volovitch, M., Dumas, J. P., Drugeon, G., and Yot, P.,** Single-stranded interruptions in cauliflower mosaic virus DNA, in *Acides Nucleiques et Synthese des proteines chez les Vegeteaux*, Colloq. No. 261, Bogorad, L. and Weil, J. H., Eds., Centre National de la Recherche Scientifique, Paris, 1976, 635.
21. **Volovitch, M., Drugeon, G., and Yot, P.,** Studies on the single-stranded discontinuities of the cauliflower mosaic virus genome, *Nucl. Acids Res.*, 5, 2913, 1978.
22. **Hull, R. and Howell, S. H.,** Structure of the cauliflower mosaic virus genome. II. Variation in DNA structure and sequence between isolates, *Virology*, 86, 482, 1978.
23. **Hull, R., Covey, S. N., Stanley, J., and Davies, J. W.,** The polarity of the cauliflower mosaic virus genome, *Nucl. Acids Res.*, 7, 669, 1979.
24. **Howell, S. H. and Hull, R.,** Replication of cauliflower mosaic virus and transcription of its genome in turnip leaf protoplasts, *Virology*, 86, 468, 1978.
25. **Hull, R. and Shepherd, R. J.,** The structure of the cauliflower mosaic virus genome, *Virology*, 79, 216, 1977.
26. **Bedbrook, J. R. and Kolodner, R.,** The structure of chloroplast DNA, *Ann. Rev. Plant Physiol.*, 30, 593, 1979.
27. **Bogenhagen, D., Gillum, A. M., Martens, P. A., and Clayton, D. A.,** Replication of mouse L-cell mitochondrial DNA, *Cold Spring Harb. Symp. Quant. Biol.*, 43, 253, 1978.
28. **Guilley, H., Richards, K. E., and Jonard, G.,** Observations concerning the discontinuous DNAs of cauliflower mosaic virus, *EMBO J.*, 2, 277, 1983.
29. **Covey, S. N., Turner, D. S., and Mulder, G.,** A small DNA molecule containing covalently-linked ribonucleotides originates from the large intergenic region of the cauliflower mosaic virus genome, *Nucl. Acids Res.*, 11, 251, 1983.

30. **Turner, D. S. and Covey, S. N.,** A putative primer for the replication of cauliflower mosaic virus by reverse transcription is virion-associated, *FEBS Lett.*, 165, 285, 1984.
31. **Hull, R. and Covey, S. N.,** Characterisation of cauliflower mosaic virus DNA forms isolated from infected turnip leaves, *Nucl. Acids Res.*, 11, 1881, 1983.
32. **Ménissier, J., Lebeurier, G., and Hirth, L.,** Free cauliflower mosaic virus supercoiled DNA in infected plants, *Virology*, 117, 322, 1982.

32a. **Marco, Y. and Howell, S. H.,** Intracellular forms of viral DNA consistent with a model of reverse transcriptional replication of the cauliflower mosaic virus genome, *Nucl. Acids Res.*, 12, 1517, 1984.

33. **Ménissier, J., de Murcia, G., Lebeurier, G., and Hirth, L.,** Electron microscopic studies of the different topological forms of cauliflower mosaic virus DNA: knotted encapsidated DNA and nuclear minichromosome, *EMBO J.*, 2, 1067, 1983.
34. **Guilfoyle, T. J.,** Transcription of the cauliflower mosaic virus genome in isolated nuclei from turnip *Brassica rapa* cultivar Just Right leaves, *Virology*, 107, 71, 1980.
35. **Olszewski, N., Hagen, G., and Guilfoyle, T. J.,** A transcriptionally active, covalently closed minichromosome of cauliflower mosaic virus DNA isolated from infected turnip leaves, *Cell*, 29, 395, 1982.
36. **Hohn, T., Hohn, B., Lesot, A., and Lebeurier, G.,** Restriction map of native and cloned cauliflower mosaic virus DNA, *Gene*, 11, 21, 1980.
37. **Howarth, A. J., Gardner, R. C., Messing, J., and Shepherd, R. J.,** Nucleotide sequence of naturally-occurring deletion mutants of cauliflower mosaic virus, *Virology*, 112, 678, 1981.
38. **Balàzs, E., Guilley, H., Jonard, G., and Richards, K.,** Nucleotide sequence of DNA from an altered virulence isolate D/H of the cauliflower mosaic virus, *Gene*, 19, 239, 1982.
39. **Delseny, M. and Hull, R.,** Isolation and characterisation of faithful and altered clones of cauliflower mosaic virus isolates Cabb B-J1, CM4-184, and Bari 1, *Plasmid*, 9, 31, 1983.
40. **Volovitch, M., Modjtahedi, N., Chouikh, Y., and Yot, P.,** Conserved and non-conserved restriction sites are interspersed in the DNA of cauliflower mosaic virus, *Biochem. Int.*, 3, 225, 1981.
41. **Franck, A., Guilley, H., Jonard, G., Richards, K., and Hirth, L.,** Nucleotide sequence of cauliflower mosaic virus DNA, *Cell*, 21, 285, 1980.
42. **Gardner, R. C., Howarth, A. J., Hahn, P., Brown-Leudi, M., Shepherd, R. J., and Messing, J.,** The complete nucleotide sequence of an infectious clone of cauliflower mosaic virus by M13mp7 shot-gun sequencing, *Nucl. Acids Res.*, 9, 2871, 1981.
43. **Hull, R.,** Cauliflower mosaic virus DNA as a possible gene vector for higher plants, in *Plant Sciences*, Panopoulos, N. J., Ed., Praeger, New York, 1981, 99.
44. **Richards, K. E., Guilley, H., and Jonard, G.,** Further characterisation of the discontinuities in cauliflower mosaic virus DNA, *FEBS Lett.*, 134, 67, 1981.
45. **Hull, R. and Covey, S. N.,** Does cauliflower mosaic virus replicate by reverse transcription?, *Trends Biochem. Sci.*, 8, 119, 1983.
46. **Pfeiffer, P. and Hohn, T.,** Involvement of reverse transcription in the replication of cauliflower mosaic virus: a detailed model and test of some aspects, *Cell*, 33, 781, 1983.
47. **Tezuka, N. and Taniguchi, T.,** Structural protein of cauliflower mosaic virus, *Virology*, 48, 297, 1972.
48. **Kelly, D. C., Cooper, J., and Walkey, D. G. A.,** Cauliflower mosaic virus structural proteins, *Microbios*, 10, 239, 1974.
49. **Brunt, A. A., Barton, R. J., Tremaine, J. H., and Stace-Smith, R.,** The composition of cauliflower mosaic virus protein, *J. Gen. Virol.*, 27, 101, 1975.
50. **Hull, R. and Shepherd, R. J.,** The coat proteins of cauliflower mosaic virus, *Virology*, 70, 217, 1976.
51. **Al Ani, R., Pfeiffer, P., and Lebeurier, G.,** The structure of cauliflower mosaic virus. II. Identity and location of the viral polypeptides, *Virology*, 93, 188, 1979.
52. **Hahn, P. and Shepherd, R. J.,** Phosphorylated proteins in cauliflower mosaic virus, *Virology*, 107, 295, 1980.
53. **Hahn, P. and Shepherd, R. J.,** Evidence for a 58-kilodalton polypeptide as precursor of the coat protein of cauliflower mosaic virus, *Virology*, 116, 480, 1982.
54. **Du Plessis, D. H. and Smith, P.,** Glycosylation of the cauliflower mosaic virus capsid polypeptide, *Virology*, 109, 403, 1981.
55. **Shepherd, R. J., Richins, R., and Shalla, T. A.,** Isolation and properties of the inclusion bodies of cauliflower mosaic virus, *Virology*, 102, 389, 1980.
56. **Shockey, M. W., Gardner, C. O., Jr., Melcher, U., and Essenberg, R. C.,** Polypeptides associated with inclusion bodies from leaves of turnip infected with cauliflower mosaic virus, *Virology*, 105, 575, 1980.
57. **Al Ani, R., Pfeiffer, P., Whitchurch, O., Lesot, A., Lebeurier, G., and Hirth, L.,** A virus-specified protein produced upon infection by cauliflower mosaic virus, *Ann. Virol. (Inst. Pasteur)*, 131E, 33, 1980.
58. **Covey, S. N. and Hull, R.,** Transcription of cauliflower mosaic virus DNA. Detection of transcripts, properties and location of the gene encoding the virus inclusion body protein, *Virology*, 111, 463, 1981.

59. **Odell, J. T. and Howell, S. H.,** The identification, mapping, and characterisation of mRNA for P66, a cauliflower mosaic virus-coded protein, *Virology,* 102, 349, 1980.
60. **Xiong, C., Balazs, E., Lebeurier, G., Hindelong, C., Stoeckel, M. E., and Porte, A.,** Comparative cytology of two isolates of cauliflower mosaic virus, *J. Gen. Virol.,* 61, 75, 1982.
61. **Daubert, S. D., Richins, R. D., and Shepherd, R. J.,** Genes and gene products of cauliflower mosaic virus, *Abstr. 5th Int. Congr. Virol.,* Strasbourg, August 2 to 7, 1981, 207.
62. **Woolston, C. J., Covey, S. N., Penswick, J. R., and Davies, J. W.,** Aphid transmission and a polypeptide are specified by a defined region of the cauliflower mosaic virus genome, *Gene,* 23, 15, 1983.
62a. **Woolston, C. J.,** A Molecular Basis for Aphid Transmission of Cauliflower Mosaic Virus, Ph.D. thesis, University of East Anglia, Norwich, U.K., 1983.
63. **Xiong, C., Müller, S., Lebeurier, G., and Hirth, L.,** Identification by immunoprecipitation of cauliflower mosaic virus *in vitro* major translation product with a specific serum against viroplasm protein, *EMBO J.,* 1, 971, 1982.
64. **Daubert, S., Richins, R., Shepherd, R. J., and Gardner, R. C.,** Mapping of the coat protein gene of cauliflower mosaic virus by its expression in a procaryotic system, *Virology,* 122, 444, 1982.
65. **Armour, S. L., Melcher, U., Pirone, T. P., Lyttle, D. J., and Essenberg, R. C.,** Helper component for aphid transmission encoded by region II of cauliflower mosaic virus DNA, *Virology,* 129, 25, 1983.
66. **Verma, I. M.,** The reverse transcriptase, *Biochem. Biophys. Acta,* 473, 1, 1977.
67. **Toh, H., Hayashida, H., and Migata, T.,** Sequence homology between retroviral reverse transcriptase and putative polymerases of hepatitis B virus and cauliflower mosiac virus, *Nature (London),* 305, 827, 1983.
67a. **Volovitch, M., Modjtahedi, N., Yot, P., and Brun, G.,** RNA-dependent DNA polymerase activity in cauliflower mosaic virus-infected plant leaves, *EMBO J.,* 3, 309, 1984.
68. **Howell, S. H., Walker, L. L., and Walden, R. M.,** Rescue of *in vitro* generated mutants of cloned cauliflower mosaic virus genome in infected plants, *Nature (London),* 293, 483, 1981.
69. **Gardner, R. C.,** Plant viral vectors: CaMV as an experimental tool, in *Genetic Engineering of Plants,* Kosuge, T., Meredith, C. P., and Hollaender, A., Eds., Plenum Press, New York, 1983, 121.
70. **Odell, J. T., Dudley, R. K., and Howell, S. H.,** Structure of the 19S RNA transcript encoded by the cauliflower mosaic virus genome, *Virology,* 111, 377, 1981.
71. **Covey, S. N., Lomonossoff, G. P., and Hull, R.,** Characterisation of cauliflower mosaic virus DNA sequences which encode major polyadenylated transcripts, *Nucl. Acids Res.,* 9, 6735, 1981.
72. **Guilley, H., Dudley, R. K., Jonand, G., Balazs, E., and Richards, K. E.,** Transcription of cauliflower mosaic virus DNA: detection of promotor sequences, and characterisation of transcripts, *Cell,* 30, 763, 1982.
73. **Dudley, K. R., Odell, J. T., and Howell, S. H.,** Structure and 5′-termini of the large and 19S RNA transcripts encoded by the cauliflower mosaic virus genome, *Virology,* 117, 19, 1982.
74. **Howell, S. H.,** Ultraviolet mapping of RNA transcripts encoded by the cauliflower mosaic virus genome, *Virology,* 112, 488, 1981.
75. **Breathnach, R. and Chambon, P.,** Organisation and expression of eucaryotic split genes coding for proteins, *Annu. Rev. Biochem.,* 50, 349, 1981.
76. **Benoist, C., O'Hare, K., Breathnach, R., and Chambon, P.,** The ovalbumin gene-sequence of putative control regions, *Nucl. Acids Res.,* 8, 127, 1980.
77. **Khoury, G. and Gruss, P.,** Enhancer elements, *Cell,* 33, 313, 1983.
78. **Messing, J., Geraghty, D., Heidecker, G., Hu, N-T., Kuidl, J., and Rubenstein, I.,** Plant gene structure, in *Genetic Engineering of Plants,* Kosuge, T., Meredith, C. P., and Hollaender, A., Eds., Plenum Press, New York, 1983, 211.
79. **Pedersen, K., Devereux, J., Wilson, D. R., Sheldon, E., and Larkins, B. A.,** Cloning and sequence analysis reveal structural variation among related zein genes in maize, *Cell,* 29, 1015, 1982.
80. **Broglie, R., Coruzzi, G., Lamppa, G., Keith, B., and Chua, N-H.,** Structural analysis of nuclear genes coding for the precursor to the small subunit of wheat ribulose-1,5-bisphosphate carboxylase, *Biotechnology,* 1, 55, 1983.
80a. **Berry-Lowe, S. L., McKnight, T. D., Shah, D. M., and Meagher, R. B.,** The nucleotide sequence, expression, and evolution of one member of a multi-gene family encoding the small subunit of ribulose-1,5-biphosphate carboxylase in soybean, *J. Mol. Appl. Genet.,* 1, 483, 1982.
81. **Shah, D. M., Hightower, R. C., and Meagher, R. B.,** Complete nucleotide sequence of a soybean actin gene, *Proc. Natl. Acad. Sci. U.S.A.,* 79, 1022, 1982.
82. **Wiborg, O., Hyldig-Nielsen, J. J., Jensen, E. O., Palndan, K., and Marcker, K. A.,** The structure of an unusual leghaemoglobin gene from soybean, *EMBO J.,* 2, 449, 1983.
83. **Triesman, R. and Kamen, R.,** Structure of polyoma virus late nuclear RNA, *J. Mol. Biol.,* 148, 273, 1981.
84. **Kozak, M.,** Comparison of initiation of protein synthesis in procaryotes, eucaryotes and organelles, *Microbiol. Rev.,* 47, 1, 1983.

85. **Varmus, H. E.,** Form and function of retroviral proviruses, *Science*, 216, 812, 1982.
86. **Varmus, H. E.,** Retroviruses, in *Mobile Genetic Elements*, Shapiro, J. A., Ed., Academic Press, New York, 1983, 411.
87. **Hull, R. and Covey, S. N.,** Replication of cauliflower mosaic virus DNA, *Sci. Prog. Oxf.*, 68, 403, 1983.
88. **Finnegan, D. J., Rubin, G. M., Young, M. W., and Hogness, D. S.,** Repeated gene families in *Drosophila melanogaster*, *Cold Spring Harb. Symp. Quant. Biol.*, 42, 1053, 1978.
89. **Kugimiya, W., Ikenaga, H., and Saigo, K.,** Close relationship between the long terminal repeats of avian leukosis-sarcoma virus and *copia*-like moveable genetic elements of *Drosophila*, *Proc. Natl. Acad. Sci. U.S.A.*, 80, 5193, 1983.
90. **Ikenaga, M. and Saigo, K.,** Insertion of a movable genetic element, 297, into the TATA box of the H3 histone gene in *Drosophila melanogaster*, *Proc. Natl. Acad. Sci. U.S.A.*, 79, 4143, 1982.
91. **Kulguskin, V. V., Ilyin, Y. V., and Georgier, G. P.,** Mobile dispersed genetic element MDG1 of *Drosophila melanogaster:* nucleotide sequence of long terminal repeats, *Nucl. Acids Res.*, 9, 3451, 1981.
92. **Tchurikov, N. A., Ilyin, Y. V., Skryabin, K. G., Ananiev, E. V., Bayer, A. A., Krager, A. S., Zelentsova, E. S., Kulguskin, V. V., Lynbomirskaya, N. V., and Georgier, G. P.,** General properties of mobile dispersed genetic elements in *Drosophila melanogaster*, *Cold Spring Harb. Symp. Quant. Biol.*, 45, 655, 1981.
93. **Will, B. M., Bayer, A. A., and Finnegan, D. J.,** Nucleotide sequence of terminal repeats of 412 transposable elements of *Drosophila melanogaster*, *J. Mol. Biol.*, 153, 897, 1981.
94. **Schever, G., Tschudi, C., Perera, J., Delius, H., and Pirrotta, V.,** *B104*, a new dispersed repeated gene family in *Drosophila melanogaster* and its analogies with retroviruses, *J. Mol. Biol.*, 157, 435, 1982.
95. **Farabaugh, P. J. and Fink, G. R.,** Insertion of the eucaryotic transposable element Ty1 creates a 5-base pair duplication, *Nature (London)*, 286, 352, 1980.
96. **Brown, S. D. M.,** A mouse dispersed repeat sequence showing remarkable similarities to the long terminal repeats of retroviruses, *Gene*, 23, 95, 1983.
97. **Streeck, R. E.,** A multicopy insertion sequence in the bovine genome with structural homology to the long terminal repeats of retroviruses, *Nature (London)*, 298, 767, 1982.

97a. **Kuff, E. L., Feenstra, A., Lenders, K., Smith, L., Hawley, R., Hozumi, N., and Shulman, M.,** Intracisternal A-particle genes as movable elements in the mouse genome, *Proc. Natl. Acad. Sci. U.S.A.*, 80, 1992, 1982.

97b. **Keshet, E. and Itin, A.,** Patterns of genomic distribution and sequence heterogeneity of a murine 'retrovirus-like' multigene family, *J. Virol.*, 43, 50, 1982.

98. **Temin, H.,** Origin of retroviruses from cellular moveable genetic elements, *Cell*, 21, 599, 1980.
99. **Shiba, T. and Saigo, K.,** Retrovirus-like particles containing RNA homologous to the transposable element *copia* in *Drosophila melanogaster*, *Nature (London)*, 302, 119, 1983.
100. **Heine, C. W., Kelley, D. C., and Avery, R. J.,** The detection of retrovirus-like entities in Drosophila melanogaster cell cultures, *J. Gen. Virol.*, 49, 385, 1980.
101. **Donehower, L. A., Huang, A. L., and Mager, G. L.,** Regulatory and coding potential of the mouse mammary tumor virus long terminal redundancy, *J. Virol.*, 37, 226, 1981.
102. **Sutcliffe, J. G., Shinnick, T. M., Verma, I. M., and Lerner, R. A.,** Nucleotide sequence of Moloney leukemia virus: 3′ end reveals details of replication, analogy to bacterial transposons, and an unexpected gene, *Proc. Natl. Acad. Sci. U.S.A.*, 77, 3302, 1980.
103. **Van Bereran, C., Van Straaten, F., Galleshaw, J. A., and Verma, I. M.,** Nucleotide sequence of the genome of a murine sarcoma virus, *Cell*, 27, 97, 1981.
104. **Schwartz, D. E., Tizand, R., and Gilbert, W.,** Nucleotide sequence of Rous sarcoma virus, *Cell*, 32, 853, 1983.
105. **Hishinuma, F., DeBona, P. J., Astrin, S., and Shalka, A. M.,** Nucleotide sequence of accepter site and termini of integrated avian endogenous provirus *ev*1: integration creates a 6bp repeat of host DNA, *Cell*, 23, 155, 1981.
106. **Shibuya, M. and Hanafusa, N.,** Nucleotide sequence of Fujinami sarcoma virus: evolutionary relationship of its transforming gene with transforming genes of other sarcoma viruses, *Cell*, 30, 787, 1982.
107. **Shimotohno, K., Mizutani, S., and Temin, H. M.,** Sequence of retrovirus provirus resembles that of bacterial transposable elements, *Nature (London)*, 285, 550, 1980.

107a. **Canaday, J., Guillemont, P., and Weil, J-H.,** The nucleotide sequences of the initiator transfer RNAs from bean cytoplasm and chloroplasts, *Nucl. Acids Res.*, 8, 999, 1980.

107b. **Ghosh, H. P., Ghosh, K., Simsek, M., and RajBhandary, U. T.,** Nucleotide sequence of wheat germ cytoplasmic initiator methionine transfer ribonucleic acid, *Nucl. Acids Res.*, 10, 3241, 1982.

108. **Galibert, F., Mandant, E., Fitoussi, F., Tiollaris, P., and Charnay, P.,** Nucleotide sequence of the hepatitis B virus genome (subtype agw) cloned in *E. coli*, *Nature (London)*, 281, 646, 1979.
109. **Galibert, F., Chen, T. N., and Mandart, E.,** Nucleotide sequence of a cloned woodchuck hepatitis virus genome: comparison with the hepatitis B virus sequence, *J. Virol.*, 41, 51, 1982.

110. **Summers, J. and Mason, W. S.,** Replication of the genome of a hepatitis B-like virus by reverse transcription of an RNA intermediate, *Cell,* 29, 403, 1982.
111. **Proudfoot, N. J. and Brownlee, G. G.,** 3′ non-coding region sequences in eucaryotic messenger RNA, *Nature (London),* 263, 211, 1976.
112. **Robinson, W. S.,** The genome of hepatitis B virus, *Annu. Rev. Microbiol.,* 31, 357, 1977.
113. **Summers, J., Smoke, J. M., and Snyder, R.,** A virus similar to human hepatitis B virus associated with hepatitis and hepatoma in woodchucks, *Proc. Natl. Acad. Sci. U.S.A.,* 75, 4533, 1978.
114. **Mason, W. S., Seal, G., and Summers, J.,** A virus of Pekin ducks with structural and biological relatedness to human hepatitis B virus, *J. Virol.,* 36, 829, 1980.
115. **Marion, P. L., Oshiro, L., Regnery, D. C., Sculland, G. H., and Robinson, W. S.,** A virus in Beechey ground squirrels which is related to hepatitis B virus of man, *Proc. Natl. Acad. Sci. U.S.A.,* 77, 2941, 1980.
116. **Gerlich, W. H. and Robinson, W. S.,** Hepatitis B virus contains protein attached to the 5′ terminus of its complete strand, *Cell,* 21, 801, 1980.
117. **Saedler, H., Bonas, U., Deumling, B., Gupta, M., Hahlbrock, K., Harrison, B. J., Kreuzaler, F., Peterson, P. A., Reif, J., Schwarz-Sommer, Zs., Shepherd, N., Sommer, H., Ubben, D., and Wienand, U.,** Transposable elements in plants, in *Genetic Rearrangement,* Chater, K. F., Cullis, C. A., Hopwood, D. A., Johnston, A. W. B., and Woolhouse, H. W., Eds., Croom Helm, London, 1983, 107.
118. **Daubert, S., Shepherd, R. J., and Gardner, R. C.,** Insertional mutagenesis of the cauliflower mosaic virus genome, *Gene,* 25, 201, 1983.
119. **Dixon, L. K., Koenig, I., and Hohn, T.,** Mutagenesis of cauliflower mosaic virus, *Gene,* 25, 189, 1983.
120. **Guilfoyle, T. J., Olszewski, N., Hagen, G., Kuzj, A., and McClure, B.,** Transcription and replication of the cauliflower mosaic virus genome: studies with isolated nuclei, nuclear lysates, and extranuclear cell-free leaf extracts, in *Plant Molecular Biology,* UCLA Symp., Goldberg, R., Ed., Alan R. Liss, New York, 1983, 117.
121. **Mount, S. M.,** A catalogue of splice junction sequences, *Nucl. Acids Res.,* 10, 459, 1982.
122. **Covey, S. N. and Turner, D. S.,** unpublished results.
123. **Condit, C. and Meagher, R. B.,** Multiple, discrete 35S transcripts of cauliflower mosaic virus, *J. Mol. Appl. Genet.,* 2, 301, 1983.
124. **Condit, C., Hagen, T. J., McKnight, T. D., and Meagher, R. B.,** Characterisation and preliminary mapping of cauliflower mosaic virus transcripts, *Gene,* 25, 101, 1983.
125. **Olszewski, N. E. and Guilfoyle, T. J.,** Nuclei purified from cauliflower mosaic virus-infected turnip leaves contain subgenomic, covalently closed circular cauliflower mosaic virus DNAs, *Nucl. Acids Res.,* 11, 8901, 1983.
126. **Rollo, F. and Covey, S. N.,** Cauliflower mosaic virus DNA persists as supercoiled forms in cultured turnip cells, *J. Gen. Virol.,* 66, 603, 1985.
127. **Hull, R.,** personal communication.
128. **Plant, A. L., Covey, S. N., and Grierson, D.,** unpublished results.
128a. **Sieg, K. and Gronenborn, B.** unpublished data.
129. **Hull, R.,** A model for the expression of CaMV nucleic acid, *Plant Mol. Biol.,* in press.
130. **Hull, R. and Zuker, M.,** unpublished observations.
131. **Darlix, J-L., Zuker, M., and Spahr, P-F.,** Structure-function relationship of Rous sarcoma virus leader RNA, *Nucl. Acids Res.,* 10, 5183, 1982.
132. **Covey, S. N.,** unpublished observations.
133. **Brisson, N., Paszkowski, J., Penswick, J., Gronenborn, B., Potrykus, I., and Hohn, T.,** Expression of a bacterial gene in plants using a viral vector, *Nature (London),* 310, 511, 1984.

Chapter 5

REPLICATION OF CAULIMOVIRUSES IN PLANTS AND PROTOPLASTS*

A. J. Maule

TABLE OF CONTENTS

* *Abbreviations:* nt(s), nucleotide(s); bp, base pairs; kb, kilobase pairs (1000 nts); dNTP, deoxynucleotide triphosphate; ddTTP, dideoxy thymidine triphosphate; cccDNA, covalently closed circular DNA; ocDNA, open circular DNA; OR, open coding region; ds, double stranded.

I. INTRODUCTION

Caulimoviruses constitute a unique group of ds DNA plant viruses which collectively infect dicotyledonous hosts from 17 plant families. The group has received a lot of attention in recent years for two reasons. First, the group provides a useful model for the study of gene organization and expression in plants (see Chapter 4), and second, as ds DNA, the viruses represent potential gene vectors with important implications for the genetic manipulation of higher plants. The type member of the group, cauliflower mosaic virus (CaMV), has been the most extensively studied and consequently will also feature most strongly in the review. Comparisons between CaMV and other caulimoviruses will undoubtedly provide a powerful tool in future when trying to characterize strategies of virus multiplication.

The general properties of caulimoviruses have been reviewed previously[1-5] and the detailed structural organization of the viral genome has been discussed in the previous chapter. However, the mechanisms involved in virus multiplication, and DNA replication in particular, are only just beginning to be resolved. Since the publication of the complete sequence of the Strasbourg isolate of CaMV,[6] the years 1982 to 1984 have revealed several pieces of information crucial to the understanding of CaMV multiplication: first, the discovery of a supercoiled form of CaMV DNA present in nuclei as a transcriptionally active minichromosome;[7-11] second, identification of fragments of CaMV DNA which have ribonucleotides attached,[12,13] most particularly a small DNA covalently linked to a tRNA molecule;[14] and third, the collation of these and other details to produce a model for viral DNA replication based upon reverse transcription.[12,15-18] The proposed model presents a challenge to be tested at the structural and enzymological level and tested for its significance for the other caulimoviruses in a wide range of hosts. Already, a coding region on the CaMV genome has been suggested as a potential gene for a virus-coded DNA polymerase enzyme[15,19,20] and a virus-specific enzyme capable of copying heteropolymeric RNA templates has been identified.[20,21]

With an increased understanding of virus multiplication we will not only be able to predictably change the characteristics of the host/virus interaction, but to use those characteristics to advantage in the genetic selection and manipulation of economically important hosts.

II. EXPERIMENTAL SYSTEMS

A. Whole Plants

Typically, members of the caulimovirus group have a narrow host range, often limited to just a single family or genus (Table 1). However, under experimental conditions, a broader host range has been recorded for some of the viruses. Hence, CaMV has members of the Cruciferae as hosts although infections of *Nicotiana clevelandii*[33] and *Datura stramonium*[34] have been reported. Furthermore, using sensitive techniques, local accumulations of virus have been identified on the inoculated leaves of cucumber, soybean, peanut, lettuce, cotton, tomato, and spinach.[35] The most extensively used host in studying the multiplication of CaMV has been *Brassica rapa* (turnip), a decision which seems to have originated from its recommendation as the highest yielding host for virus purification.[36]

For most of the work relating to CaMV multiplication, whole turnip plants have been used. Clearly, there are advantages in having large amounts of readily accessible infected material which is fairly reproducible in its response to infection. However, it is important to note that the extent to which a virus multiplies in the tissues of its host does not solely depend upon the competence of individual cells, but also factors such as tissue specificity, antagonism from host defences, efficiency of virus spread, host physiology, and environmental conditions. The progressive nature of infections in intact tissues will mean that at

Table 1
THE CAULIMOVIRUS GROUP

Virus	Abbreviation Members	Host-range family	Ref.[a]
Cauliflower mosaic	CaMV	Cruciferae (Solanaceae)[b]	3,4
Carnation etched ring	CERV	Caryophyllaceae	3,22
Dahlia mosaic	DaMV	Compositae (Amaranthaceae, Chenopodiaceae, Solanaceae)	3,23
Figwort mosaic	FMV	Scrophulariaceae (Solanaceae)	3,24
Mirabilis mosaic	MMV	Nyctaginaceae	3,25
Strawberry vein banding	SVBV	Rosaceae	3,26
Thistle mottle	ThMV	Compositae	24
Soybean chlorotic mottle	SoyCMV	Leguminoseae	27
	Possible Members		
Blueberry red ringspot	BRRV	Ericaceae	28
Cassava vein banding	CVBV	Euphorbiaceae	29
Cestrum virus	(CV)[c]	Solanaceae	30
Petunia vein clearing	PVCV	Solanaceae	31
Plantago virus 4	PlV4	Plantaginaceae	32

[a] Where possible, review reference given.
[b] Families in parentheses are laboratory infections.
[c] This virus has not yet been named.

best studies will reveal the presence of all the components associated with virus replication and spread, possibly in quantities greater than those obtained from single-cell systems. However, owing to the asynchrony characteristic of progressive infection, the correlation of particular molecular species with the kinetics of infection in individual cells is not possible.

The experimental transmission of CaMV particles to the host can be achieved either mechanically or through using aphids, or just mechanically for viral DNA. The successful inoculation of turnip plants with CaMV virion DNA, and linear, or circularized cloned CaMV DNA[37-39] has provided a mechanism for examining viral gene structure and for testing the infectivity of in vitro modified CaMV DNA. It was demonstrated that the infectivity of cloned CaMV DNA was dependent upon its being excised from the recombinant plasmid at the restriction endonuclease site used for cloning;[38] restriction enzymes giving longer cohesive 3′ or 5′ ends released DNA with greater infectivity. Evidently, in these circumstances a very early event in infection is the re-ligation of linear molecules. An exception to this rule is seen when plasmids contain nested CaMV genomes. Here excision is unnecessary, infection occurring after intragenomic recombination.[40] Full-length clones of CERV, FMV, and MMV have also been shown to be infectious for their respective hosts.[24] Infections with cloned DNA produce progeny virion DNA with the single-strand discontinuities characteristic of virus particles.

Unfortunately, the nature of whole plant infections means that in vitro constructs can only be tested for their ability to initiate a fully competent and systemic infection. Constructs deficient in some aspect of virus replication or virus spread would not be detected. Most modifications of CaMV DNA result in a loss of infectivity for whole plants.[41-43] Attempts to develop a system for complementation of defective mutants have shown that pairs of molecules with potential for complementation preferentially undergo recombination to produce wild-type virus and systemic infection.[41,44,45] Such a phenomenon provides a good experimental system for examining the functional overlap between essential genes occurring in different viruses and isolates, although this has not been exploited.

B. Protoplasts

Biochemical mapping of defective mutants would be most usefully achieved using DNA-inoculated protoplasts to search for defective stages in the replication cycle. Infections of turnip protoplasts with CaMV DNA have been reported[46-48] although only a small proportion of the protoplasts became infected, with each producing less virus than those infected with CaMV virions. A more efficient system would almost certainly be required for the analysis of macromolecules following inoculation with defective mutants.

Infections of protoplasts from turnip[49-52] and other *Brassica* species[52] with CaMV particles have been demonstrated, and the accumulation of virus-specific antigen,[50-52] transcripts,[49] and DNA [52] measured. In all cases, a cycle of virus replication occurred in 2.5 to 4 days, although in rape protoplasts there was a discrepancy between maximal DNA and virion accumulation.[52]

The major advantage of protoplasts over intact tissue when studying virus replication is that they are infected in a synchronous fashion and theoretically show virus multiplication synchronously. Thus it is possible to study the kinetics of viral DNA replication (Section VI), transcription, virus-specific protein synthesis, etc. It should be noted however, that the production of individual molecules may take only minutes, and hence cannot be synchronous within a virus replication cycle taking days.

The concept of CaMV DNA as a gene vector for plants would ideally involve the introduction of manipulated DNA into single cells from which whole plants could be regenerated. Difficulties experienced when inoculating protoplasts with DNA necessitates that this be followed by a selection for the inserted gene, prior to regeneration. Alternatively, DNA manipulated to a size within virion packaging limits could be introduced into protoplasts as virus particles where the >90% infection efficiencies possible using turnip protoplasts[50,52] would be beneficial. Unfortunately, turnip mesophyll protoplasts have not been completely regenerated to whole plants, whereas protoplasts from rape,[53] cauliflower,[54] and cabbage[55] have. CaMV infection of protoplasts from the related *Moricandia arvensis,*[52] a plant not known to be susceptible to CaMV, may indicate that it is possible to have nonhost species acting as recipients for manipulated DNA.

C. Callus

Recently, the multiplication of CaMV in rapidly dividing pieces of callus[56] and in protoplast-derived callus clones[57] from infected tissue, has been examined. Here the ability to replicate viral DNA varies considerably between clones. The observation that even in the most active callus the proportion of unencapsidated viral DNA is much higher than in infected leaves suggests a fundamental difference in the control of virus synthesis between the two systems and is important both for the understanding of the controlling steps and for its implications for genetic manipulation.

D. Environmental Conditions

The only systematic study of environmental parameters in relation to caulimovirus replication has examined the effect of temperature on multiplication of CaMV in turnip plants and protoplasts.[51] The optimum temperature for virus multiplication was around 20°C, temperatures around 33°C giving a complete cessation of growth. Effects at the higher temperature could be reversed by a return to 20°C. It was not clear which stages in virus replication were temperature sensitive, although inclusion body formation was inhibited to a lesser degree.

In vitro-infected turnip protoplasts are widely tolerant of nutritive culture conditions for CaMV synthesis,[52] although in the presence of complex media, the extent of synthesis became more reproducible. This probably represents an influence upon host physiology rather than demonstrating a specific requirement for efficient CaMV replication.

The formation of undifferentiated callus from pieces of infected leaf tissue cultured in the presence of high auxin concentrations, and the associated fall in levels of accumulated CaMV virions may indicate that indirect environmental control of virus synthesis may operate in this system.[56]

III. SYSTEMIC SPREAD AND SYMPTOMATOLOGY

For most isolates of CaMV, mechanical inoculation of young (3-week) turnip plants with virus particles is followed (4 to 7 days) by the appearance of chlorotic local lesions[58] before systemic symptoms begin to appear (10 to 15 days). Normal growth of the plant in the intervening period results in leaves which fail to show any symptoms and do not accumulate virus.[59] First systemic symptoms may again be of the chlorotic lesion type, often occurring asymmetrically with regard to left or right, and proximal or distal regions of the leaf. All subsequent leaves show the more severe systemic symptoms of vein clearing and chlorotic mottle (Figure 1).

The observation that protoplasts from systemically infected leaves are 90[51] to 100%[60] infected with CaMV and that these protoplasts contain 10- to 100-fold more virus than infected protoplasts from inoculated leaves, indicates a greater compatibility or susceptibility for CaMV in younger cells. This association might be anticipated, since the synthesis of viral DNA in older nonmitotic cells would logically require a major diversion of purine and pyrimidine intermediates. It is surprising therefore that, following isolation from fully differentiated turnip leaves, protoplasts infected in vitro synthesize as much virus[51] or viral DNA[52] as is present in the younger tissues, estimated as 3×10^5 particles or 10^4 to 10^5 genome equivalents, respectively.

Two recent findings may explain this phenomenon. First, isolated turnip mesophyll protoplasts extensively incorporate ^{3}H-thymidine into host DNA.[59] Whether constituting *de novo* host DNA synthesis, or DNA repair following the trauma of protoplast isolation is not known but in either case a mobilization of DNA precursors would be required. Second, and perhaps of greater significance is the ability of turnip protoplasts to divert nucleosides from RNA to DNA synthesis. Measured as the incorporation of ^{3}H-uridine into host DNA,[59] this probably represents the activity of scavenging pathways known to exist in eukaryotic systems. The large numbers of infected cells present in systemically infected turnip leaves (90 to 100%), and the fact that even when very small the leaves show full systemic symptoms, both indicate that CaMV is replicating in the very young tissues close to the meristematic apex of the host plant. Whether the virus invades the apical meristem itself is not known. However, poor replication of CaMV in rapidly dividing callus tissue derived from infected turnip leaves,[62] the elimination of CaMV in a majority of protoplast-derived clones from infected turnip,[57] and removal of CERV from *Dianthus* spp. by meristem tip culture[63] suggest that this is unlikely. Rapid spread of the virus through newly formed vascular tissue may be sufficient to infect all leaf cells prior to leaf emergence. Nevertheless, it must also be considered that replication in meristematic tissues can occur but with the production of very little stable virus. Protoplast-derived callus clones maintaining CaMV do so without extensive production of virus particles.[56,57] *In situ* hybridization of CaMV-specific DNA or RNA in sections of host meristems should resolve this question.

For all the caulimoviruses, spread from cell to cell and systemically is probably by virus particles. Virus particles of CERV,[64] DaMV,[23] and PVCV[31] occur free in host cytoplasm and, moreover, have been observed in plasmodesmata in DaMV-infected zinnia tissues[65,66]

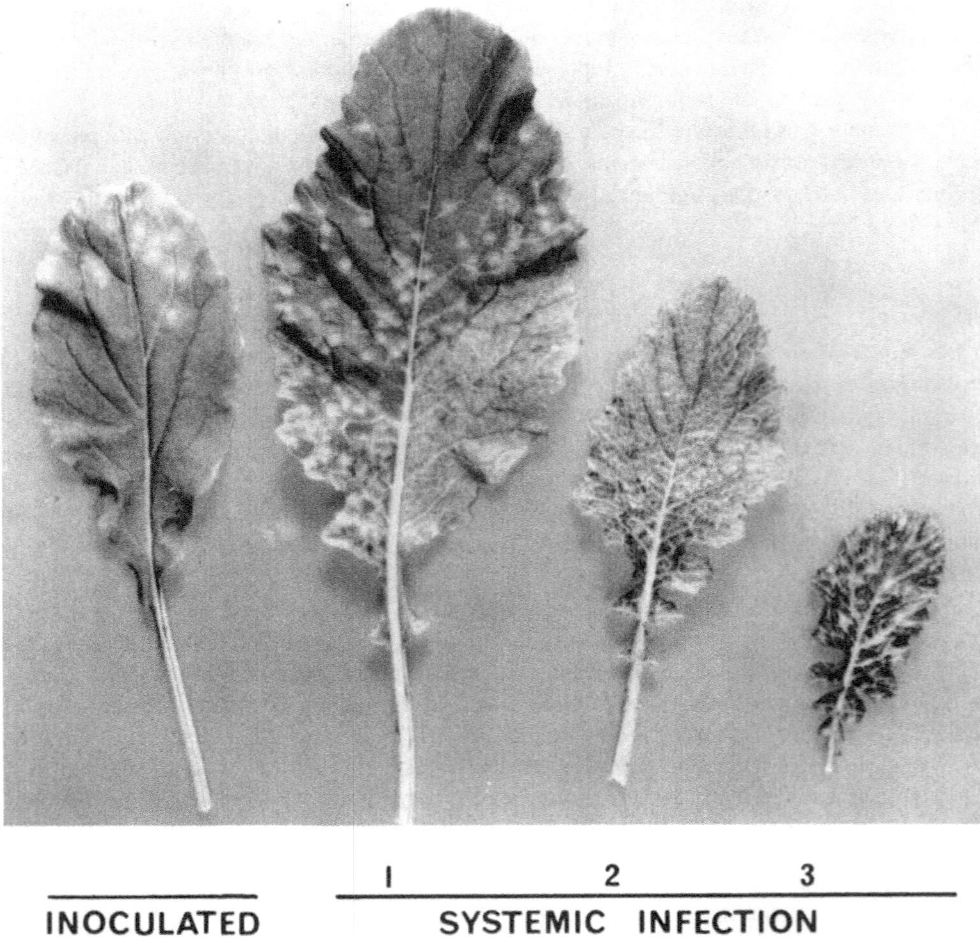

FIGURE 1. Symptoms on leaves of "Just Right" turnip after inoculation with CaMV Cabb B-JI. Symptoms are apparent on the inoculated leaf 4 to 7 days after inoculation. Systemic infection is seen as intermediate symptoms in the first infected leaf (1), and consistent vein clearing symptoms on all subsequent leaves (2,3).

CERV-infected *Dianthus barbatus*[64] and *Saponaria vaccaria,*[67] and MMV-infected *Mirabilis jalapa,*[25] and also in the vascular tissues of petunia infected with PVCV[31] and zinnia infected with DaMV.[23]

Alternative infectious agents could be inclusion bodies, or parts thereof, or viral DNA, but the problem of size for the former and the sensitivity to enzyme degradation for the latter make them improbable alternatives. Furthermore, if naked viral DNA were responsible for systemic spread then it would have to be in a form different from encapsidated infectious DNA, since this is barely detectable in a free form in infected tissue or protoplasts (see Section V).

IV. INTRACELLULAR SITES FOR REPLICATION

The intracellular site(s) for caulimovirus multiplication have still not been fully resolved, although the physical association of virus with inclusion bodies (or viroplasms) in the cytoplasm (Figure 2) and the accumulation of viral DNA in the nucleus[7,68,69] shows roles for both in a complex replication cycle.

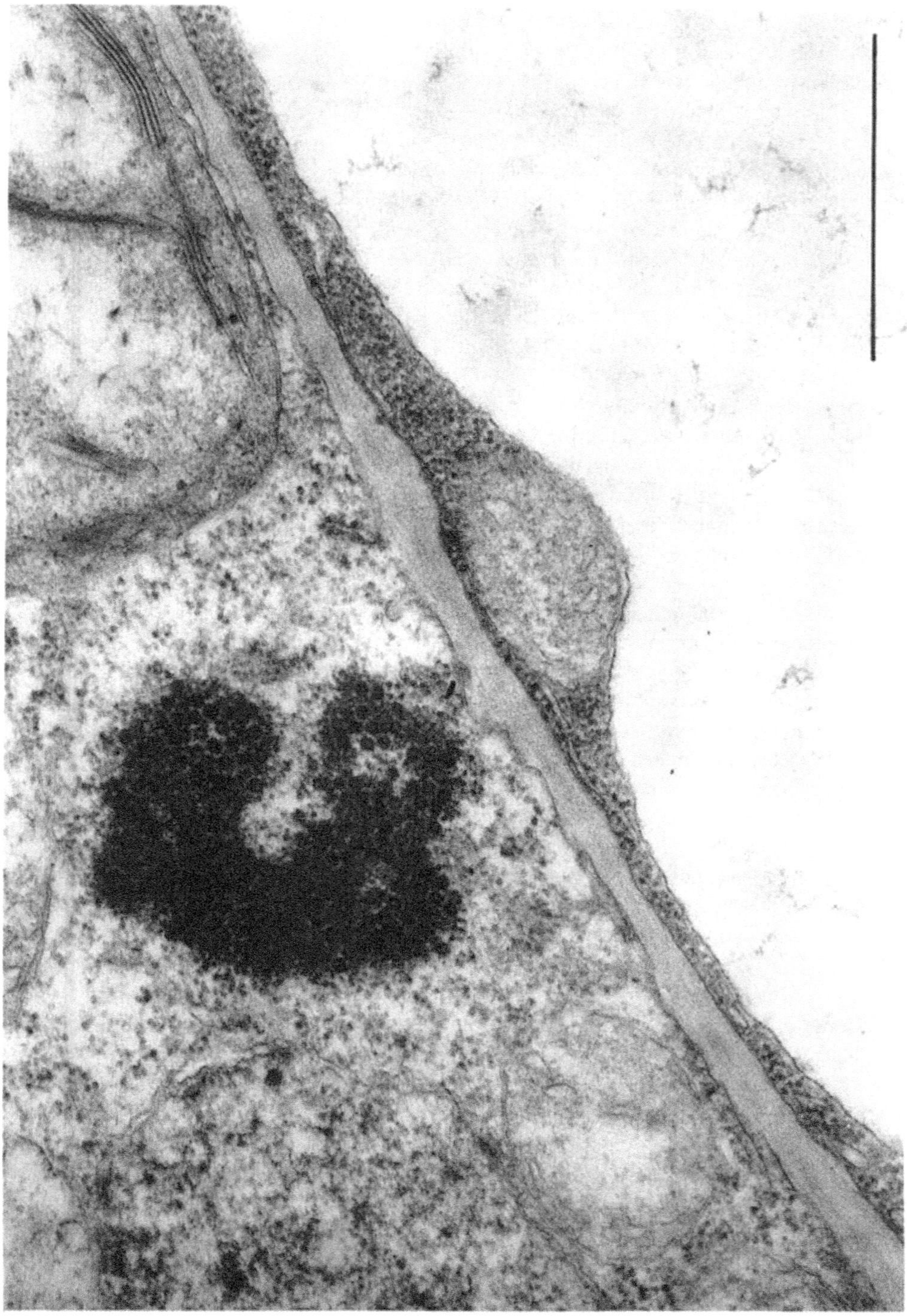

FIGURE 2. Electron micrograph showing the accumulation of CaMV in inclusion bodies in infected turnip cells. Bar represents 1 μm.

Kamei et al.[70] presented early evidence based upon incorporation of ^{3}H-thymidine that inclusion bodies were the sites of replication for CaMV, and this was later corroborated.[71] Furthermore, it has been shown that they are able to synthesize CaMV-specific DNA in a cell-free system.[72] Early ultrastructural[73] and immunochemical[74] evidence supported this with CaMV particles only being observed inside inclusion bodies, and not free in the cytoplasm or in nuclei. However, for other caulimoviruses the evidence is less clear. There is only a loose physical association between either PVCV,[31] CERV,[64] or DaMV[23] and inclusion bodies in infected cells, and in direct contrast to CaMV,[73] CERV occurs free in cytoplasm prior to the formation of inclusions.[64] Moreover, PVCV,[31] CERV,[64,67] BRRV,[28] FMV,[75] and very recently, CaMV[76] particles have been observed free inside nuclei of infected cells, albeit not abundantly.

Typically, caulimovirus inclusion bodies appear as electron-dense amorphous masses of variable size[77] with virus particles embedded or in vacuole-like areas (Figure 2). They are composed of protein and may[78,79] or may not[66,73] have free DNA present. With CaMV, the major protein constituent is of about 60,000 mol wt[80-83] which is coded for by gene VI of the virus genome[81,83,84] (Chapter 4). Whatever role this protein may play in the replication of the virus, it is clearly essential, since most insertions (or deletions) of DNA into the viral genome which disturb gene VI are not tolerated and infectivity is lost[41,42] and even when infectivity is retained, only attenuated sypmtoms are apparent.[43]

For CaMV strain D/H in turnip, electron microscopy revealed inclusion bodies which were markedly less dense than other strains although the accumulation of virions remained unaffected.[85] Similar observations made for CaMV multiplying in turnip protoplasts led to the conclusion that the inclusion matrix was not essential for virus multiplication.[51]

The implications for inclusion bodies present in chloroplasts[77] and of inclusion bodies of different appearance and without virus particles[82] remain unclear.

Active participation of nuclei in the virus replication process is shown by their ability to incorporate labeled precursors into viral DNA,[69] although it cannot be excluded that nuclei preparations may have been contaminated with inclusion bodies. Nuclei also accumulate a covalently closed and supercoiled version of the complete viral genome[7,8,10,11,68,69,86] and subgenomic supercoils[8] which are not present in mature virus particles. Prior to deproteination with phenol this 8kb supercoiled DNA exists as a minichromosome structure[7,9] which is transcriptionally active.[8,10,17,68] Transcription from the supercoiled template produces the 19S mRNA for inclusion body protein and a 35S RNA species, around which has been built a model for viral DNA replication which depends upon reverse transcription (see Section VII).

Discussing the model, Pfeiffer and Hohn[17] postulate as to the division of responsibility for viral replication between the nucleus and cytoplasm. They give the nucleus a role in the production of supercoiled molecules and transcription, and assign reverse transcription and virus assembly to inclusion bodies. However, without applying techniques which will separate nuclei and inclusion bodies, such considerations must remain speculative. No account is taken of the possible occurrence of virus particles in nuclei. Should this scheme prove to be correct, then it will demonstrate for the first time in plants the processing of DNA molecules in a nonorganellar cytoplasmic compartment. The s/sDNA geminiviruses appear to replicate predominantly in the nucleus (see Chapter 6). It is interesting that DNA replication with the retroviruses of animals is generally believed to occur in the cytoplasm. This subject has been comprehensively reviewed.[87]

In this section, I have considered CaMV to be typical of the caulimovirus group, but it cannot be discounted that differences in inclusion body structure and synthesis and the distribution of virus particles may be virus and/or host specific. Even so, it must have been with some foresight that Lawson and Hearon[67] proposed that for the replication of CERV separate roles could be assigned to the nucleus and the cytoplasm.

V. STRUCTURE OF CaMV DNA AND ITS TRANSCRIPTS

A detailed description of CaMV genome organization has been given in the previous

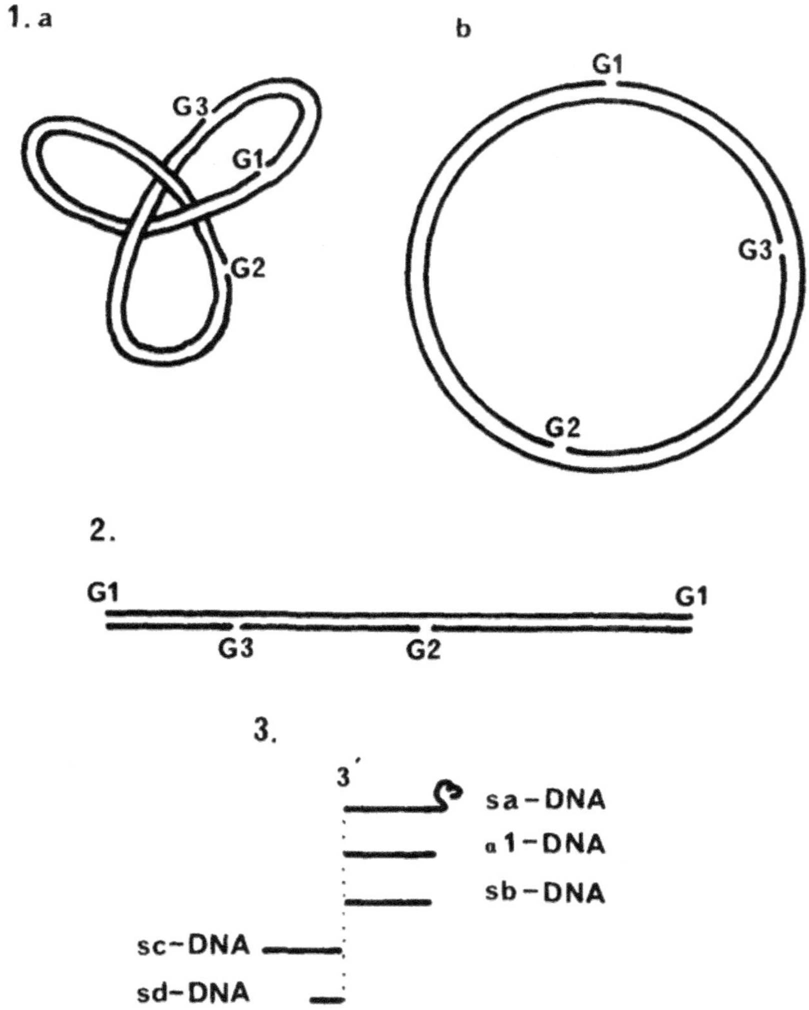

FIGURE 3. CaMV encapsidated DNA. (1) Open circular DNA present in the form of (a) topological knots, (b) open circles with three single strand discontinuities. (2) 8kb Linear DNA derived from open circular DNA linearized predominantly at G1. (3) Small DNAs of (+) and (−) strand polarity, one of which (sa-DNA) has a $tRNA^{met}_{i}$ at its 5′ end.

chapter and, except for those features implicated in the replication of viral DNA, will only be summarized here.

A. Virion DNA (Figure 3)

In all cases, DNA isolated from virus particles exists either as an open circular structure or as linear DNA. The open circles which exist in a variety of topological forms, result from single-strand discontinuities or gaps in the ds DNA which occur at specific sites. Although the number of discontinuities varies between viruses (see Chapter 4), there is always one strand (α[−]strand) which contains just a single gap (G1), the other having between one and three discontinuities (G2, G3, G4). These delimit single-strand DNA fragments termed β, γ, δ. The biological significance of the discontinuities has not been fully resolved but evidence against these being important for infection comes from the infectivity of excised, cloned CaMV DNA, lacking gap structures.[37-39] However, the con-

Table 2
DNA PRIMING SITES FOR DNA REPLICATION

(−) Strand synthesis

Virus	tRNA Priming binding site
RSV[a]	TGGTGACCCCGACGTGAT..............
*ev*1	TGGTGACCCCGACGTGAT..............
SNV	TGGGGGCTCGTCCGGGAT..............
MoMLV,MSV	TGGGGGCTCGTCCGGGAT..............
MMTV	TGGCGCCCGAACAGGGAC..............
CaMV[b]	TGGTATCAGAGCCA-G................
	++++++++++++++ +
	ACCAUAGUCUCGGUCC................
	3, tRNA$^{Met}_{i}$ primer 5'

↑ Initiation site for (-) DNA synthesis

(+) Strand synthesis

Virus	(+) Strand primer sequence
RSV	......................AGGGAGGGGGA
*ev*1	......................AGGGAGGGGGA
SNV	......................AGTGGGG
MoMLV,MSV	..................AGAAAAAGGGGGG
MMTV	..............AAAAAGAAAAAAAGGGGGA
CaMV[b] Gap 2	..................AGAGGGGAGGAGG
CaMV Gap 3	......................AAGAGTGGGGG

Initiation site for (+) DNA synthesis ↑

[a] RNA tumor viruses:[104] Rous sarcoma virus (RSV); endogenous chicken provirus, *ev*1 locus; spleen necrosis virus (SNV); Moloney strains of murine leukemia and murine sarcoma viruses (MLV, MSV); mouse mammary tumor virus (MMTV).

[b] CaMV DNA sequence shown in (+) RNA sense.

served nature of these structures, particularly the presence of only a single discontinuity in one strand, the [−] strand for CaMV, indicates that they must play an important role in the multiplication of the caulimoviruses.

Sequence analysis[6,88] has revealed that these discontinuities are not gaps but are sites of sequence overlap which could form limited triple-stranded regions. For each discontinuity in CaMV the 5′ nucleotide is in a fixed position[89] while the 3′ nucleotide is in a variable position resulting in short overlaps of between 8 and 43 nucleotides.[88] Recently, it has been reported that the 5′ extremity of the interrupted strand is sometimes a ribonucleotide.[12] It appears to be important that for CaMV, at least, the AT-rich genome has immediately upstream from G2 and G3, purine and G-rich regions of 13 and 10 nucleotides, respectively[6] (Table 2). The role these sites may have as primers for (+)strand DNA synthesis will be discussed later (see Section VI.)

The linear form of CaMV DNA is not infectious[90] and arises by nicking of the circular form[91-93] predominantly at the sites with single-strand interruptions.[61,90] This probably occurs by both mechanical cleavage and S_1-type enzymatic cleavage.

Further, small DNA species have been identified in DNA extracted from CaMV virions.[10,12,13] Covey et al.[13] identified small quantities of 625, 575, and 200 nucleotide fragments (termed sb-, sc-, and sd-DNA) present in virion DNA and mapped their position relative to the (−)strand discontinuity. In all cases, either the 5′ or 3′ terminus mapped close to the 5′ extremity of the (−)strand.

These species migrated as a single band on nondenaturing gels and accordingly demonstrated different polarities. sb DNA has the same polarity as the (−)strand and is of opposite polarity to sc- and sd-DNA, and probably corresponds with a separately identified small DNA (termed α 1)[12] of 587 nucleotides. The 5′ end of α 1 is coterminal with the extreme 5′ end of the (−)strand and its 3′ end is two nucleotides downstream from the initiation site of the large transcript.[94,95] A proportion of α 1 molecules possess a ribonucleotide at their 5′ terminus.[12] Recently, a larger molecule of 680 nucleotides (termed sa DNA), initially identified in DNA and poly(A)+ RNA fractions from CaMV-infected turnip leaves,[13] has been found associated with DNA from mature virions.[14] This has the same polarity as sb-DNA and α 1 and has RNA which is probably a $tRNA_i^{met}$ bound at its 5′ end.

The structural association that these small molecules have with the infectious open circular DNA or virus particles has not been fully resolved. Clearly the differences in polarity mean that they cannot only hybridize with each other but also with larger DNA species. Hence they have only been observed separately[13] if at all[12] under denaturing conditions. Comparing DNA extracted from virion preparations before and after RNase or DNase treatment, it has been observed[61,96] that sa-DNA is positioned on the particle such that its RNA, but not its DNA moiety is susceptible to enzyme digestion. Similarly, following inoculation of protoplasts with virus particles, sa-DNA is rapidly degraded to a smaller species which is itself released from virus particles in advance of the large genomic species.[61] Whether sa-DNA is comparatively loosely attached to virus particles as a consequence of incomplete encapsidation, and whether the presence of $tRNA_i^{met}$ "outside" the virus particle is important in the earliest stages of infection are both intriguing questions.

B. Unencapsidated or "Free" Viral DNA (Figure 4)

The insensitivity of caulimovirus particles to disruption by phenol means that for this virus, at least, a convenient means is available for examining viral DNA species not encapsidated in vivo. This approach has been used most extensively for CaMV although some information relating to CERV, FMV, MMV, and ThMV is also available.[75] Hull and Covey[97] estimate that "free" or unencapsidated CaMV DNA comprises about 10% of total CaMV DNA present in infected turnip tissue, which is higher than a similar estimate for in vitro infected protoplasts.[61] Since there are equivalent accumulations of viral DNA per cell in both systems,[52] it would appear that the fidelity of viral DNA replication in protoplasts after a single cycle of virus multiplication is greater than in persistently infected leaf mesophyll cells. When this DNA extracted from either infected turnip leaves[13,18,86,97] or protoplasts[61] is analyzed on agarose gels without denaturation, a number of different molecular species are revealed. Some of these have been partially or fully characterized and their relationship with either transcription or DNA replication discussed. However, it is difficult to place a meaningful biological interpretation on the occurrence of these molecules based solely upon a biochemical analysis of structure. With the exception of supercoiled viral DNA, a biological role has not been positively assigned to any other "free" DNA species. Difficulty in distinguishing between real and defective intermediates in virus multiplication could lead to erroneous conclusions.

1. Supercoiled CaMV DNA

Supercoiled CaMV DNA has been identified in total DNA extracts from infected turnip plants[9,86,97] and protoplasts,[61] from subcellular fractions enriched for nuclei and virus inclusion bodies,[17] and from nuclei preparations.[7,8,10,11]

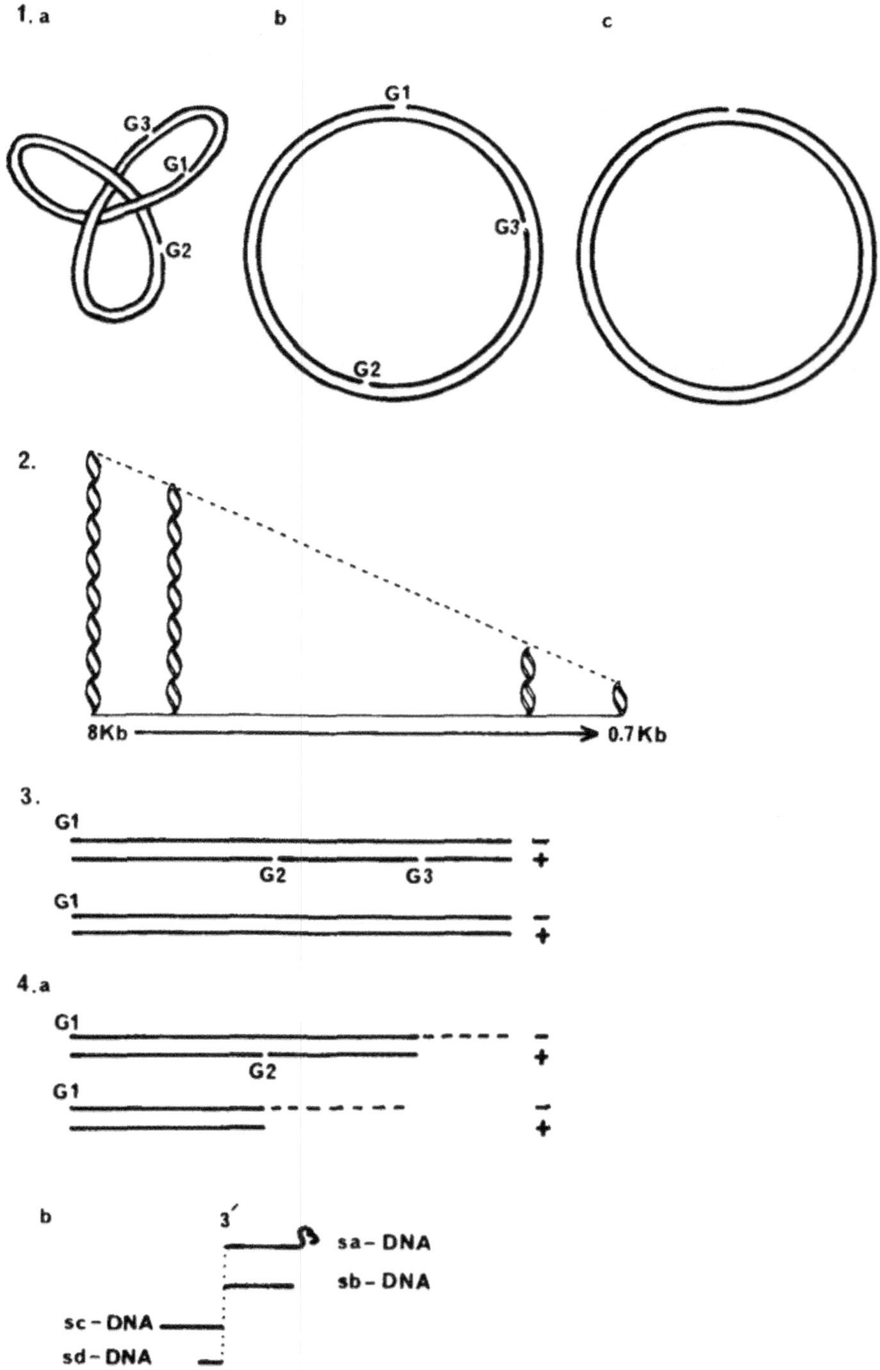

FIGURE 4. CaMV unencapsidated DNA. (1) Open circular DNA present as topological knots (a), and open circles (b) with three single-strand discontinuities, and (c) open circles with a single discontinuity. (2) Supercoiled DNA ranging in size frm 8kb to 0.7kb. Other predominant species of slightly less than 8kb and 2kb are indicated. (3) Linear DNA (8kb) arising by linearization of the open circular forms at G1. (4) Linear DNA (<8kb) comprised of (a) two groups of truncated linear molecules with variable length extensions of the (−) strands and (b) a range of small molecules also found in encapsidated DNA one of which (sa-DNA) has a $tRNA^{met}_{i}$ bound to its 5′-end.

The molecule of full genomic size, which is not found in virions, occurs in vivo as a nuclear minichromosome.[7,9] Electron microscopy showed it to have a compaction ratio of 2.9 and an average of 41.0 nucleosomes per molecule, giving a DNA repeat of 196bp. This value is very similar to that for host chromatin.[9] It is not clear where the minichromosome resides in the nucleus.

The discovery of supercoiled CaMV DNA of less than genomic length present in nuclei[8,10] may have implications in the search for specific transcripts for protein open coding regions (OR)I to V and VII (see Chapter 4) and in elucidating the mechanism by which supercoiled molecules arise. The authors suggest that these subgenomic supercoils result from intragenomic recombination. From their behavior on sucrose gradients, it appears that these molecules exist as nucleoprotein complexes, possibly also as minichromosomes.

A direct comparison between the New Zealand and Cabb-BJI isolates of CaMV in the same host showed a considerable difference in the amount of accumulated 8kb supercoiled DNA,[97] possibly signifying that its concentration is dependent upon some virus-specified function.

Guilfoyle[68] demonstrated that CaMV-specific transcription occurred in the nucleus, and accumulated evidence (previous chapter) now indicates that it is the minichromosome which is active in transcribing at least the 35S RNA.

2. Open Circular DNA

oc CaMV DNA has been observed in unencapsidated DNA from infected turnip tissue,[86,97] protoplasts,[61] and nuclei preparations[7] and has been attributed to the oc form with site-specific discontinuities found in virion DNA[86,97] or random nicking of the supercoiled DNA.[7,8,11] When the unencapsidated oc extracted from infected protoplasts was characterized by 2-D electrophoresis in agarose gels,[61] it was clear that it contained only a single discontinuity. The triple-gapped oc was not detectable. A single-gapped molecule also constituted a major proportion of unencapsidated oc DNA present in CaMV-infected turnip tissues.[98] It is conceivable that this molecule may represent an intermediate in the formation of supercoiled DNA.

3. Linear DNA (8kb)

As with virion DNA, genomic length linear DNA present in preparations of unencapsidated CaMV DNA[9,61,86,97] probably results from cleavage of oc molecules at single-strand sites. However, 2-D electrophoresis[97] shows that cleavage is preferentially at G1 yielding ss DNA of 8, 3.8, 2.6, and 1.6kb upon denaturation. Although in accord with infected protoplasts containing an oc molecule with a single gap, this molecule appears to be completely double stranded.

4. Linear DNA (<8kb)

A range of linear DNA forms of <8kb can be detected in unencapsidated DNA from CaMV-infected turnip plants[13,18,97] and protoplasts.[61] Some of these correspond with molecules predicted from more than one single-strand cleavage of virion DNA.

Two other unusual molecular forms have been characterized.[97] These are the major forms of unencapsidated DNA from plants. Both comprise ds DNA molecules with single-strand extensions. The smaller form is abundant in unencapsidated DNA from both CaMV strain Cabb-BJI-infected tissue[97] and protoplasts[61] and maps on the CaMV genome as a double-stranded molecule between G1 and G2 with a variable length extension of the (−)strand from the position of G2 to G3.[97] The larger form is a molecule with similar structure being double-stranded anticlockwise from G1 to G3 with a variable extension of the (−)strand round to G1. In synchronously infected protoplasts, the latter form has not been detected. Recently, similar forms of unencapsidated DNA have been identified in squirrel livers infected with ground squirrel hepatitis virus, where the (+) strand was always present as double-stranded DNA and the (−) strand occurred in excess.[97] This observation paralleled that for duck hepatitis B virus,[100] which replicates asymmetrically by reverse transcription of an RNA intermediate.[101]

Small DNA species present in DNA from virus particles (Section V) have also been observed in "free" CaMV DNA[13] and in much greater relative concentrations, and in nuclei lysates from infected turnips.[10] Their function in vivo is not certain but they and the double-stranded molecules with single-strand extensions have been implicated in the model for CaMV-DNA replication based upon reverse transcription (Section VII).

5. Other Caulimoviruses

For CERV, FMV, and MMV,[75] two-dimensional electrophoresis of unencapsidated DNA from their respective hosts revealed a complex spectrum of DNA molecules, some of which were similar in their relative migration to molecules characterized for CaMV. On this basis, supercoiled DNA, open circular DNA, low-molecular-weight DNAs, and various forms of linear DNA have been identified. It is significant when evaluating some proposed mechanisms of viral DNA replication (Section VII.A) that those viruses containing three gaps in the (+)strand (MMV; FMV) have an equal number of molecular forms which are double stranded with (−)strand extensions of variable length.

C. CaMV-Specific RNA

Details regarding the origin and structure of CaMV-RNA transcripts have been discussed in detail in the previous chapter and only the salient features relating to mechanisms of DNA replication will be mentioned here.

Poly(A) + RNA isolated from infected turnip leaves contains two abundant CaMV-specific RNAs.[94,95,102,103] The smaller of these is the 1.9kb transcript of ORVI and is the messenger RNA for the inclusion body protein. The larger RNA (35S RNA) is a transcript of the complete DNA (−)strand and contains all the open coding regions found in the DNA genome. Individual mRNAs for the other ORs have not been found.

The 3′ ends of both RNAs are coterminal within the large intergenic region approximately 400 nucleotides 3′ to the (−)strand gap of virion DNA. The 5′ terminus of 35S RNA is also within the large intergenic region and maps about 180 nucleotides upstream of its own 3′ end. 35S RNA is therefore longer than the DNA (−)strand from which it is transcribed and has a 180-nucleotide terminal repeat.

It has been noted[15-17] that 35S RNA has several features in common with the genomic RNA of animal retroviruses. These RNAs have closely clustered protein coding regions with no introns, they are transcribed asymetrically from only one DNA strand, and have direct repeats at each end. Downstream from the 3′ side of the terminal repeat is an 18-nucleotide sequence complementary to the 3′ end of a tRNA, a molecule which is believed to act as a primer for reverse transcription. For CaMV, a similar complementarity has been identified between the first 14 bases at the 3′ end of a $tRNA_i^{met}$ from wheat germ or bean cytoplasm, and 14 bases on the 35S RNA approximately 600 nucleotides downstream from its 5′ end. If it is taken that there is a close homology between the $tRNA_i^{met}$ from widely divergent plant species, then a natural RNA primer for reverse transcription exists.

VI. KINETICS OF CaMV REPLICATION

Accumulation of CaMV in in vitro-inoculated turnip protoplasts, assessed as radiolabel incorporated into virus particles[49] or immunochemical staining for virus particles,[50,52] showed a cycle of replication to take between 72 and 96 hr. The pattern of accumulation of total viral DNA was similar reaching a maximum 65 to 96 hr postinoculation.[52,61]

Attempts to characterize the kinetics of formation of unencapsidated CaMV DNA in turnip protoplasts has been complicated by the presence of large amounts of inoculum-derived DNA at the early times after inoculation.[61] However, a sequence of changes in unencapsidated DNA can be identified. Within 1 hr after inoculation, virion-associated sa DNA has been processed to a lower molecular weight form presumably equivalent to sb DNA. This occurs before the virus particle relaxes sufficiently to permit phenol extraction of genomic virion DNA. By 24 hr postinoculation, all the encapsidated DNA has been released and most has been degraded by single-strand nuclease activity to 8kb and smaller linear DNA fragments. Degradation continues such that very little of the inoculum remains by 72 hr.[52]

Novel synthesis of viral DNA was first detectable by 24 hr as incorporation of ^{3}H-thymidine into 8kb linear DNA[59] and by the formation of open circular DNA with one continuous strand; i.e., with less than the three gaps characteristic of virion DNA.[61] This latter species has been detected as early as 5 hr after inoculation and increases in concentration to reach a maximum at the end of the virus replication cycle (65 to 96 hr) when it represents a major species in unencapsidated DNA. It is not known whether this species present early in infection is repaired inoculum DNA or the progeny open circular DNA typical of unencapsidated DNA late in infection. Supercoiled CaMV DNA not previously identified in turnip tissue infected with the CaMV isolate Cabb B-JI,[97] the lower molecular weight ds DNA molecule with variable extensions of the (−) strand (Section V), and an unidentified species of faster migration, all increase with time in parallel with the accumulation of mature virions.[61]

The degradation product of sa DNA (sb DNA) is detectable throughout the infection, while sa DNA itself reappears after 24 to 48 hr and increases continuously with time. A further species migrating between sa- and sb-DNA on neutral gels shows similar increases.

While the time taken for a "single cycle" of virus replication in in vitro-infected protoplasts can be estimated as 65 to 96 hr, more complex experiments would be necessary in order to measure the kinetics of the individual components of the cycle. Pulse-chase experiments have not yet been attempted with caulimovirus-infected protoplasts. Using an alternative approach with protoplasts from CaMV-infected leaves,[49] it was suggested that the time taken for viral DNA synthesis and encapsidation was only 10 to 15 hr, and that by comparison with in vitro-infected protoplasts, the early stages in the infection cycle (e.g., virus entry, uncoating, construction of the machinery for DNA synthesis) were relatively slow. However no account was taken of the deleterious effects and the artificial nature of protoplast inoculation procedures.

VII. MECHANISMS OF VIRAL DNA REPLICATION

A. Replication by Reverse Transcription

The hypothesis that hepatitis B viruses (HBV) are replicated by a process involving reverse transcription[101] and recognition of the structural similarities between HBV and CaMV with regard to the DNA genome coincided with a search for evidence for a similar mechanism operating with CaMV. Based upon the identification of potential (+) and (−) strand priming sites from the CaMV sequence (Table 2), recognition of similarities between CaMV-35S RNA and retrovirus genomes[15-17] identification of a piece of CaMV DNA still linked to its $tRNA_i^{met}$ primer[15] and equivalent to retrovirus "strong stop" DNA,[14] and some enzymological data[10,17,20,21] a model for reverse transcription of CaMV DNA has emerged. At the time of writing, the evidence in favor of the model was strong. With the identification of a virus-specific enzyme active in producing CaMV DNA on an endogenous template and capable of copying a heteropolymeric RNA template,[20,21] it must now only require the isolation of bona fide reverse transcriptase to provide conclusive evidence for the principle features of the model. Further support would be obtained if it could be demonstrated that it was possible to initiate a competent infection with 35S RNA alone. It should be noted that while offering

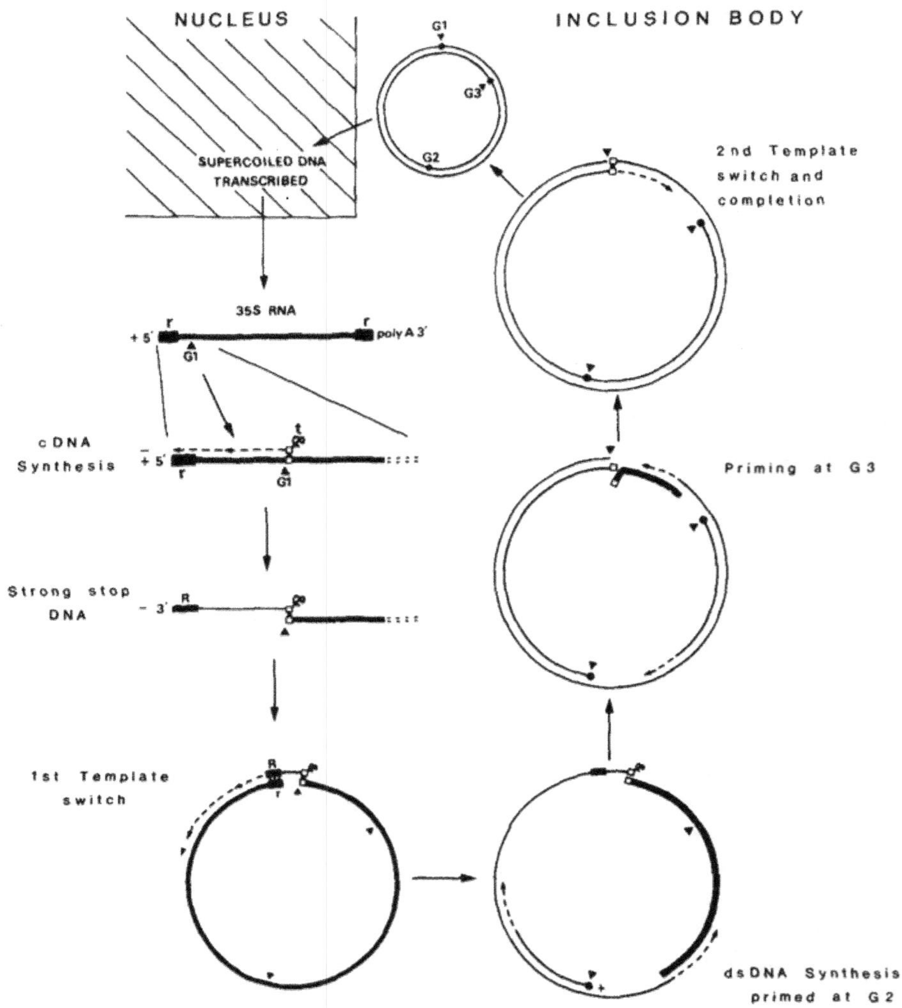

FIGURE 5. A model for the replication cycle of CaMV DNA. Supercoiled DNA, produced from gapped virion DNA, is first trancsribed into 35S RNA (thick line) which has a terminal direct repeat (r) of 180 nucleotides. Host cell $tRNA_i^{met}$ molecule (t) then base-pairs with a complementary sequence (open square) in the RNA immediately to the right of the G1 position. Reverse transcription proceeds towards the 5′ end of 35S RNA, removing the RNA template. The 5′ r sequence is copied into minus-strand DNA (R). The DNA R sequence base-pairs with the 35S RNA 3′ r sequence thus circularizing the molecule (bottom left). Minus-strand synthesis continues and passes the DNA plus-strand primer sequences at G2 (bottom right) and G3 (middle right). Plus strands are synthesized on the DNA minus-strand template. Both plus and minus strands copy the tRNA primer and tRNA primer complementary sequence in 35S RNA, respectively. These tRNA sequences in the DNA strands base-pair to effect a second jump of template for the transcriptase to complete plus-strand synthesis (upper right). (From Hull, R. and Covey, S. N., *Sci. Prog. (Oxford)*, 68, 420, 1983. With permission.)

a good base for further work, the model will almost certainly be a simplification of a more complex process.

The Model[15-17] *(Figure 5)*

1. 35S RNA is transcribed in the nucleus from CaMV minichromosome and polyadenylated.
2. At the correct intracellular site (?) 35S RNA becomes a template for reverse tran-

scription using a $tRNA_i^{met}$ primer. The 3′ end of the primer hybridizes to 14 nucleotides about 600 nucleotides downstream form the 5′ end of 35S RNA (Table 2).

3. Reverse transcription copies the 35S RNA to its 5′ end, producing a small DNA molecule (sa-DNA) including the terminally repeated sequence, equivalent to "strong stop" DNA. α1 DNA[12] and sb-DNA[13] are presumably equivalent molecules with most or all the ribonucleotides removed or lost.
4. Reverse transcription then jumps to the 3′ end of 35S RNA which has the terminal repeat sequence, and (−) strand synthesis continues while the copied RNA template (but not the 14 nucleotides hybridized to $tRNA_i^{met}$ primer) is degraded behind the leading DNA strand, probably by RNase H activity.
5. On passing nucleotide 4218 (G2) and then nucleotide 1632 (G3), priming for (+) strand synthesis occurs at the purine- and G-rich regions (+strand). Primers at these sites (Table 2) may either be residual RNA not degraded owing to the resistance of rG:dC base pairs to RNase H attack, or some unidentified RNA with the consensus sequence[17]

$$\begin{array}{c} \text{U} \\ \downarrow \\ 3'\text{AGAPuPuPuAGGGGGPu} \end{array}$$

Ribonucleotides positioned at the 5′ extremities of the β and γ-strands[12] are probably residues from (+) strand primers.

6. Synthesis of the β (+) strand is extended beyond the 5′ end of the (−) strand to copy the $tRNA_i^{met}$ primer sequence following strand displacement of the 5′ end of 35S RNA. The $tRNA_i^{met}$ primer is degraded.
7. Reverse transcription of 35S RNA continues and copies the $tRNA_i^{met}$ binding sequence, resulting in a repeat sequence that can hybridize with the 3′ end of the (+) strand.
8. A second switch of templates occurs and (−) and (+) strand synthesis continues to completion. The triple-stranded structures at G1, G2, and G3 are formed by displacement into and beyond the 5′ primer sites by the extending α, β, and γ strands. The formation of single-strand (+) strands by a process of strand displacement has also been shown for Rous sarcoma virus replication.[105]

This model offers explanations for the structure, number, and distribution of the gaps in the caulimoviruses generally (Section V) and for some of the unencapsidated DNA forms observed in CaMV-infected tissue and protoplasts. Of the small DNAs, sa-, sb-, and α 1 DNA have deen accounted for. sc- and sd-DNA are of opposite polarity to the (−) strand and may originate from minor (+) strand priming sites on 35S RNA[17] or by a loop-back mechanism.[10] The accumulation of sa-DNA in infected tissues and protoplasts may reflect the inefficiency of the first template switch. Whether the encapsidation of sa-DNA and other small DNAs of (+) and (−) polarity occurs adventitiously, or whether it occurs in preparation for some biological event early in infection, is unknown. The ds DNA fragments with 3′ extensions of the (−) strand (Section V) accumulate and consequently may represent defective rather than true replication intermediates.[17] They could arise by defective priming for (+) strand synthesis at G3 for the smaller form, or for the larger, cleavage of the $tRNA_i^{met}$ primer from the 5′ of the (−) strand. Nuclease activity would then degrade the single-strand extension to a variable length. It is curious that in protoplasts producing a "single cycle" of virus replication that only the smaller form accumulates.[61] Perhaps there is a greater fidelity of DNA replication in this system than in persistently infected cells in intact tissue, and the accumulation of defectives depends only upon the efficiency of (+) strand priming rather than enzymic degradation. The presence of a pyrimidine (T) at nt1627

(Table 2) could produce a more labile primer-binding site for the (+) strand synthesis. In accordance with these forms resulting from a process of reverse transcription, CaMV strain CM4-184 which has only one gap in the (+) strand also has only a single ds DNA with 3′ extension in unencapsidated DNA.[18]

B. Replication of Supercoiled DNA

Overshadowed by the interest in the reverse transcription model, little attention has been given to the origin and accumulation of supercoiled DNA in the nucleus. It is assumed that supercoiled DNA originates from repair and twisting of the gapped virion DNA and accordingly, an open circle with less than three gaps can be identified early in the infection of turnip protoplasts with CaMV (Section VI). Pfeiffer and Hohn[17] propose a replication cycle for CaMV whereby progeny virion DNA with gaps, synthesized in inclusion bodies, is transported back to the nucleus for repair to increase the pool of supercoiled DNA. A similar mechanism operates with avian retroviruses where linear duplex DNA synthesized in the cytoplasm is transported back to the nucleus prior to circularization and integration.[87] A number of factors provide reason to question this hypothesis for CaMV and to consider possible independent mechanisms of replication for supercoiled DNA, although it should be stated that there is no evidence for independent replication of unintegrated retroviral DNA in the nucleus. However, the biological role of supercoiled DNA is different in the two situations. Supercoiled retroviral DNA is used solely for integration after which transcription occurs, whereas supercoiled CaMV DNA is the transcription template.

The mechanism proposed by Pfeiffer and Hohn[17] would require virion DNA to be transported from widely dispersed inclusion bodies through the cytoplasm to the nucleus. However, fully gapped virion DNA is not present in large quantities in preparations of "free" DNA from infected tissue,[97] or detectable as "free" DNA in infected protoplasts.[61] Preparations of "free" DNA do contain relatively large amounts of open circular DNA but this contains only one single-strand discontinuity. This structure would not result from the proposed model for reverse transcription, and there is no evidence for the occurrence of intermediate two-gapped molecules produced during repair of virion DNA. However, it cannot be discounted that the accumulated open circles represent molecules defective for the repair of one gap, or that for unknown reasons the repair of one of the gaps is a rate-limiting step in the production of supercoiled molecules. The relative concentrations of 8kb supercoiled molecules have been compared between different isolates of CaMV replicating in turnip[97] and have been shown to vary greatly. Finally, preparations of nuclei, purported to contain no inclusion bodies, incorporate labeled precursor into open circular viral DNA.[69]

Sequence analysis[106,107] of the CaMV-Cabb S isolate has revealed the distribution of potential hairpin structures capable of forming cruciforms, which may be important in the regulation of DNA replication.[108] Unlike some other eukaryotic viruses, CaMV does not show a preponderance of these structures in the intergenic regions, the largest inverted repeats occurring at the beginning[106] and the end[107] of coding region V. There are no obvious similarities between these structures and the 13bp inverted repeat close to an AT-rich region on SV40 DNA considered to be the origin of replication.[109]

In the absence of further data it is not easy to postulate which mechanism could operate in the synthesis of CaMV supercoiled DNA. From its early appearance in infected protoplasts it is possible that oc DNA with a single gap may be involved. Similar open circles are produced following D-loop bidirectional replication of mitochondrial DNA. This mechanism produces a RNA primer linked to a 500- to 600-nt fragment of DNA (7S mitochondrial DNA) reminiscent of some small CaMV DNA fragments. The two daughter strands are synthesized asynchronously and have different origins of replication. Whatever origin could be functional in supercoiled CaMV DNA, it appears that it is not recognized as such in yeast, since the complete CaMV genome, or *Eco*RI fragments, fails to supply an origin of replication when cloned into the YIp5 plasmid.[110] However, origins of replication present on two other eukaryotic DNA viruses, SV40[111] and polyoma virus,[112] are also not recognized in the yeast system.

In discussing subgenomic supercoiled CaMV DNA, Olszewski and Guilfoyle[8] suggest that they may represent redundant products of defective CaMV replication, or since they accumulate, may be capable of replication themselves possibly with intact virus genomes as helpers.

The concept of independent replication of supercoiled CaMV DNA is attractive in offering an explanation for the maintenance of CaMV DNA replication without the accumulation of virus DNA in rapidly growing protoplast-derived callus clones from infected turnip.[57]

Clearly the possibility of the independent replication of supercoiled CaMV DNA remains an open question. The answer will not only further the understanding of CaMV multiplication but may have important implications for the study of host DNA replication.

C. Enzymology and Inhibitors of Replication

The model for reverse transcription for CaMV replication has provided an open invitation for the enzymologist, and one would anticipate that in the coming years this area will receive much attention.

In the study of the enzymology it is important that what is being characterized is shown to be infection specific, and is cases where DNA polymerase activity is functioning on endogenous template, is shown to be virus specific by hybridization.

Plants have been shown to contain several DNA-polymerizing enzymes which are broadly similar to polymerases α, β, and γ characterized for animal systems,[113] and which have also been correspondingly termed polymerase B, C, and A.[114] Each DNA polymerase shows different specificities with homo- and heteropolymeric templates and different responses to specific inhibitors and environmental conditions. Retroviral reverse transcriptases show some properties in common with several host DNA polymerases while remaining distinct from all of them. In the search for virus-specific polymerase it is valuable to note two points; first, DNA polymerase γ (A) may show different sensitivities to inhibitors when using RNA or DNA as a template[113] and second, a free host reverse transcriptase has recently been demonstrated from healthy embryonic tissues of the Japanese quail[115] and may well exist in the uninfected tissues of caulimovirus hosts.

To date two main approaches have been used to study CaMV-replicative activities. First, subcellular fractions enriched for nuclei and inclusion bodies,[10,72] "enzyme complexes" leached from them,[17] and preparations of dense replication complexes pelleted through 60% sucrose[21] have all been shown to incorporate labeled precursors into host and viral DNA using endogenous templates. Second, relatively crude preparations of enzymes extracted from subcellular fractions have been characterized by their ability to function on different artificial templates.[10,20,21] For nuclei preparations,[10] incorporation of dNTPs into CaMV DNA was "largely resistant" to aphidicolin, a specific inhibitor of DNA polymerase α and was also resistant to RNase and DNase. Presumably, enzymes are not readily taken up by isolated nuclei. The activity was, however, inhibited by high salt concentrations, possibly indicative of polymerases dissimilar to DNA polymerase γ. Enzymically active preparations leached from subcellular fractions enriched for nuclei and inclusion bodies could themselves be separated into fractions directing CaMV-specific RNA and DNA synthesis.[10,17] Pfeiffer and Hohn[17] concluded from inhibitor, RNase, and salt sensitivity tests that the DNA-synthesizing activity involved an enzyme similar to DNA polymerase A (itself similar to the animal DNA polymerase γ) and that this enzyme was active in copying both a DNA and RNA template. However, it is not clear whether these properties related to the CaMV-specific activity, and without formal indentification of the enzyme template as RNA it cannot be taken that this represents virus-specific "reverse transcriptase". Two consistent findings

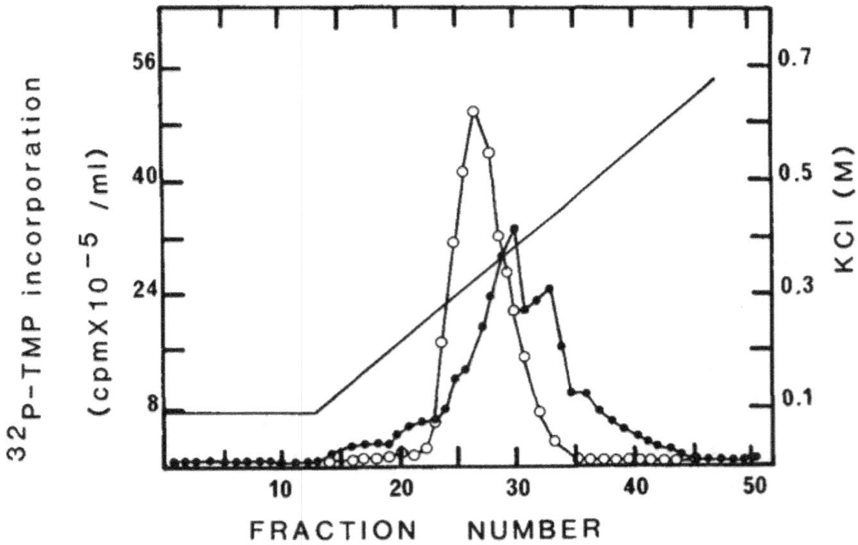

FIGURE 6. Heparin Sepharose chromatography of DNA polymerase preparations. From 100 g of healthy (○-○) and CaMV-infected (●-●) turnip tissue. DNA polymerase or "reverse transcriptase"[10] was assayed using a poly rA:oligo dT_{12-18} template/primer. Heparin Sepharose columns were eluted with a linear gradient of KCl (0.1 to 1.0 *M*). (Data taken from Guilfoyle, T., Olszewski, N., Hagen, G., Kuzj, A., and McClure, B., in *Plant Molecular Biology*, UCLA Symp., Goldberg, R., Ed., Alan R. Liss, New York, 1983, 17. With permission.)

do emerge from these studies; first, the DNA polymerizing activities leached from subcellular fractions are partially sensitive to RNase, and second, they occur in gradient fractions containing open circular and linear CaMV DNA but very little of the supercoiled molecule. At this time, it is not clear where on these gradients hypothetical intermediates in the replication of supercoiled DNA would occur.

Using the artificial template/primer polyA/oligo dT_{12-18}, the activity of DNA polymerase γ (A) has been measured in extracts from healthy and CaMV-infected turnip[10,116] and cabbage.[20] No major virus-specific increase in polymerase γ-type activity was observed, although following separation on heparin Sepharose[10] (Figure 6) or phosphocellulose[20] (Figure 7) columns, an infection-specific change in the elution profile was recorded. This was particularly so in the latter case where the workers found a distinct new enzyme species with an additional and unique activity when given poly rC/oligo dG as a template/primer. Furthermore, it was found that extracts from infected leaves contained DNA polymerase activity stimulated by heteropolymeric mRNAs, provided oligo dT_{12-18} was also present, and that this activity occurred even in the presence of actinomycin D.[20] The question as to whether this enzyme constituted a virus-coded reverse transcriptase was addressed in their discussion. Suffice it to say that the poly rC/oligo dG directed activity is characteristic of many reverse trancriptases but without further characterization of the infection-specific protein involved it remains premature to consider it as virus coded. Similar observations[21] indicated that replication complexes isolated from CaMV-infected tissue or protoplasts which were active in copying endogenous viral templates were also capable of reverse transcribing primed cowpea mosaic virus RNA.

As yet there have been no reports of reverse transcriptase activity in preparations of CaMV particles, as occurs with retroviruses.[87] This may reflect the difference in the nature of the viral genome. The extreme stability of caulimovirus particles will doubtlessly hinder the identification of any encapsidated enzyme.

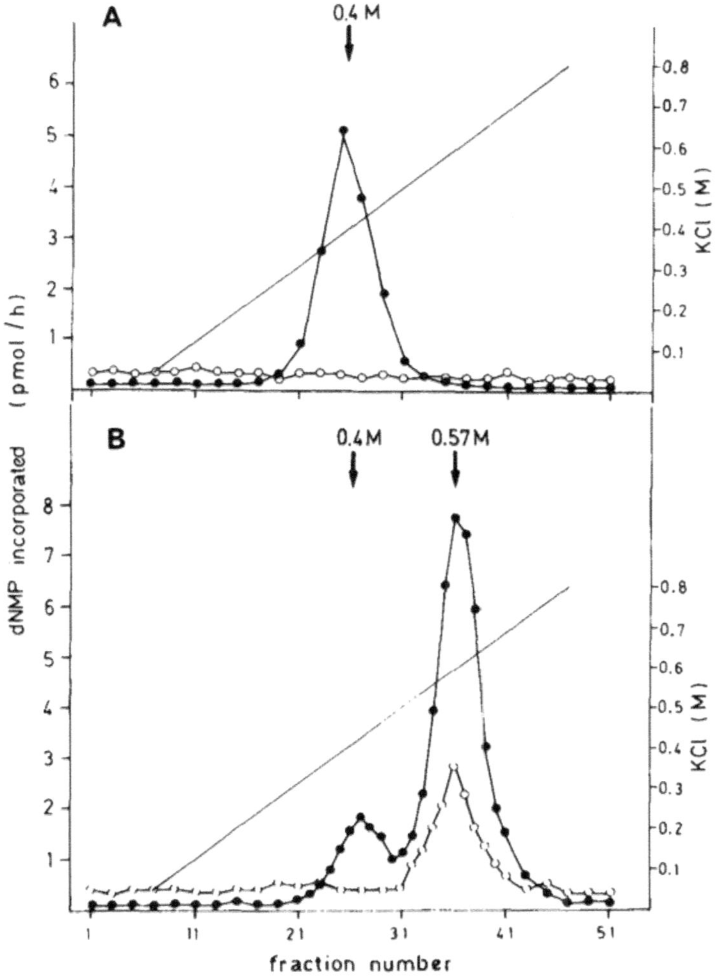

FIGURE 7. Phosphocellulose chromatography of DNA polymerase preparations from healthy (A) or CaMV-infected (B) cabbage leaves. Activity was measured using poly rA:oligo $dT_{12\text{-}18}$ (●-●) or poly rC:oligo $dG_{12\text{-}18}$ (○-○) as template/primers. Columns were eluted with a linear gradient of KCl; vertical arrows indicate the KCl concentrations at which different peaks of activity were eluted. (From Volovitch, M., Modjtahedi, N., Yot, P., and Brun, G., *EMBO J.*, 3, 311, 1984. With permission.)

Recently a computer-assisted analysis of the amino acid composition of retrovirus reverse transcriptase, and all the unassigned open coding regions of HBV and CaMV has revealed a remarkable sequence similarity between the N-terminal conserved region of retrovirus pol gene products and the largest OR of HBV and CaMV[19,20] (Figure 8). Coincidentally, this is the largest OR and has the potential to code for a protein of about 80K, a size similar to reverse transcriptase from murine leukemia virus.

It is to be expected that other enzyme activities will be found associated with the multiplication cycle of CaMV. With the exception of DNA-dependent RNA polymerase II, identified by α-amanitin sensitivity as the enzyme transcribing RNA from CaMV minichromosome,[7] other activities have not yet been identified. One can, however, predict several essential activities: first, a DNA polymerase for repair of the gaps prior to formation of

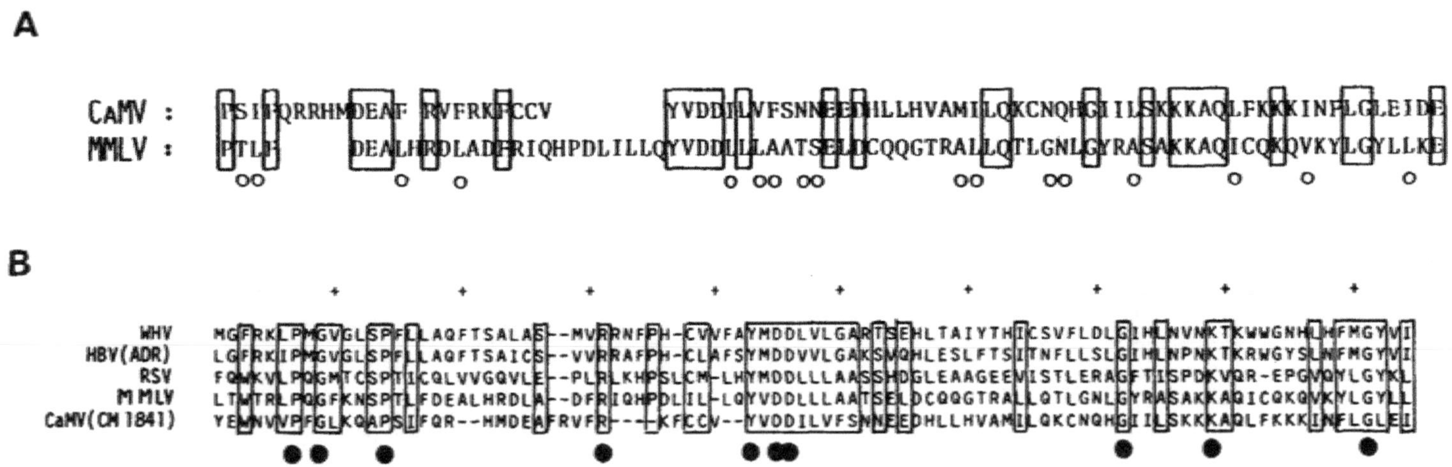

FIGURE 8. (A) Alignment of CaMV-strain CM1841 putative polymerase with Moloney murine leukemia virus (MMLV) polymerase. CaMV ORV amino acids (upper line) from position 380 to position 454 have been aligned with MMLV polymerase amino acids (lower line) from position 316 to 395. Gaps are indicated by blanks and identical amino acids are boxed. Amino acids belonging to the same group (basic, acidic, polar uncharged, unpolar) are indicated by open circles. (From Volovitch, M., Modjtahedi, N., Yot, P., and Brun, G., *EMBO J.*, 3, 314, 1984. With permission.) (B) Alignment of polymerase and putative polymerase among five different viruses. WHV (wood chuck hepatitis virus) residue positions 544-634; HBV (adr strain) 510-600; RSV, 141-230; MMLV, 303-393; CaMV (strain CM1841), 368-453. Gaps have been inserted to increase similarity. Boxed residues indicate identities or favored amino acid substitutions among four or five viruses, where favored substitutes are defined as pairs of residues falling in the groups A,S,T,P and G; N,D,E and Q; H,R and K; M,L,I and V; F,Y and W. Positions that are invariant among the five viruses are designated by ○. Amino acid nomenclature is as recommended by IUPAC-IUB. (Reprinted by permission from Toh, H., Hayashida, H., and Miyato, T., *Nature*, 305, 829. Copyright (c) 1983 Macmillan Journals Limited.)

supercoiled DNA. This enzyme may be nuclear associated or, since it would be fulfilling a very early function in infection, could conceivably be inside virus capsids. Evidence against this has been published,[117] although under the conditions used CaMV particles would probably have remained intact. A DNA polymerase activity is found associated with 28-nm cores in preparations of HBV Dane particles.[118] Second, there has been no report of CaMV-specific RNase H activity. With the monomeric and multimeric reverse transcriptases of retroviruses, RNase H activity is an integral function of the same molecule.[87] By further analogy with retroviruses, it has been assumed that the activities synthesizing (+) and (−) strands of CaMV DNA are located on the same enzyme. Experimental evidence in favor of this is not yet available, although it has been demonstrated that subcellular fractions from infected protoplasts are capable of synthesizing both strands on an endogenous template.[21]

Unfortunately, there has been little work done on the in vivo inhibition of CaMV replication to support any proposed model of DNA replication. It has been reported that actinomycin D,[119] 5-bromodeoxyuridine, 5-fluorodeoxyuridine, and formycin B inhibited CaMV multiplication in leaves, whereas blasticidin S and 2-thiouracil did not.[120] However, the lack of specificity of these inhibitors provides little support for any particular mechanism of replication.

More useful and specific inhibitors include aphidicolin for DNA polymerase α and ddTTP for polymerase γ. Aphidicolin has been shown to be inhibitory to the replication of avian sarcoma virus in vivo. Its action prevents the accumulation of unintegrated circular DNA and integrated proviral DNA, although synthesis of unintegrated linear DNA remains unaffected.[121,122] It was proposed that an enzyme similar to DNA polymerase α was involved in the process of circularization of unintegrated DNA. Interestingly, it had been noted previously[123] that total retroviral DNA synthesis in vitro was unaffected by aphidicolin; a similar observation has been made for CaMV DNA synthesis in vitro.[10] The circularization of unintegrated retroviral DNA is also sensitive to cyloheximide[124] providing circumstantial evidence that early viral protein synthesis is required for supercoiled viral DNA formation.

A number of inhibitors have been shown to inhibit reverse transcriptase, but for most of them their inhibitory action extends to other DNA and/or RNA polymerases. One inhibitor, retrostatin, isolated from *Streptomyces retrostaticus*, is claimed to be effective against avian myeloblastosis virus reverse transcriptase in vitro and ineffective against prokaryotic DNA-directed DNA polymerase and eukaryotic DNA-directed RNA polymerase.[125] The action of this inhibitor has not yet been tested against CaMV reverse transcriptase.

VIII. VIRUS MULTIPLICATION CYCLE

A. In a Single Cell (Figure 9)

The infecting virus particle(s) enters the cell and is uncoated at an unknown site. Virion DNA in the nucleus has the gaps repaired; the molecule becomes supercoiled and forms a minichromosome. The minichromosome is replicated, possibly independently, to produce a transcriptionally active pool synthesizing 19S and 35S RNA species. 19S and at least some 35S RNA is polyadenylated and transported to the cytoplasm for translation to inclusion body and other virus-coded proteins. At an uncertain site, but probably the inclusion bodies, 35S RNA is used as a template for reverse transcription giving rise to fully gapped virion DNA which is promptly encapsidated *in situ*.

B. In Whole Plants

Virus occurring free or released from inclusion bodies moves from cell to cell through plasmodesmata and systemically in the vascular tissue. Systemic spread to most or all of the cells of the host leaf has occurred before leaf emergence.

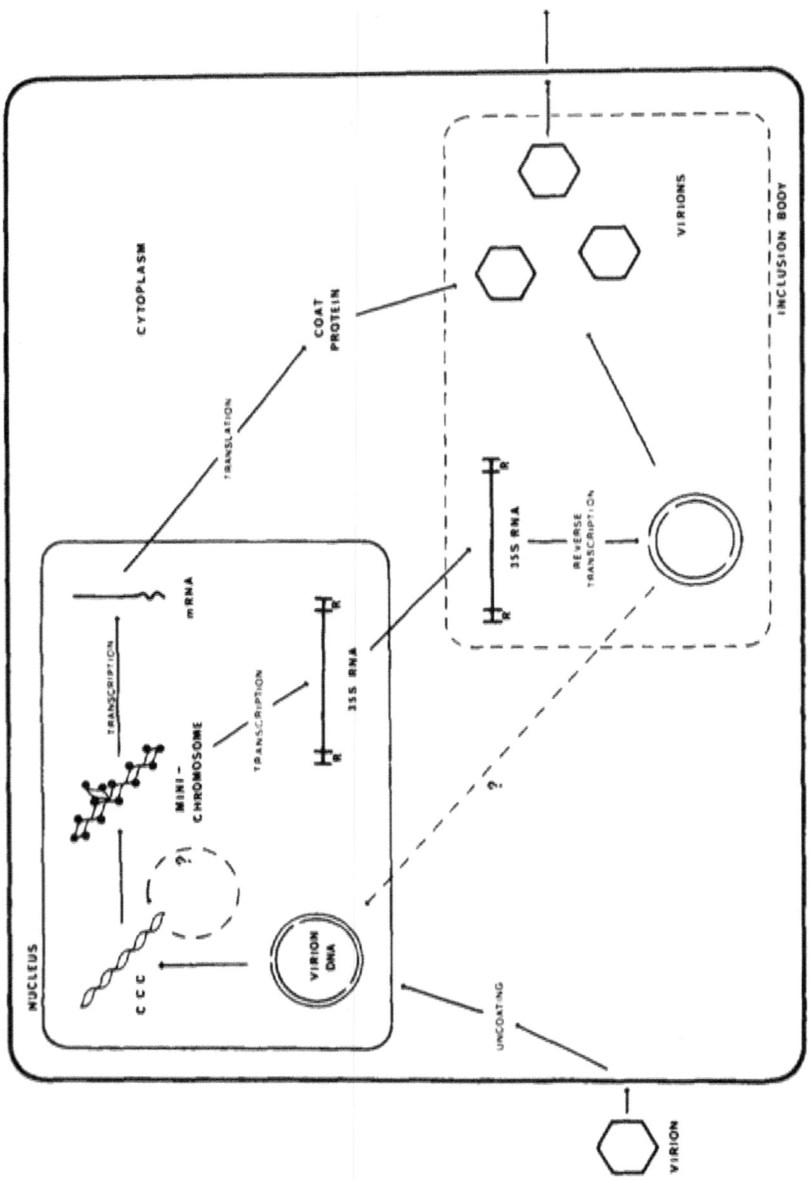

FIGURE 9. Virus multiplication cycle in a single cell.

XI. FUTURE PROSPECTS

Great advances in the understanding of the mechanisms of caulimovirus multiplication have been made in the last few years, and the discovery of reverse transcriptase in plants will probably be noted as one of the landmarks in plant molecular biology. After this advance we are left with the difficult task of elucidating the fine details of caulimovirus multiplication. The enzymology has obvious attractions and will continue to receive attention. There are other important areas that also need attention. At the single-cell level, we know little of the earliest stages of infection (uptake and uncoating), how the pool of supercoiled material is generated, the structure of the functional mRNAs for OR I-V, possible processing events for viral proteins, and the nature of the environment formed within inclusion bodies which favors reverse transcription and encapsidation. In whole plants, the relationship between viral DNA replication and the cell cycle would help us appreciate and predict how the plant might process any incoming foreign DNA, and studying the biochemistry of the host/virus interaction may help us understand symptom expression.

Many of the advances made so far have been achieved using intact infected plants. In the future, the use of synchronous protoplast infections will provide a powerful tool for the identification of transient events or minor components. However, as with many other biological systems, and virus infections of animals and plants in particular, greatest progress will be made by examining engineered peturbations of the normal processes. Thus use could be made of inhibitor studies, specific mutations, and particularly, temperature-sensitive mutations.

In the previous chapter the readers attention was drawn to the advantages of site-directed mutagenesis and the use of DNA transfection systems. Unfortunately, efficient DNA transfections of plant protoplasts have not yet been achieved using established methods which induce uptake of macromolecules. Considering the importance of the work, perhaps more time should be devoted to trying fresh approaches to this problem. The use of nonlethal or conditional mutants would avoid the need for DNA transfection systems and will probably provide the key to a further understanding of CaMV and the processes controlling its multiplication. Caulimovirus research is a fast-moving and exciting subject which will continue to provide important and fundamental contributions to the fields of plant molecular biology and plant pathology.

NOTE ADDED IN PROOF

Recently, activity gel analysis has identified a DNA polymerase of 75kd associated with CaMV-infected tissues.[126] This size is close to that expected for the CaMV gene V product. Also, a similar analysis of proteins from purified CaMV preparations[127] has indicated that CaMV particles contain a DNA polymerase of 76kd, although it is not yet known what the biological function of this enzyme is.

ACKNOWLEDGMENTS

I am grateful to Drs. J. W. Davies, S. N. Covey and R. Hull, and Margaret Boulton for helpful discussions. I would also like to thank Drs. R. Hull, S. N. Covey, C. J. Woolston, and J. Beggs, and Mr. C. M. Thomas for providing unpublished data.

REFERENCES

1. **Shepherd, R. J.,** DNA viruses of higher plants, *Adv. Virus Res.*, 20, 305, 1976.
2. **Shepherd, R. J.,** DNA plant viruses, *Annu. Rev. Plant Physiol.*, 30, 405, 1979.
3. **Shepherd, R. J.,** Caulimoviruses, in *Handbook of Plant Virus Infections and Comparative Diagnosis*, Kurstak, E., Ed., Elsevier/North Holland Biomedical Press, Amsterdam, 1981, 847.
4. **Shepherd, R. J.,** Cauliflower mosaic virus, in *CMI/AAB Descriptions of Plant Viruses*, No 243, 1981.
5. **Hull, R.,** Caulimovirus Group, in *CMI/AAB Descriptions of Plant Viruses*, No 295, 1983.
6. **Franck, A., Guilley, H., Jonard, G., Richards, K., and Hirth, L.,** Nucleotide sequence of cauliflower mosiac virus DNA, *Cell*, 21, 285, 1980.
7. **Olszewski, N., Hagen, G., and Guilfoyle, T. J.,** A transcriptionally active covalently closed minichromosome of cauliflower mosaic virus DNA isolated from infected turnip leaves, *Cell*, 29, 395, 1982.
8. **Olszewski, N. E. and Guilfoyle, T. J.,** Nuclei purified from cauliflower mosaic virus-infected turnip leaves contain subgenomic covalently closed circular cauliflower mosaic virus DNAs, *Nucl. Acid Res.*, 11, 8901, 1983.
9. **Menissier, J., de Murcia, G., Lebeurier, G., and Hirth, L.,** Electron microscopic studies of the different topological forms of the cauliflower mosaic virus DNA: knotted encapsidated DNA and nuclear minichromosome, *EMBO J.*, 2, 1067, 1983.
10. **Guilfoyle, T., Olszewski, N., Hagen, G., Kuzj, A., and McClure, B.,** Transcription and replication of the cauliflower mosaic virus genome: studies with isolated nuclei, nuclear lysates and extranuclear cell-free leaf extracts, in *Plant Molecular Biology*, UCLA Symp., Goldberg, R., Ed., Alan R. Liss, New York, 1983, 117.
11. **Guilfoyle, T., Olzewski, N., and Hagen, G.,** The cauliflower mosaic minichromosome, in *Structure and Function of Plant Genomes*, Ciferri, O. and Dure, L., III, Eds., Plenum Publishing, New York, 1983.
12. **Guilley, H., Richards, K. E., and Jonard, G.,** Observations concerning the discontinuous DNAs of cauliflower mosaic virus, *EMBO J.*, 2, 277, 1983.
13. **Covey, S. N., Turner, D., and Mulder, G.,** A small DNA molecule containing covalently linked ribonucleotides from the large intergenic region of the cauliflower mosaic virus genome, *Nucl. Acid Res.*, 11, 251, 1983.
14. **Turner, D. S. and Covey, S. N.,** A putative primer for the replication of cauliflower mosaic virus by reverse transcription is virion-associated, *FEBS Lett.*, 165, 285, 1984.
15. **Hull, R. and Covey, S. N.,** Does cauliflower mosaic virus replicate by reverse transcription?, *Trends Biochem. Sci.*, 8, 119, 1983.
16. **Hull, R. and Covey, S. N.,** Replication of cauliflower mosaic virus DNA, *Sci. Prog. (Oxford)*, 68, 403, 1983.
17. **Pfeiffer, P. and Hohn, T.,** Involvement of reverse transcription in the replication of cauliflower mosaic virus: a detailed model and test of some aspects, *Cell*, 33, 781, 1983.
18. **Marco, Y. and Howell, S. H.,** Intracellular forms of viral DNA consistent with a model of reverse transcriptional replication of the cauliflower mosaic virus genome, *Nucl. Acids Res.*, 12, 1517, 1984.
19. **Toh, H., Hayashida, H., and Miyata, T.,** Homology of reverse transcriptase of retrovirus with putative polymerase gene products of hepatitis B virus and cauliflower mosaic virus, *Nature (London)*, 305, 827, 1983.
20. **Volovitch, M., Modjtahedi, N., Yot, P., and Brun, G.,** RNA-dependent DNA polymerase activity in cauliflower mosaic virus-infected plant leaves, *EMBO J.*, 3, 309, 1984.
21. **Thomas, C. M., Hull, R., Bryant, J. A., and Maule, A. J.,** Isolation of a fraction from cauliflower mosaic virus-infected protoplasts which is active in the synthesis of (+) and (−) strand viral DNA, and reverse transcription of primed RNA templates, *Nucleic Acids Res.*, submitted
22. **Lawson, R. H., Hearon, S. S., and Civerolo, E. L.,** Carnation etched ring virus, in *CMI/AAB Descriptions of Plant Viruses*, No. 182, 1977.
23. **Brunt, A. A.,** Dahlia mosaic virus, in *CMI/AAB Descriptions of Plant Viruses*, No. 51, 1971.
24. **Donson, J. and Hull, R.,** Physical mapping and molecular cloning of caulimovirus DNA, *J. Gen. Virol.*, 64, 2281, 1983.
25. **Brunt, A. A. and Kitajima, E. W.,** Intracellular location and some properties of Mirabilis mosaic virus, a new member of the cauliflower mosaic group of viruses, *Phytopathol. Z.*, 76, 265, 1973.
26. **Frazier, N. W. and Converse, R. H.,** Strawberry vein binding virus, in *CMI/AAB Descriptions of Plant Viruses*, No. 219, 1980.
27. **Iwaki, M., Isogawa, Y., Tsuzuki, H., and Y.,** Soybean chlorotic mottle, a new caulimovirus on Soybean, *Plant Dis.*, in press.
28. **Kim, K. S., Ramsdell, D. C., Gillett, J. M., and Fulton, J. P.,** Virions and ultrastructural changes associated with Blueberry red ringspot disease, *Phytopathology*, 71, 673, 1981.

29. **Kitajima, E. W. and Costa, A. S.**, Particulas esferoidais associadas virus do mosaico das nervuras da manidioca, *Bragantia*, 25, 211, 1966.
30. **Ragozzino, A.**, Una virosi del*Cestrum parqui* L'Erit (Fam. Solanaceae) in Campania, *Ann. Fac. Sci. Agrar. Univ. Napoli*, 8, 3, 1974.
31. **Leseman, D. and Casper, R.**, Electron microscopy of petunia vein clearing virus, an isometric plant virus associated with specific inclusions in Petunia cells, *Phytopathology*, 63, 1118, 1973.
32. **Hammond, J.**, Viruses occuring in *Plantago* species in England, *Plant Pathol.*, 30, 237, 1981.
33. **Hills, G. J. and Campbell, R. N.**, Morphology of Broccoli necrotic yellows virus, *J. Ultrastruct. Res.*, 24, 134, 1968.
34. **Lung, M. C. Y. and Pirone, T. P.**, *Datura stramonium*, a local lesion host for certain isolates of cauliflower mosaic virus, *Phytopathology*, 62, 1473, 1972
35. **Melcher, U., Brannan, C., Gardner, C. O., and Essenberg, R. C.**, Leaf skeleton hybridisation assesses cauliflower mosaic virus host range, in *Proc. 7th Beltsville Symp. Genet. Eng. Appl. Agric.*, Abstr. No. 21, 1982.
36. **Hull, R., Shepherd, R. J., and Harvey, J. D.**, Cauliflower mosaic virus: an improved purification procedure and some properties of the virus particles, *J. Gen. Virol.*, 31, 93, 1976.
37. **Howell, S. H., Walker, L. L., and Dudley, R. K.**, Cloned cauliflower mosaic virus DNA infects turnips (*Brassica rapa*), *Science*, 208, 1265, 1980.
38. **Lebeurier, G., Hirth, L., Hohn, T., and Hohn, B.**, Infectivities of native and cloned DNA of cauliflower mosaic virus, *Gene*, 12, 139, 1980.
39. **Delseny, M. and Hull, R.**, Isolation and characterisation of faithful and altered clones of the genomes of cauliflower mosaic virus isolates Cabb B-JI, CM4-184 and Bari I, *Plasmid*, 9, 31, 1983.
40. **Walden, R. M. and Howell, S. H.**, Uncut recombinant plasmids bearing nested cauliflower mosaic virus genomes infect plants by intragenomic recombination, *Plant Mol. Biol.*, 2, 927, 1983.
41. **Howell, S. H., Walker, L. L., and Walden, R. M.**, Rescue of *in vitro* generated mutants of cloned cauliflower mosaic virus genome in infected plants, *Nature (London)*, 293, 483, 1981.
42. **Dixon, L. K., Koenig, I., and Hohn, T.**, Mutagenesis of cauliflower mosaic virus, *Gene*, 25, 189,1983.
43. **Daubert, S., Shepherd, R. J., and Gardner, R. C.**, Insertional mutagenesis of the cauliflower mosaic virus genome, *Gene*, 25, 201, 1983.
44. **Lebeurier, G., Hirth, L., Hohn, B, and Hohn, T.**, *In vivo* recombination of cauliflower mosaic virus DNA, *Proc. Natl. Acad. Sci. U.S.A.*, 79, 2932, 1982.
45. **Walden, R. and Howell, S. H.**, Intergenomic recombination events among pairs of defective cauliflower mosaic virus genomes in plants, *J. Mol. Appl. Gen.*, 1, 447, 1982.
46. **Yamaoka, N., Furusawa, I., and Yamamoto, M.**, Infection of turnip protoplasts with cauiflower mosaic virus DNA, *Virology*, 122, 503, 1982.
47. **Sakai, F. and Shohara, Y.**, Transfection of turnip protoplasts by cauliflower mosaic virus DNA, in *Plant Tissue Culture, 1982, Proc. 5th Int. Congr. Plant Tissue and Cell Culture*, Fujiwara, A., Ed., Japanese Association for Plant Tissue Culture, Maruzen Co. Ltd., Tokyo, 1982, 669.
48. **Yamaoka, N., Furusawa, I., Yamamoto, M., Yamamoto, F., and Furusawa, M.**, A simple method for introducing cauliflower mosaic virus DNA into plant cells by pricking, in *Proc. 4th Int. Cong. Plant Pathol.*, Abstr. 494, Melbourne, Australia, 1983.
49. **Howell, S. H. and Hull, R.**, Replication of cauliflower mosaic virus and transcription of its genome in turnip leaf protoplasts, *Virology*, 86, 468, 1978.
50. **Furasawa, I., Yamaoka, N., Okuno, T., Yamamoto, M., Kohno, M., and Kunoh, H.**, Infection of turnip protoplasts with cauliflower mosaic virus, *J. Gen. Virol.*, 48, 431, 1980.
51. **Yamaoka, N., Morita, T., Furusawa, I., and Yamamoto, M.**, Effect of temperature on the multiplication of cauliflower mosaic virus, *J. Gen. Virol.*, 61, 283, 1982.
52. **Maule, A. J.**, Infection of protoplasts from several *Brassica* species with cauliflower mosaic virus following inoculation using polyethylene glycol, *J. Gen. Virol.*, 64, 2655, 1983.
53. **Kartha, K. K., Michayluk, M. R., Kao, K. N., Gamborg, O. L., and Constabel, F.**, Callus formation and plant regeneration from mesophyll protoplasts of rape plants (*Brassica napus* L. cv. Zephyr), *Plant Sci. Lett.*, 3, 265, 1974.
54. **Vatsya, B. and Bhaskaran, S.**, Plant regeneration from cotyledonary protoplasts of cauliflower (*Brassica oleracea* L. var. *botrytis* L.), *Protoplasma*, 113, 161, 1982.
55. **Bidney, D. L., Shepard, J. F., and Kaleikan, E.**, Regeneration of plants from mesophyll protoplasts of *Brassica oleracea*, *Protoplasma*, 117, 89, 1983.
56. **Rollo, F. and Covey, S. N.**, Cauliflower mosaic virus persists as supercoiled forms in cultured turnip cells, *J. Gen. Virol.*, 66, 603, 1985.
57. **Paszkowski, J., Shinshi, H., Koenig, I., Lazar, G. B., Hohn, T., Mondak, V., and Potrykus, I.**, Proliferation of cauliflower mosaic virus in protoplast derived clones of turnip (*Brassica rapa*), in *Proc. 6th Int. Protoplast Symp.*, Potrykus, I., Harms, C. T., Hinnen, A., Hütter, R., King, P. J., and Shillito, R. D., Eds., Berkhäuser-Verlag, Basel, 1983, 139.

58. **Tomlinson, J. A. and Shepherd, R. J.**, Studies on mutagenesis and cross protection of cauliflower mosaic virus, *Ann. Appl. Biol.*, 90, 223, 1978.
59. **Maule, A. J.**, unpublished data, 1983.
60. **Maule, A. J., Hull, R., and Donson, J.**, The application of spot hybridization to the detection of DNA and RNA viruses in plant tissues, *J. Virol. Meth.*, 6, 215, 1983.
61. **Maule, A. J.**, Partial characterisation of different classes of viral DNA and kinetics of DNA synthesis in turnip protoplasts infected with cauliflower mosaic virus, *Plant Mol. Biol.*, in press.
62. **Hagen, T. J., Taylor, D. B., and Meagher, R. B.**, Rocket immunoelectrophoresis assay for cauliflower mosaic virus, *Phytopathology*, 72, 239, 1982.
63. **Paludan, N.**, Carnation etched ring virus. Survey, infection trials, thermotherapy and meristem culture, *Tidsskr. Planteavl*, 74, 75, 1970.
64. **Rubio-Huertos, M., Serafino, C., Fujisawa, I., and Matsui, C.**, Electron micriscopy of the formation of carnation etched ring virus intracellular inclusion bodies, *J. Gen. Virol.*, 15, 257, 1972.
65. **Kitajima, E. W. and Lauritis, J. A.**, Plant virions in plasmodesmata, *Virology*, 37, 681, 1969.
66. **Kitajima, E. W., Lauritis, J. A., and Swift, H.**, Fine structure of zinnia leaf tissues infected with dahlia mosaic virus, *Virology*, 39, 240, 1969.
67. **Lawson, R. H. and Hearon, S. S.**, Ultrastructure of carnation etched ring virus-infected *Saponaria vaccaria* and *Dianthus caryophyllus*, *J. Ultrastruct. Res.*, 48, 201, 1974.
68. **Guilfoyle, T. J.**, Transcription of the cauliflower mosaic virus genome in isolated nuclei from turnip leaves, *Virology*, 107, 71, 1980.
69. **Ansa, O. A., Bowyer, J. W., and Shepherd, R. J.**, Evidence for replication of cauliflower mosaic virus DNA in plant nuclei, *Virology*, 121, 147, 1982.
70. **Kamei, T., Rubio-Huertos, M., and Matsui, C.**, Thymidine-^{3}H uptake by X bodies associated with cauliflower mosaic virus, *Virology*, 37, 506, 1969.
71. **Favali, M. A., Bassi, M., and Conti, G. G.**, A quantitative autoradiographic study of intracellular sites for replication of cauliflower mosaic virus, *Virology*, 53, 115, 1973.
72. **Modjtahedi, N., Volovitch, M., Sossountzov, L., Habricot, Y., Bonneville, J. M., and Yot, P.**, Cauliflower mosaic virus-induced viroplasm support viral DNA synthesis in a cell-free system, *Virology*, 133, 289, 1984.
73. **Conti, G. G., Vegetti, G., Bassi, M., and Favali, M.**, Some ultrastructural and cytochemical observations on chinese cabbage infected with cauliflower mosaic virus, *Virology*, 47, 694, 1972.
74. **Day, M. F. and Venables, D. G.**, The transmission of cauliflower mosaic virus by aphids, *Aust. J. Biol. Sci.*, 14, 187, 1961.
75. **Donson, J.**, Comparative Studies on Caulimoviruses, Ph.D. thesis, University of East Anglia, Norwich, U.K., 1983.
76. **Francki, R. I. B., Hatta, T., and Milne, R. G.**, Caulimovirus group, in *Atlas of Plant Viruses*, Vol. 1, CRC Press, Boca Raton, Fla., 1985, 17.
77. **Shalla, T. A., Shepherd, R. J., and Peterson, L. J.**, Comparative cytology of nine isolates of cauliflower mosaic virus, *Virology*, 102, 381, 1980.
78. **Martelli, G. P. and Castellano, M. A.**, Light and electron microscopy of the intracellular inclusions of cauliflower mosaic virus, *J. Gen. Virol.*, 13, 133, 1971.
79. **Woolston, C.**, personal communication, 1983.
80. **Al Ani, R., Pfeiffer, P., and Lebeurier, G.**, The structure of cauliflower mosaic virus. II. Identity and location of the viral polypeptides, *Virology*, 93, 188, 1979.
81. **Odell, J. T. and Howell, S. H.**, The identification, mapping and characterisation of mRNA for P66, a cauliflower mosaic virus coded protein, *Virology*, 102, 349, 1980.
82. **Shepherd, R. J., Richins, R., and Shalla, T. A.**, Isolation and properties of the inclusion bodies of cauliflower mosaic virus, *Virology*, 102, 389, 1980.
83. **Covey, S. N. and Hull, R.**, Transcription of cauliflower mosaic virus DNA. Detection of transcripts, properties and location of the gene encoding the virus inclusion body protein, *Virology*, 111, 463, 1981.
84. **Xiong, C., Muller, S., Lebeurier, G., and Hirth, L.**, Identification by immunoprecipitation of cauliflower mosaic virus *in vitro* major translation product with a specific serum against viroplasm protein, *EMBO J.*, 1, 971, 1982.
85. **Xiong, C., Balazs, E., Lebeurier, G., Hindelang, C., Stoeckel, M. E., and Porte, A.**, Comparative cytology of two isolates of cauliflower mosaic virus, *J. Gen. Virol.*, 61, 75, 1982.
86. **Mennisier, J., Lebeurier, G., and Hirth, L.**, Free cauliflower mosaic virus supercoiled DNA in infected plants, *Virology*, 117, 322, 1982.
87. **Varmus, H. and Swanstrom, R.**, Replication of Retroviruses, in *Molecular Biology of Tumour Viruses, part 2, RNA Tumour Viruses*, 2nd. ed., Weiss, R., Teich, N., Varmus, H., and Coffin, J., Eds., Cold Spring Harbor Laboratory, Cold Spring Harbor, New York, 1982, 369.
88. **Richards, K. E., Guilley, H., and Jonard, G.**, Further characterisation of the discontinuities in cauliflower mosaic virus, *FEBS Lett.*, 134, 67, 1981.

89. **Hull, R., Covey, S. N., Stanley, J., and Davies, J. W.,** The polarity of the cauliflower mosaic virus genome, *Nucl. Acids Res.*, 7, 669, 1979.
90. **Hull, R. and Shepherd, R. J.,** The structure of cauliflower mosaic virus genome, *Virology*, 79, 216, 1977.
91. **Hull, R.,** The DNA of plant DNA viruses, in *Nucleic Acids in Plants*, Vol. 2, Hall, T. C. and Davies, J. W., Eds., CRC Press, Boca Raton, Fla., 1979, 3.
92. **Russell, G. J., Follett, E. A. C., Subak-Sharpe, J. H., and Harrison, B. D.,** The double-stranded DNA of cauliflower mosaic virus, *J. Gen. Virol.*, 11, 129, 1971.
93. **Hull, R. and Howell, S. H.,** Structure of the cauliflower mosaic virus genome. II. Variation in DNA structure and sequence between isolates, *Virology*, 86, 482, 1978.
94. **Covey, S. N., Lomonosoff, G. P., and Hull, R.,** Characterisation of cauliflower mosaic virus DNA sequences which encode major polyadenylated transcripts, *Nucl. Acids Res.*, 9, 6735, 1981.
95. **Guilley, H., Dudley, R. K., Jonard, G., Balazs, E., and Richards, K. E.,** Transcription of cauliflower mosaic virus DNA: detection of promotor sequences and characterisation of transcripts, *Cell*, 30, 763, 1982.
96. **Harmsen, J. and Covey, S. N.,** personal communication, 1983.
97. **Hull, R. and Covey, S. N.,** Characterisation of cauliflower mosaic virus DNA forms isolated from infected turnip leaves, *Nucl. Acids Res.*, 11, 1881, 1983.
98. **Hull, R.,** personal communication, 1984.
99. **Weiser, B., Ganem, D., Seeger, C., and Varmus, H. E.,** Closed circular viral DNA and asymmetrical heterogeneous forms in livers from animals infected with ground squirrel hepatitis virus, *J. Virol.*, 48, 1, 1983.
100. **Mason, W. S., Aldrich, C., Summers, J., and Taylor, J. M.,** Asymmetric replication of duck hepatitis B virus DNA in liver cells (free minus strand DNA), *Proc. Natl. Acad. Sci. U.S.A.*, 79, 3997, 1982.
101. **Summers, J. and Mason, W. S.,** Replication of the genome of a hepatitis B-like virus by reverse transcription of an RNA intermediate, *Cell*, 29, 403, 1982.
102. **Odell, J. T., Dudley, R. K., and Howell, S. H.,** Structure of the 19S transcript encoded by the cauliflower mosaic virus genome, *Virology*, 111, 377, 1981.
103. **Dudley, R.K., Odell, J. T., and Howell, S.H.,** Structure and 5′ termini of the large and 19S RNA transcripts encoded by the cauliflower mosaic virus genome, *Virology*, 117, 19, 1982.
104. **Varmus, H. E.,** Form and function of retroviral proviruses, *Science*, 216, 812, 1982.
105. **Taylor, J. M., Cywinski, A., and Smith, J. K.,** Discontinuities in the DNA synthesised by an avain retrovirus, *J. Virol.*, 48, 654, 1983.
106. **Hohn, T., Richards, K., and Lebeurier, G.,** Cauliflower mosaic virus on its way to becoming a useful plant vector, *Curr. Top. Microbiol. Immunol.*, 96, 193, 1981.
107. **Müller, U. R. and Fitch, W, M.,** Evolutionary selection for perfect hairpin structures in viral DNA, *Nature (London)*, 298, 582, 1982.
108. **Sims, J., Capon, D., and Dressler, D.,** dnaG (Primase)-dependent origins of DNA replication, *J. Biol. Chem.*, 254, 12615, 1979.
109. **Gutai, M. W. and Nathans, D.,** Evolutionary variants of Simian Virus 40: nucleotide sequence of a conserved SV40 DNA segment containing the origin of viral DNA replication as an inverted repetition, *J. Mol. Biol.*, 126, 259, 1978.
110. **Beggs, J. and Hull, R.,** personal communication, 1983.
111. **Zakian, V. A.,** Origin of replication from *Xenopus laevis* mitochondrial DNA promotes high-frequency transformation of yeast, *Proc. Natl. Acad. Sci. U.S.A.*, 78, 3128, 1981.
112. **Simanis, V. and Beggs, J.,** personal communication, 1983.
113. **Bryant, J.,** Biochemical aspects of DNA replication with particular reference to plants, *Biol. Rev.*, 55, 237, 1980.
114. **Castroviejo, M., Tharaud, D., Tarrago-Litvak, L., and Litvak, S.,** Multiple deoxyribonucleic acid polymerases from quiescent wheat embryos. Purification and characterisation of three enzymes from the soluble cytoplasm and one from purified mitochondria, *Biochem. J.*, 181, 183, 1979.
115. **Mondal, H. and Hofschneider, P. H.,** Demonstration of free reverse transcriptase in the nuclei of embryonic tissues of the Japanses quail, *Biochem. Biophys. Res. Commun.*, 116, 303, 1983.
116. **Thomas, C., Bryant, J., Dunhasm, V., and Hull, R.,** Deoxyribonucleic acid polymerase activities in healthy and infected turnip plants, *Biochem. Soc. Trans.*, 11, 367, 1983.
117. **Hull, R.,** Genetic engineering in plants: the possible use of cauliflower mosaic virus DNA as a vector, in *Plant Cell Cultures: Results and Perspectives*, Sala, F., Parisi, B., Cella, R., and Ciferri, O., Eds., Elsevier/North-Holland Biomedical Press, Amsterdam, 1980, 219.
118. **Caplan, P. M., Greenman, R. L., Gerin, J. L., Purcell, R. H., and Robinson, W. S.,** DNA polymerase associated with human hepatitis B antigen, *J. Virol.*, 12, 995, 1973.
119. **Tezuka, N., Taniguchi, T., and Matsui, C.,** Inhibition of cauliflower mosaic virus multiplication by actinomycin D., *Virology*, 43, 717, 1971.

120. **Tezuka, N. and Taniguchi, T.,** Effects of some chemicals on the multiplication of cauliflower mosaic virus, *Ann. Phytopathol. Soc. Jpn.*, 39, 259, 1973.
121. **Hagino-Yamagishi, K., Kano, K., and Mano, Y.,** Effects of aphidicolin on retrovirus DNA synthesis *in vivo, Biochem. Biophys. Res. Commun.*, 102, 1372, 1981.
122. **Hsu, T. W. and Taylor, J. M.,** Effect of aphidicolin on avian sarcoma virus replication, *J. Virol.*, 44, 493, 1982.
123. **Pedrali-Noy, G. and Spadari, S.,** Effect of aphidicolin on human and viral DNA polymerase, *Biochem. Biophys. Res. Commun.*, 88, 1194, 1979.
124. **Yang, W. K., Den-Mei, Y., and Kiggans, J. O., Jr.,** Covalently closed circular DNAs of murine type C retrovirus: depressed formation in cells treated with cycloheximide early after infection, *J. Virol.*, 36, 181, 1980.
125. **Nishio, M., Kuroda, A., Suzuki, M., Ishimaru, K., Nakamura, S., and Nomi, R.,** Retrostatin, a new specific enzyme inhibitor against avain myoblastosis virus transcriptase, *J. Antibiotics*, 36, 761, 1983.
126. **Pfeiffer, P., Laquel, P., and Hohn, T.,** Cauliflower mosaic virus replication complexes: characterisation of the associated enzymes and of the polarity of the DNA synthesised *in vitro, Plant Mol. Biol.*, 3, 261, 1984.
127. **Menissier, J., Laquel, P., Lebeurier, G., and Hirth, L.,** A DNA polymerase is associated with cauliflower mosaic virus, *Nucleic Acids Res.*, 12, 8769, 1984.

Chapter 6

STRUCTURE AND FUNCTION OF THE DNA GENOME OF GEMINIVIRUSES

John Stanley and Jeffrey W. Davies

TABLE OF CONTENTS

I. INTRODUCTION

The geminiviruses comprise a unique group of viruses characterized by their twinned particles. Since the earliest reports of such structures,[1,2] the numbers of both recognized and probable members of the group have risen steadily, based on particle morphology, the cytopathological effects induced within the nuclei of infected tissue which are often in the form of distinctive fibrillar rings,[3-6] and the structure of the nucleic acid encapsidated within the twinned particles. The observation that bean golden mosaic virus (BGMV) contained a genome of DNA[7] distinguished it from the majority of characterized plant viruses, and subsequent work showing the DNA of this virus to be predominently in the form of single-stranded circles[8] set it apart from the only other known group of plant viruses to have a DNA genome, the caulimoviruses, which contain double-stranded DNA (reviewed in Reference 9). A genome of single-stranded circular DNA has now been shown to be a characteristic of all geminiviruses so far investigated in depth.[10-12]

Comprehensive reviews of the geminiviruses are available[13-15] in which may be found descriptions of the members and the consequences of their infection in plants. Although information of relevence to the forthcoming discussion is displayed in Table 1 it is not the purpose of this article to reiterate this material in detail but to comment on specific aspects of virus structure and replication in the light of recent information concerning the fine structure of the genome. In this respect, most of the data will concern those members of the group most intensely studied at the molecular level, namely cassava latent virus (CLV, also known as African cassava mosaic virus (ACMV)[15a]) for which the complete nucleotide sequence is now available[16] and also BGMV and tomato golden mosaic virus (TGMV) for which possible intermediates in the replication cycle have been studied. However, as there is at present a general lack of supportive data, much of the discussion concerning genome organization (Section IV), its replication (Section V), and transcription (Section VI) will be of necessity speculative and the conclusions drawn will be based on comparisons with other more widely characterized systems.

II. RELATIONSHIPS BETWEEN GROUP MEMBERS

The members of the geminiviruses can be subdivided into two main groups as shown in Table 1, those transmitted by the whitefly *Bemisia tabaci* and those that are leafhopper transmitted. The latter group contains the three members known to infect monocotyledons, namely, chloris striate mosaic virus (CSMV), maize streak virus (MSV), and wheat dwarf virus (WDV). Whether or not this division into subgroups will be reflected in terms of genome structure remains to be seen, but available evidence concerning serological studies on the various members suggests that relationships do exist throughout the group. Thus, a close serological relationship has been demonstrated between the whitefly-transmitted viruses BGMV, CLV, and TGMV,[34,35] although these viruses must be considered distinct on the basis of their unique host ranges. However, while the whitefly-transmitted viruses squash leaf curl virus (SLCV) and tobacco leaf curl virus (TLCV) show no serological relationship to BGMV,[25,26] and TLCV shows no relationship to CLV,[26,34] SLCV is apparently closely related to CLV[25] although again must be considered a distinct virus in terms of host range. Of the leafhopper-transmitted viruses, bean summer death virus (BSDV) and tobacco yellow dwarf virus (TYDV) are closely related strains of the same virus.[28] A distant relationship has been proposed between beet curly top virus (BCTV) and TYDV[28] and, more surprisingly from the grouping of Table 1, a faint serological relationship has been detected between the leafhopper-transmitted virus BCTV and the whitefly-transmitted viruses CLV[34] and TGMV,[35] although there is now some doubt about this.[121] It has previously been noted that such observations, together with the failure in some instances to demonstrate any serological

Table 1
MEMBERS OF THE GEMINIVIRUS GROUP

Virus[a]	Host range	Insect vector	Coat protein (mol wt)	DNA conformation[c]	DNA (mol wt)	Ref.
		Whitefly				
Abutilon mosaic	Malvaceae	*Bemisia tabaci*	—	–	—	17,18
Bean golden mosaic	Leguminosae	*Bemisia tabaci*	27,400	+	8.0×10^5	8,19-21
Cassava latent	Euphorbiaceae	*Bemisia tabaci*	34,000	+	8.0×10^5	10,22,23
Euphorbia mosaic	Euphorbiaceae	*Bemisia tabaci*	—	–	—	24
Mungbean yellow mosaic	Leguminosae	*Bemisia tabaci*	—	–	—	6
Squash leaf curl	Cucurbitaceae	*Bemisia tabaci*	—	–	—	25
Tobacco leaf curl	Solanaceae Caprifoliaceae Compositae	*Bemisia tabaci*	—	–	—	26
Tomato golden mosaic	Solanaceae	*Bemisia tabaci*	28,000	+	7.9×10^5	12,27
		Leafhopper				
Bean summer death	Leguminosae	*Orosius argentatus*	—	–	—	28
Beet curly top[b]	Chenopodiaceae Compositae Cruciferae Leguminosae Solanaceae	*Circulifer tenellus*	—	–	—	1,29
Chloris striate mosaic	Gramineae	*Nesocultha obscura*	28,000	+	7.1×10^5	11,30
Maize streak	Gramineae	*Cicadulina mbila*	28,000	+	7.1×10^5	10,22
Tobacco yellow dwarf	Solanaceae	*Orosius argentatus*	27,500	–	—	28,31
Wheat dwarf	Gramineae	*Psammotettix alienus*	—	–	—	32

[a] Potential members, not yet officially recognized, are included (see References 13, 15, and 33).
[b] Only example host families are shown. The host range is much wider.
[c] (+) Single-stranded circular DNA has been observed, (–) data not yet available.

relationship between virus members within a single subgroup,[34,35] in addition to the fact that completely unrelated viruses may be transmitted by a common vector,[14] illustrate the basic problem of virus classification on a basis of transmission vector.

Recent evidence concerning the leafhopper-transmitted viruses CSMV and MSV has suggested that fundamental differences exist with respect to genome structure when compared to the whitefly-transmitted viruses so far examined in detail (see following section). A more appropriate subdivision of the geminivirus group may well be made between members infecting either monocotyledons or dicotyledons. However, with respect to those members that infect monocotyledons, even a distant serological relationship has yet to be demonstrated.[30,32]

III. THE BIPARTITE NATURE OF THE GENOME

A. Initial Evidence

DNA extracted from geminivirus particles has been observed by electron microscopy to be in the form of either circular or linear molecules. The available evidence suggests that the circular molecules represent the infective form of the DNA while the linear forms arise from DNA breakage and probably play no active role in virus replication.[10] The estimated molecular weights of the circular DNA (Table 1) represent a DNA size of between 2200 and 2500 nucleotides which, when compared with other eukaryotic DNA viruses, appears unprecedentedly small. This observation, together with the earlier theory[22] (now believed incorrect[19,30]) that the twinned capsid morphology reflected a functionally divided genome, led to the suggestion that the geminivirus genome might be bipartite. Although a functionally divided genome has not been reported for any other virus group containing a genome of DNA, there are many such cases within RNA-containing viruses (for example, see Chapter 4). The initial evidence that the geminivirus BGMV may have a bipartite genome came from observations on the cumulative size of restriction enzyme fragments produced from the digestion of both virus DNA and a double-stranded DNA form, the latter found in infected leaf tissue and proposed as being a replicative intermediate.[36] Using restriction enzymes *Hae* III and *Hha* I (both of which recognize single-stranded as well as double-stranded DNA[37,38]) the gel-fractionated fragments showed a total size approximating twice that expected for a single species of DNA of molecular weight 8.0×10^5. Although such results may be explained in terms of co-infection of different virus types, appropriately divergent strains or even polymorphism within the strain itself, further evidence was presented concerning the dilution kinetics of infection of leaf mesophyll protoplasts to suggest that two components were responsible for the infection process.[36] Similar conclusions were reached for TGMV, both by restriction enzyme analysis of double-stranded replicative form extracted from infected tissue,[39] and subsequently by the molecular cloning of two distinct components of the genome in the *E. coli* plasmid pAT153.[40]

B. Direct Proof of the Bipartite Genome

To demonstrate unequivocally a multicomponent genome, infectivity studies must be carried out to show that all of the proposed components contribute information essential for virus multiplication. This has proved difficult for geminiviruses due to the problem of separation of the two putative similar-sized DNA molecules from the mixture found in naturally occuring virus populations. This problem may be alleviated, however, by the use of recombinant DNA technology to produce faithful full-length clones of each component. This was first demonstrated for CLV[41] with the use of bacteriophage M13 vectors. The restriction enzyme maps of the two components of the genome shown in Figure 1 were deduced from the nucleotide sequence of the DNAs (discussed in Section IV). To produce full-length copies of the genomic DNAs suitable for infectivity studies, it is necessary to use a single-cutting restriction enzyme that allows cloning into an appropriate vector without destroying the cloning site. Suitable enzymes for this purpose in the case of CLV are *Mlu* I, *Bcl* I, and *Cla* I for DNA 1 and *Pst* I, *Cla* I, *Eco* RI, and *Sal* I for DNA 2. A double-stranded form of the virus DNA was produced by primed synthesis in vitro using DNA polymerase. DNA 1 was cloned at the *Mlu* I site in a modified version of M13 mp8 (clone pJS092) and DNA 2 at the *Pst* I site in M13 mp9 (clone pJS094).[42]

When the cloned DNAs of CLV were excised from the bacteriophage M13 vectors as full-length linear double-stranded DNAs and mechanically inoculated onto *Nicotiana benthamiana*, both severe stunting and leaf curling symptoms concomitant with the development of cytopathological structures characteristic of CLV infection occurred when both components of the genome were present. Furthermore, infective geminate particles containing circular single-stranded DNA indistinguishable from that of CLV DNA could be isolated from the plants (Figure 2). The coat protein encapsidating the virus DNA shows the same mobility on SDS-containing polyacrylamide gels and an identical immunoprecipitation response to that of the original virus stock. Therefore the infection process to produce mature virus from a mixture of cloned DNA components is indistinguishable from that using virus

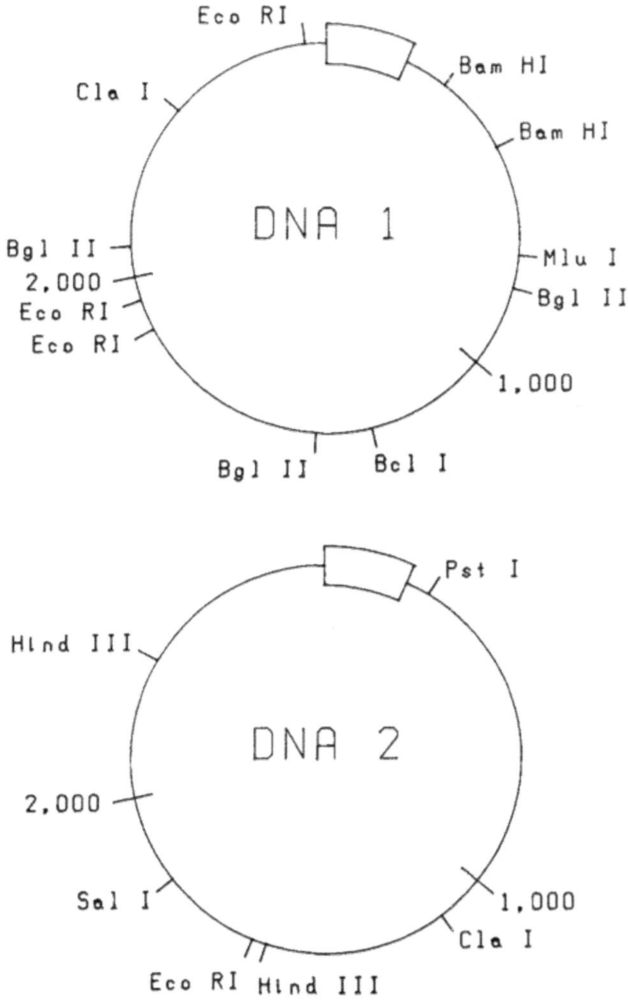

FIGURE 1. Restriction enzyme recognition sites in cassava latent virus DNAs 1 and 2, based on the deduced nucleotide sequence. Noncutters include *Hind* III, *Pst* I, and *Sal* I in DNA 1 and *Bam* HI, *Mlu* I, *Bgl* II, and *Bcl* I in DNA 2. The open box represents the region containing sequences common to both DNAs (see Section IV). (Reprinted by permission from Stanley, J., *Nature*, 305, (5935), 643. Copyright (c) 1983, Macmillan Journals Ltd.)

particles in the initial inoculum. Individually, the linearized double-stranded DNA components do not induce symptons of infection and no detectable virus DNA is transported systemically throughout the plant. As the sequence information presented in Section IV rules out the possibility that infection is a result of complementation of defective DNA molecules, these data unambiguously demonstrate the bipartite nature of the CLV genome. The possibility remains however, that independent replication of one DNA component takes place within infected cells but the DNA remains localized in the absence of coat protein or some other virus-specific transport factor[43,44] and consequently, undetected in the inoculated leaves. Independent replication of one component of the bipartite genomes of some plant RNA viruses has been demonstrated.[45-47] However, before this possibility can be investigated for the geminiviruses, a method for the efficient introduction of the cloned DNA into host cells

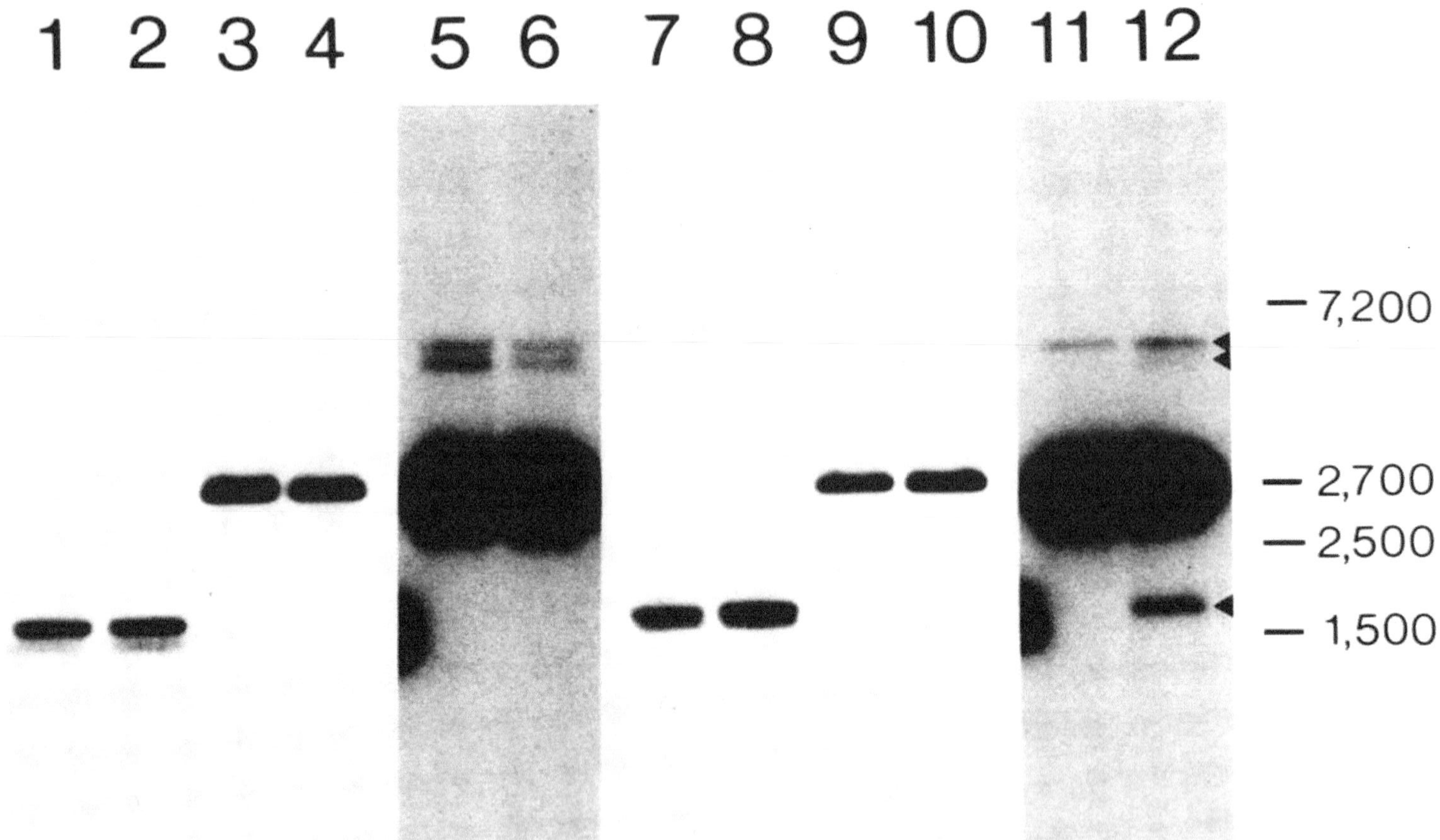

FIGURE 2. The production of cassava latent virus DNA in *N. benthamiana* as shown by Southern blotting after electrophoresis in a 1.4% agarose gel. Hybridization was with nick-translated probes specific to DNAs 1 (1 to 6) and 2 (7 to 12). The nucleic acid was extracted from geminate particles isolated from equal weights of leaf tissue inoculated with either cloned DNAs 1 and 2 (1, 3, 7, and 9) or virus DNA (2, 4, 8, and 10). DNA samples either remained untreated (1, 2, 7, and 8) or were denatured by treatment with glyoxal (3, 4, 9, and 10). 5, 6, 11, and 12 represent longer exposure times of 3, 4, 9, and 10, respectively. (Reprinted by permission from Stanley, J., *Nature*, 305 (5935), 643. Copyright (c) 1983, Macmillan Journals Ltd.)

using for example, a suitable protoplast system or cell suspension culture derived from callus tissue must be developed.

The full-length copies of both CLV components, when intact within the bacteriophage M13 vectors, are infectious when mechanically transmitted to *N. benthamiana*.[122] This is surprising considering that the cloning site may well occur within a gene or disrupt a control sequence essential to virus replication. The *Mlu* I cloning site within DNA 1 (at nucleotide 734) is believed to lie within the coat protein gene (see Section IV). Although a detailed analysis has not yet been undertaken, preliminary studies suggest that the bacteriophage M13 segments are excised from the chimeric DNA during the infection cycle.

C. Advantages of a Bipartite Genome for Science and in Nature

The bipartite nature of the genome of TGMV has now been demonstrated in much the same manner as for CLV by cloning the two components, denoted DNAs A and B, in the *E. coli* plasmid pAT153.[48] Whereas the starting material to produce the infective clones in the case of CLV was single-stranded DNA isolated from purified virus particles, from which a double-stranded form was synthesized in vitro, that for TGMV was a preparation of covalently closed circular double-stranded DNA isolated directly from infected cells. Sequence homologies between the viruses suggest that DNAs A and B of TGMV correspond to DNAs 1 and 2, respectively, of CLV.[48] A detailed comparison of the two viruses, which have been shown to be closely related serologically,[35] awaits the completion of the nucleotide sequence of TGMV. In addition to their unique natural hosts (CLV, *Manihot esculenta*;[49] TGMV, *Lycopersicon esculentum*), both viruses share common hosts in *Nicotiana* species. With the availability of cloned, potentially infective DNA components, pseudorecombination experiments may produce a viable hybrid genome with which the contribution of each DNA component to virus replication may be studied. Such experiments may also give an insight into the mechanisms controlling virus host range.

Although the evolutionary significance of a multicomponent genome is not fully understood, it presumably serves to increase the potential for genetic reassortment by pseudorecombination events which may be evolutionary advantageous to the virus. Furthermore, splitting the genetic information into discrete packages may provide a level of control over the production of related gene products, effectively dividing the genome into functional operons. Thus, gene products required at a similar time during the infection cycle might be expected to be found on the same component of the genome.

D. Encapsidation of the DNA Components

Observations on the buoyant density of both CSMV[30] and BGMV[19] and chemical analysis of the DNA and protein content of BGMV[19] suggest that a single DNA molecule is individually encapsidated within a geminate particle. A minimum of two such geminate particles is therefore required to encompass the entire genome and so instigate the infection cycle. Recent observations on the encapsidated single-stranded DNA of CLV has shown that DNA with an apparent size much greater than either DNA 1 or 2 is tightly associated with the geminate particles[41] (see also Figure 2). Although the DNA is to be found in relatively small amounts, this observation suggests that the geminate structures are capable of encapsidating DNA molecules larger than the individual genome components. There is no information available however, concerning the assembly of both DNA and coat protein to produce mature geminivirus particles or any reasonable explanation as to why the DNA is encapsidated in such bizarre twinned morphological structures.

E. The Bipartite Genome as a Common Feature of the Geminiviruses

A bipartite genome has been unequivocally demonstrated for CLV and TGMV, and other evidence strongly suggests a similar situation prevails for BGMV. In addition, in all three cases, each component of the genome has been shown to be represented in similar amounts within infected tissue[36,39] (see also Figure 2). These three geminiviruses are transmitted in nature by the whitefly *Bemisia tabaci* and all infect dicotyledons (Table 1). Recent evidence concerning the leafhopper-transmitted geminiviruses CSMV[50] and MSV,[50a] both of which infect monocotyledons, suggests that either a single DNA component exists or the second component is represented in such relatively low amounts within infected tissue that it has so far avoided detection. The possibility remains therefore, that a bipartite genome is not a common feature of all members of the geminivirus group. Those viruses capable of infecting monocotyledons may have involved entirely different multiplication strategies. If this proves to be the case they will have to be considered as a subgroup within the geminiviruses or possibly even as an entirely separate virus group. Furthermore, if these viruses do have a genome consisting of a single circle of DNA of approximately 2500 to 2600 nucleotides,[50,50a] they represent the smallest known viral genomes characterized so far.

IV. ORGANIZATION OF THE GENOME

A. Nucleotide Sequence of the DNA Components

The complete nucleotide sequence of the genome of one member of the geminivirus group, CLV (West Kenyan isolate 844), is now available.[16] The sequence was established by a combination of two methods. First, the virus DNA was digested with restriction enzymes known to act on single-stranded DNA (*Hae* III[37] and *Hha* I[38]), the fragments end-labeled using polynucleotide kinase or terminal transferase and sequenced using the chemical degradation technique of Maxam and Gilbert.[51] The second approach involved primed DNA polymerase-directed synthesis in vitro of a complementary strand using the single-stranded DNA as a template.[52] The resultant double-stranded DNA was digested with various restriction enzymes and the fragments cloned into appropriate M13 vectors.[53] Sequences of the inserted DNA were established using the dideoxy-termination procedure.[54] As both strategies used single-stranded DNA extracted from virus particles it is not altogether surprising that a certain level of polymorphism was observed within the sequences, as has been seen for other RNA[55,56] and DNA[57] plant viruses. However, as infective full-length copies of both components of the genome have now been cloned into M13 vectors (as discussed in the previous section), the unambiguous sequences of representative viable DNA molecules may be established. This has been achieved by subcloning restriction enzyme fragments from the full-length copies into M13 vectors and deriving the sequences, shown in Figure 3, using the dideoxy-termination procedure. The two DNAs are of similar length, DNAs 1 and 2 containing 2779 and 2724 nucleotides, respectively. The size difference of less than 2% accounts for the failure to resolve the DNA components by gel electrophoresis[10] and hence the initial difficulty in recognizing the bipartite nature of the genome. Size estimation of the circular single-stranded DNA of CLV by electron microscopy[10] produced a figure of 8.0×10^5 for the molecular weight (Table 1), corresponding to approximately 2500 nucleotides, slightly lower than that from the sequence data.

B. Intermolecular Homology

One outstanding feature that becomes apparent on comparison of the two DNA components is a region of homology that exists between them, encompassing almost 200 nucleotides which represents over 7% of each component. The extent of the homology can be seen in Figure 4. Subsequent comparison of the DNAs using computer-generated dot matrices does not reveal any further extended inter- or intramolecular homology. The 5′ termini of the homologous regions provide a convenient place to arbitrarily locate nucleotide 1 of each sequence shown in Figure 3. This region of each molecule, which appears to be completely noncoding (as discussed below), presumably carries out a function common to both com-

60 120
CTCAACTAGAGACACTCTTGAGCATCTCCTCCTATTAATTGGAGACATTATATAGGTGTCTCTAAATGGCATTCTTGTAATAAGTTGAACTTTAATTTGAATTAAAAGGCTCAAAAGGCT

180 240
CAGAACACCCAAGGGGCCAACCGTATAATATTACCGGTTGGCCCCGCCCCCTTTAATGTGGTCCCCGCGCACTACTTATGTCGGCCAATCATGATGTAGCTTTAAAGGTTATGTATTAGT

300 360
GGTGGGCCACTATATAATTGCAGGCGAAGTTGTGGCTAGTGCGCAATGTGGGATCCACTGGTGAATGAGTTTCCAGACTCGGTGCATGGGCTTAGGTGTATGCTGGCAATTAAATATTTG

420 480
CAGGCCTTAGAGGATACATACGAGCCCAGTACGTTGGGCCACGAACTGGTGAGGGATCTAGTCTCAGTTATCAGGGCTCGTAATTATGTCGAAGCGACCAGGAGATATCATCATTTCCAC

540 600
TCCAGGATCCAAGGTTCGTCGAAGACTGAACTTCGACAGCCCATACAGGAACCGTGCTACTGCCCCCACTGTCCACGTCACAAATCGAAAACGGGCCTGGGTGAACAGGCCCATGTACAG

660 720
AAAGCCCACGATGTACAGGATGTATAGAAGCCCAGACATACCTAGGGGCTGTGAAGGCCCATGTAAGGTCCAGTCGTTTGAGCAGAGGGATGATGTGAAGCACCTTGGTATCTGTAAGGT

780 840
GATTAGTGATGTGACGCGTGGGCCTGGGCTGACACACAGGGTCGGAAAGAGGTTTTGTATCAAGTCCATTTACATTCTTGGTAAGATCTGGCTGGATGAAACTATTAAGAAGCAAAATCA

900 960
CACTAATAATGTGATTTTTTACCTGCTTAGGGATAGAAGGCCGTATGGCAATGCGCCCCAAGACTTCGGGCAGATATTTAACATGTTTGATAATGAGCCCAGTACTGCAACAATTAAGAA

1020 1080
CGATTTGAGGGATAGGTTTCAGGTGTTGAGGAAATTTCATGCCACTGTTGTTGGTGGTCTATATTGCATGAAGGAGCAGGCGTTGGTGAAAAGGTTTTACAGGTTGAATCATCACGTGAC

1140 1200
ATACAATCATCAGGAGGCAGGGAAGTATGAGAATCACACAGAGAATGCTTTGCTTCTGTACATGGCATGTACTCATGCCTCCAATCCTGTATATGCGACGTTGAAAATACGTATATACTT

1260 1320
CTATGACAGTATTGGCAATTAATAAACATTGAATTTTATTTCATGAGTCAACTTGCACTCAATAGTGTTGGCAATTACATTGAACAAAACATGATCAGCAGCTCTAATTACATTGTTAAT

1380 1440
TGAGATAACATCTATATTATCCAAGTATTTAATTACTTGGTATCTAAAGACCCTTAAGAAAAGACCAGTCTGAGGCTGTAAGGTTGTCCAGATCTGGAAGTTGAGAAAACATTTGTGAAT

FIGURE 3A. The nucleotide sequence of a full-length clone of cassava latent virus DNA 1 (pJS092).

```
                                                              1500                                                        1560
CCCCAGCTCCTTCCTCAGGTTGTGATTGAATCGAACCTGGACTGTTATGATGTCCTGGTTCAGCAGGAATGGTCGTTGTTGGTGCCTGGTGATTGTGAAATACAGGGGATTGTTTATTTC

                                                              1620                                                        1680
CCAGGTATACACGCCATTCATTGCTTGAGGAGCAGTGATGACTTCCCCTGTGCGTAAATCCATGATTGGAGCAGTTGATATGGAGGTAATATGAACATCCACAGACAAGATCCACTCTCC

                                                              1740                                                        1800
TACGCCGGATGGCTCGCTTCTTGACTTGTCTGTGAGTGACTTTGATTGGAACCTGAGTAGAGTGGTTCTGTGAGGGTGATGAAGATTGCATTCTTTAATGCCCAGGCCTTTAGCGCTTCT

                                                              1860                                                        1920
TGCTTTTCCTCGGCTAGGAACTCTTTATAGGACGAGGTAGGTCCTGGATTGCAGAGGAAGATAGTGGGAATTCCACCTTTAATTTGAACGGGTTTCCCGTATTTTGTGTTGGACTGCCAG

                                                              1980                                                        2040
TCCCTCTGGGACCCCATGAATTCCTTAAAGTGCTTTAGGTAGTGGGGATCGACGTCATCAATGACGTTGTACCAGGCAGCATTATTGAAGACCTTTGGACTAAGGTCCAGGTGTCCACAC

                                                              2100                                                        2160
AGGTAATTATGTGGGCCTAAAGATCTGGCCCATATCGTCTTCCCTGTTCTGCTATCACCTTCTATGACAATACTATTAGGTCTCCATGGCCGCGCAGCGGAATCCCTAACATTATCAGCG

                                                              2220                                                        2280
ACCCATTCTTCAATTTCAACAGGAACTTGGTCAAAGGAAGAACATGGGAAGGGAGAAACATAAGGAGCTGGTGGCTCTTGGAAAATCCTATCTAAATTACTATTTAGATTATGAAACTGA

                                                              2340                                                        2400
AGTACAAAGTCCTTTGGGACTAATTCCCTAATGACATTAAGAGCTTCTGACTTACTGCCGCTGTTAAGCGCTTTGGCGTAAGCATCATTCGCTGATTGTTGCCCGCCTCTAGCAGATCGT

                                                              2460                                                        2520
CCATCGATTTGAAATTGTCCCCATTCGACGGTGTCGCCGTCCTTATCCAGATAGGACTTGACATCCGAGCTTGATTTGGCACCTTGAATGTTGGGGTGGAAACTGGTGCTACAGCTTGGG

                                                              2580                                                        2640
TGTACACAATCGAAGAGACGATTGTTCGTAATCGTGATTTTACCCTCGAATTGAATGAGGGCATGCAAGTGAGGTTCCCCATTCTGATGCAGCTCTCTACAGATTTTAATGAACTTAGGG

                                                              2700                                                        2760
TTTGATGGGAGAGAAAGTGTTTGAATGAATGACAGTAGGTGTTCTTTGGGTATAGAACACTTTGGGTATGTGAGAAAGACATTCTTGGCTTGAATTCGAAAACGAGGAGTTCTCATGTTG

ACCAAGTCAATTGGAGACA
```

FIGURE 3A. (continued)

```
                                                            60                                                          120
CTCAACTGGAGACACACTTGAGCATCTCCTCCTATTAATTGGAGACATTATATAGGTGTCTCTAAAGGGCATTCTTGTAATAAGTTGAACTTTAATTTCAAATAAAAGGCTCAAAAGGCT

                                                           180                                                          240
CAGAACACCCAAGGGGCCAACCGTATAATATTACCGGTTGGCCCCGCCCCTTTTGAAGTGTGGACCCCGCGCACTGGTTGGCTTCGTCTCCATTTGAAAGTGGGTCCTCATCTTTTTCTC

                                                           300                                                          360
TTTCCTGCAGATACAACGGTGGTCCCCGCATATAGATGTTGTCAGCTATGGATATTGCATGTGTGGCCCACATGCAAATAAATAATCTGCGGGCCAATCGAATGACAGCTTGTGATACTA

                                                           420                                                          480
AATTATATTTAGTATATTGTTCATGGCTCTGAGTGTAATAAAAGGATATATGTTAGTTAACGTTTAATATTAGCCATTCGATATTATTGATTGATAGTCGTCAATCTGACAACGACCATT

                                                           540                                                          600
AAGGACAATTAAGGCCAGTAGCGTAATTGTTTATGTATTATATATAGGGTGCGAAATACAGGAGTTGGAGAATATCATTTATTGGAAGTATATACATTACGTTGTAATGTATAGTATCAG

                                                           660                                                          720
GAAGCAGTCTAGAAATCTCCAAAGGAAATGGAACAGCAATATCACTAACAGGTATCCAATTAAGCGTAAGTATGTTGCCGGGCATACGAGACCATGTGTTAGGCGAAGATTGTTATATGA

                                                           780                                                          840
GCCAGTGGAGAGACCGTTTGGCCATAATGTTTTGTGTGAGAAGCAACATGGTGATGTGTTTAATTTGCAGCAGAACACAAGTTACACGTCGTTTGTGACGTATCCTTCCAGGGGACCATC

                                                           900                                                          960
TGGTGATGGGCGTAGTAGGGATTACATCAAGTTGCAAAGTATGTCTGTCTCGGGAGTGATACATGCGAAGGCTAATGGCAATGATGATCCTATGGAGGTGTCGCCTGTGGTTAATGGAGT

                                                          1020                                                         1080
GTTCGTGTTTAGCTTAATCATGGATACAAAGCCTTATTTACCGGCTGGTGTTCAAGGTCTTCCCACATTTGAAGAGTTGTTCGGACCATATTCCGCTTGTTACGTTAATTTAAGATTATT

                                                          1140                                                         1200
GAATAATCAACAGCATCGATATAGGGTATTGCACAGCGTGAAACGATTCGTTTCATCTTCAGGAGATACTAAGGTATCTCAATTTAGGTTTAATAAGCGATTGAGCACCAGGCGATATAC

                                                          1260                                                         1320
TATATGGGCATCATTCCATGATGGGGATCTGGTAAATGCAGGTGGAAATTATAGGAACATAAGCAAGAATGCTATTCTTGTCAGTTATGCTTTTGTATCAGAGCATGCAATGTCTTGTAA
```

FIGURE 3B. The nucleotide sequence of a full-length clone of cassava latent virus DNA 2 (pJS094).

```
                                                            1380                                                        1440
GCCATTTGTACAAATAGAAACATCCTATGTGGGATGATTTATTATTTATAAATCTTTATGGCTACTACCTCTATGTGCTATTAAGCACTTGTTTACAGTTTGTTCTATAATGCTTTCAAT

                                                            1500                                                        1560
GTCCTTTTTCGACATGGACTGGGTTTGAGAAATTGAGTCCCCTGGATCTAGGGAAGCTTCAGGTAGCTGGTGTAGCCCTCTAAGTGGGAATTCAGCCTCAGAGGTATACTGCTTGGAGGA

                                                            1620                                                        1680
TGAGTCGTCGATGTCAATTGGGTTTGGTCTTGTGTATCTCATGCTTGTATAGCGCCCAATTGTGGATTTTGTAGCCCATGTTTCTCCTGGTTGGACCGTGATGGGCCTATATTTAGGCCC

                                                            1740                                                        1800
AGTGTCATAATCTGTGCCGGGCTCATTTTGAATGAGCCTTCGAATGGGCTTAGGTTTTCCAACGGACCAGAAGTCGACACAGTCAGGCCCATAGTCTTTGGATAGTATTTTTATAGTGGG

                                                            1860                                                        1920
CTGTTTAAATTTAATGTCCGTTGAATGTTTGGCAGCTGACAGCTTAAGTTTTGCCTTAATTTGGGCAAATGTGATTCCATTCTTTACGTTTGAGTCTTCAACCTTATACAGCAGCTGCCA

                                                            1980                                                        2040
TGGTACGTTATCATCAATTGAAAAGTATGATGCAGAGAAATAATGGAGGTCCACATTACACCCTATGGGAAATGTGAACTGGGCTTGATCCTGTTGCTCATCACTGAGTCTAGTGTCCCT

                                                            2100                                                        2160
AATGGTAACAATAACAGAACCTTGTGCATTAAAGGGCACTTGATTCCTGTACTCAATGACTACATGGTCTACCTTCATGCATTTGCCCATGATACGTACCCTTGTTCTCTCAAGGGTCGA

                                                            2220                                                        2280
AGGAAATTGAAGTGTAATCGGCGATTCATCATTCGTCAGTTTGTATTCTGTTCTTGCGGAGTGAATATAATCGCTGGATATGACAGGTACAGATGTATCCATGTTCAACACTTTGAGTAT

                                                            2340                                                        2400
AAGCTTGTCAAAGGCACAACACCTGTCTCATGAGAAACATGATTAATCTATTGTAAAACTGGCCGCGCAGCGGAATGGGTTTAATTCAAGATAAATCAGAAGATTTTGTCAACAGGATAA

                                                            2460                                                        2520
ACAGAGTGTCGAAATATAGACATATATTTCTTATCCTATCGGAAAAATGCTTCTACGTGTAATCCAATTCAAACATAAATATATGCATACACATTAGTCGATGAAACATTATATATTTAG

                                                            2580                                                        2640
AGATATGTACCTTACAATAACTTGAATTGGATATCCTCTGGAGTGGAATGTGGATGGTCTTATTATCAAAGTGGTCATATAAATATGTTTTATATATTTATTAAGGAGTTATCCAACGAG

                                                            2700
TGATAGTTTACGTTGTTAGATTTGCATGTTCAAATCTAAAATAAGCAATTACATTATAGTGAGAGAAAGTTAGAGAGAGACGCT
```

FIGURE 3B. (continued)

```
      2769      2779        10        20        30        40        50        60
GACCAAGTCAATTGGAGACACTCAACTAGAGACACTCTTGAGCATCTCCTCCTATTAATTGGAGACATTATATAGGTGTC
  ******* ****  ** *       *       *
GAAAGTTAGAGAGAGACGCTCTCAACTGGAGACACACTTGAGCATCTCCTCCTATTAATTGGAGACATTATATAGGTGTC
      2714      2724        10        20        30        40        50        60

        70        80        90       100       110       120       130       140
TCTAAATGGCATTCTTGTAATAAGTTGAACTTTAATTTGAATTAAAAGGCTCAAAAGGCTCAGAACACCCAAGGGGCCAA
      *                               *  *
TCTAAAGGGCATTCTTGTAATAAGTTGAACTTTAATTTCAAATAAAAGGCTCAAAAGGCTCAGAACACCCAAGGGGCCAA
        70        80        90       100       110       120       130       140

       150       160       170         180       190       200       210
CCGTATAATATTACCGGTTGGCCCCGCCCCCTTT  AATGTGGTCCCCGCGCACTACTTATGTCGGCCAATCATGATGTA
                              *   ** *     *           **  *** ** * ***** ******
CCGTATAATATTACCGGTTGGCCCCGCCCCTTTTGAAGTGTGGACCCCGCGCACTGGTTGGCTTCGTCTCCATTTGAAAG
       150       160       170       180       190       200       210       220
```

FIGURE 4. Regions of the cassava latent virus genome containing sequences common to both DNA components. Differences between DNA 1 (upper sequence) and DNA 2 are shown by asterisks. A gap of two nucleotides has been introduced after nucleotide 174 of DNA 1 for the purposes of comparison. (Reprinted by permission from Stanley, J. and Gay, M. R., *Nature*, 301 (5897), 260. Copyright (c) 1983, Macmillan Journals Ltd.)

ponents of the genome. Thus, the bipartite genome of the geminiviruses provides an opportunity for intermolecular comparisons that would otherwise only be possible by comparing members of the same or related virus groups. It is tempting to assign the homologous region a role in DNA replication. This aspect will be discussed in more detail in Section V. Additionally, as both DNA components appear to be packaged within identical capsids, one DNA molecule within a single geminate particle,[19,30] the region common to both components may well be involved in coat protein recognition and DNA packaging. It is conceivable that such a region of the genome may contribute to the development of a self-replicating expression vector for plant genes (see volume 3 of this series).

C. Coding Capacity of the Genome

The potential polypeptide coding capacity of the genome of CLV has been investigated by positioning open reading frames within the DNA sequence, achieved by screening for nonsense codons (UAA, UAG, and UGA) within an equivalent RNA sequence. The results displayed in Figure 5 summarize the open reading frames as read from the first in-phase AUG initiation codon in both the virus strand sense (+) and its complement (−). Although smaller open reading frames exist throughout the genome, the data of Figure 5 are limited to those capable of coding for polypeptides of a minimum molecular weight of 10,000. To facilitate their identification in the following discussion, the numbered open reading frames are defined in Table 2 in terms of both molecular weight coding capacity and coordinates within the genome.

The initial conclusion to be reached from the data depicted in Figure 5 is that the genetic information appears to be distributed between both senses of the DNA strands, inferring bidirectional transcription of the genome. Assuming this to be a common feature of the geminiviruses, the transcription strategy may well show more in common with the papovaviruses SV40 and polyoma virus (reviewed in Reference 58) than with other autonomously replicating DNA-containing viruses including the other group of plant DNA viruses, the caulimoviruses, for which the evidence points to asymetric transcription of the DNA.[59] Further aspects of transcription of the geminivirus genome will be discussed in Section VI.

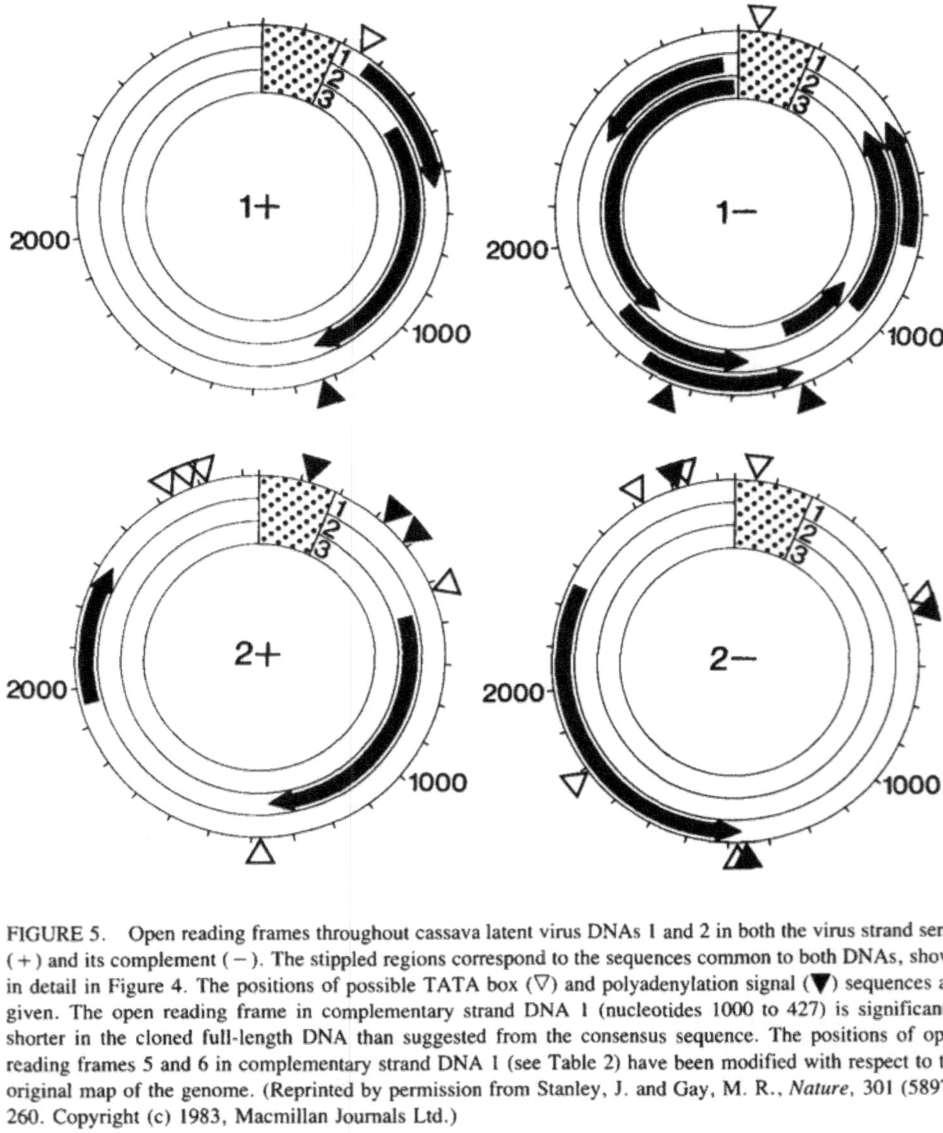

FIGURE 5. Open reading frames throughout cassava latent virus DNAs 1 and 2 in both the virus strand sense (+) and its complement (−). The stippled regions correspond to the sequences common to both DNAs, shown in detail in Figure 4. The positions of possible TATA box (▽) and polyadenylation signal (▼) sequences are given. The open reading frame in complementary strand DNA 1 (nucleotides 1000 to 427) is significantly shorter in the cloned full-length DNA than suggested from the consensus sequence. The positions of open reading frames 5 and 6 in complementary strand DNA 1 (see Table 2) have been modified with respect to the original map of the genome. (Reprinted by permission from Stanley, J. and Gay, M. R., *Nature*, 301 (5897), 260. Copyright (c) 1983, Macmillan Journals Ltd.)

Whereas most of the nucleotide sequence outside of the common region of DNA 1 can be accounted for in a theoretical role of polypeptide coding, this appears not to be the case for DNA 2. That part of the molecule across the common region encompassing nucleotides 2311 to 586, representing almost 30% of the molecule excluding the common region, contains no extended open reading frames. The function of these sequences can only be guessed at with the limited data available.

D. The Coat Protein Gene

When analyzed by SDS-containing polyacrylamide gel electrophoresis the geminivirus particles are shown to comprise a single predominant protein species, which for CLV shows a rather uncharacteristically high molecular weight of 34,000[22] (Table 1). Comparison of the theoretical amino acid composition of each of the putative polypeptides of Figure 5 with the empirical value deduced experimentally for the CLV coat protein[123] shows that only one of the open reading frames can code for a product that resembles the coat protein. The

Table 2
OPEN READING FRAMES WITHIN THE GENOME OF CASSAVA LATENT VIRUS

Open reading frame	DNA strand	DNA sense[a]	Coordinates[b]	Coding capacity (mol wt)[c]
1	1	+	286—624	13,100
2			446—1219	30,200
3		−	2756—1683	40,300
4			2722—2303	15,600
5			1771—1367	15,100
6			1623—1222	15,800
7			1229—957	10,600
8			1000—427	21,000
9			789—434	13,500
10	2	+	587—1354	29,300
11			1963—2310	13,300
12		−	2262—1369	33,700

[a] Virus strand (+), complementary strand (−).
[b] All coordinates of the open reading frames correspond to the nucleotide numbering of Figure 3. The open reading frames within the complementary (−) DNA sense are read in a counterclockwise direction to correspond with Figure 5.
[c] The molecular weights are calculated from the first in-phase AUG codon.

correlation in this case is almost exact and the putative protein in question is encoded by open reading frame 2 within virus strand DNA 1. Furthermore, the major virus-specific transcript maps across open reading frame 2 (see Section VI) and translation of the RNA in vitro produces a polypeptide that comigrates with the virus coat protein in polyacrylamide gels.[123] The theoretical coding capacity of this open reading frame in terms of molecular weight is 30,200, significantly lower than the experimental value, but much closer to those values for the coat proteins of the other geminiviruses so far investigated (Table 1). However, the theoretical value does not take into consideration any transcript splicing events that may alter the open reading frame size or posttranslation modification of the primary polypeptide that might lead to a change in the electrophoretic mobility of the coat protein when analyzed by SDS-containing polyacrylamide gel electrophoresis. Recent estimations of the coat protein size in this laboratory[124] suggest the value for the molecular weight to be between 30,500 and 31,000, comparable with that deduced from open reading frame 2. As the same isolate of CLV was used for the coat protein molecular weight determination, the reason for the discrepancy between this value and that of Table 1 is unclear. In order to delineate the nucleotide sequences responsible for coat protein coding it will be necessary to determine the C- and N-terminal amino acid sequences of the protein.

The coat protein is the only virus-specific product so far positively identified. Interestingly, dot matix analysis has demonstrated a significant degree of homology between the amino acid sequence of the coat protein as deduced from the nucleotide sequence of open reading frame 2 and that of open reading frame 10 on virus strand DNA 2.[60] As these two open reading frames can code for similar sized polypeptides (see Table 2) and are to be found in similar positions within DNAs 1 and 2, it was concluded that the bipartite genome had evolved from a single common ancestral DNA. The possibility also exists that the product of open reading frame 10, because of the homology with the coat protein, represents a minor component of the virus capsid that has so far remained undetected. One may envisage that a product such as this may be responsible for the maintenance of the geminate particle morphology.

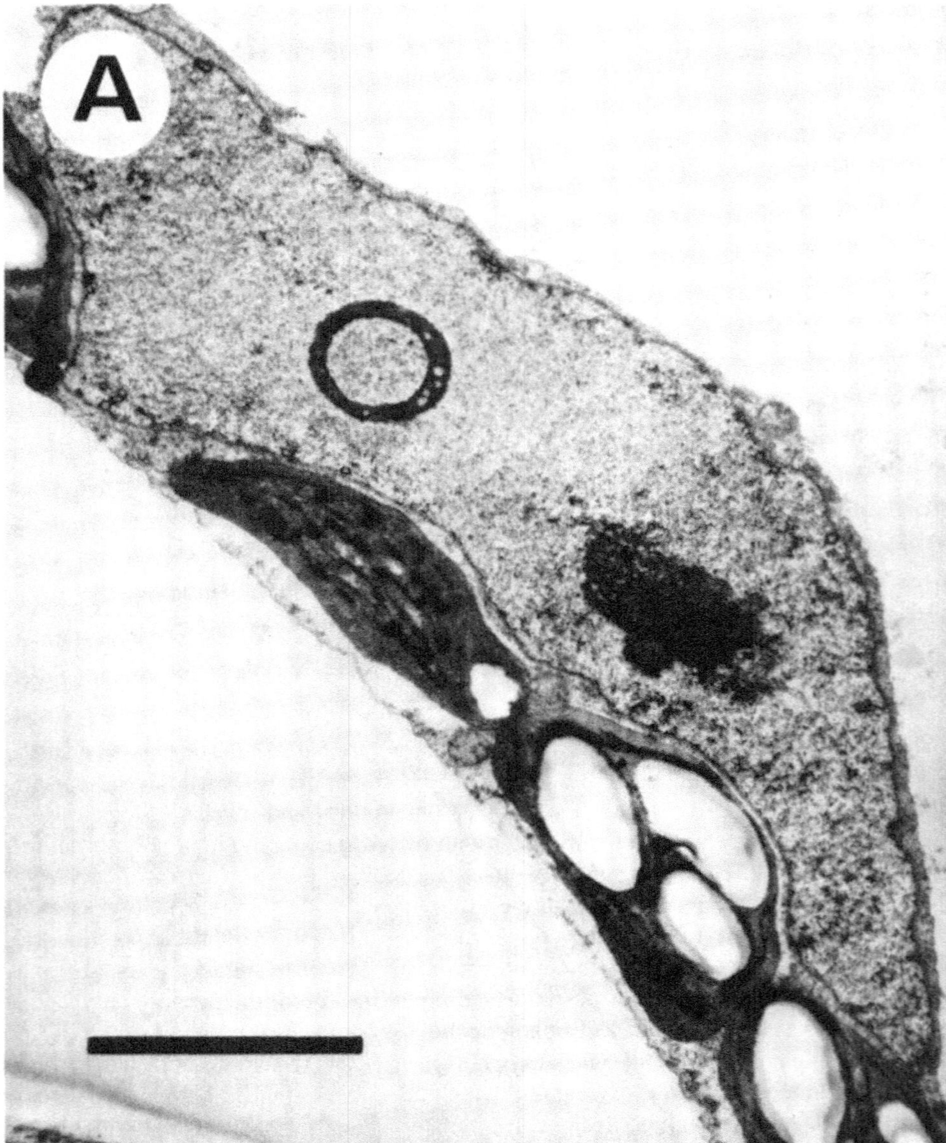

FIGURE 6. Electron micrographs of a cell from *N. benthamiana* infected with CLV showing a fibrillar ring structure within the nucleus (A, the bar represents 5000 nm) and a cell nucleus from *Z. mays* infected with MSV containing inclusions of virus-like particle arrays (B, the bar represents 500 nm).

V. REPLICATION OF THE GENOME

A. Site of DNA Replication

All geminiviruses so far studied appear to assemble in the nuclei of infected host cells, the infection cycle being associated in many cases with the reorganization of the nuclear material and formation of characteristic cytopathological structures.[3,4] Typical effects resulting from CLV infection of *N. benthamiana* (fibrillar rings) and MSV infection of *Zea mays* (arrays of virus-like particles) are shown in Figure 6. Electron-dense masses, termed

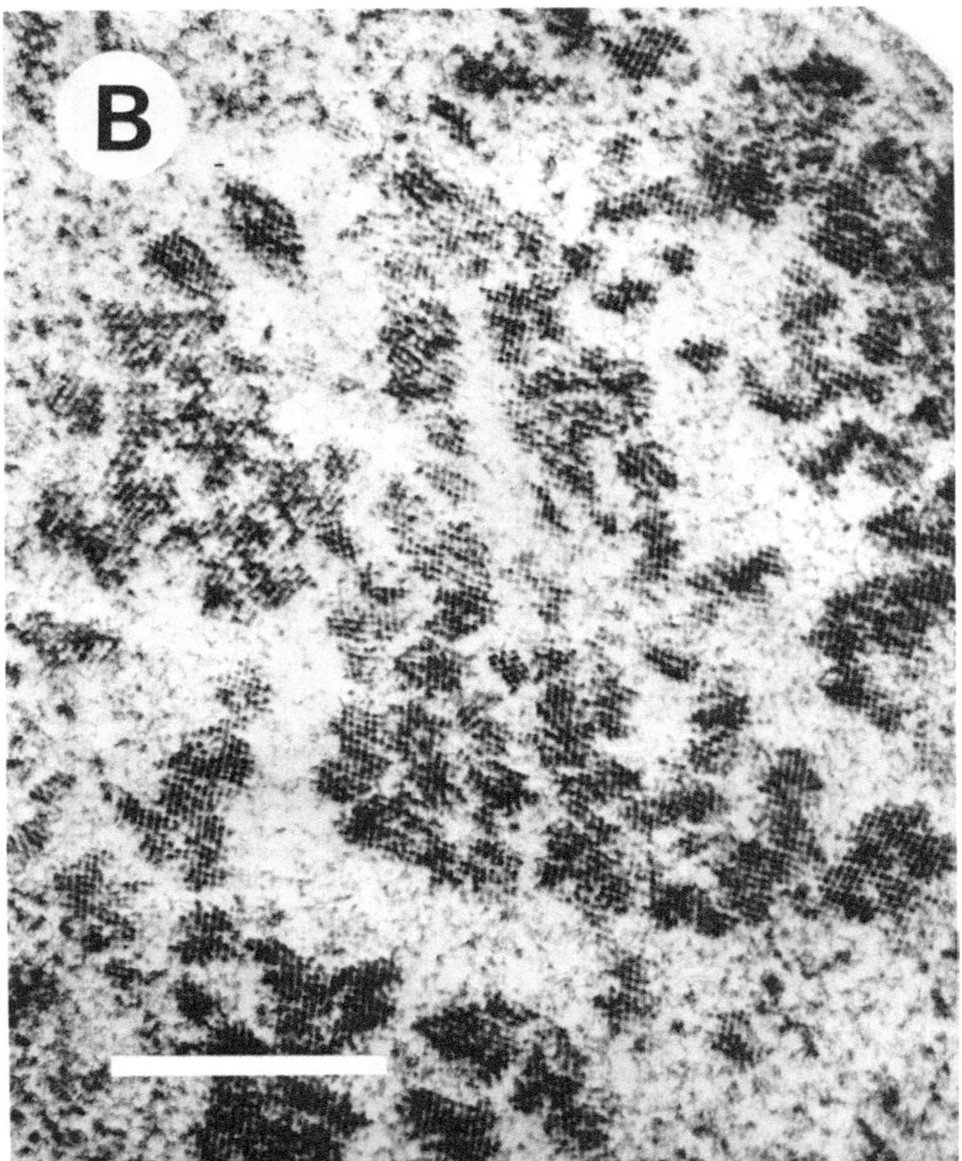

FIGURE 6B.

inclusion bodies, found in the cytoplasm of caulimovirus-infected plant tissue, have been tentatively implicated as having a role in virus assembly[61] or replication (see Chapter 5). The inclusion bodies are composed of virus particles embedded within virus-coded matrix protein.[62,63] An analogy with certain geminivirus-induced structures may be drawn although it remains to be seen if the material within the geminivirus-infected tissue is virus coded.

B. Virus-Specific DNA Intermediates in the Replicative Cycle

The search for possible replicative intermediates in *Phaseolus vulgaris* leaves, systemically infected with BGMV revealed the presence of open-circular, double-stranded DNA specific to the virus.[64] The finding prompted the suggestion that BGMV is replicated via a circular template. Similar structures were later identified together with linear and covalently closed circular (ccc) double-stranded DNA in extracts of *Nicotiana benthamiana* infected with TGMV[39] and CLV.[122] It was suggested[39] that the absence of cccDNA in BGMV-infected leaf tissue reflected either a unique replicative process, or a more rigorous extraction procedure than that employed for the isolation of the TGMV DNA. The control experiment of copurification of BGMV DNA in the presence of M13 cccDNA however suggested that BGMV cccDNA was not being nicked during the isolation procedure.[64] The fact that cccDNA was isolated from CLV-infected *N. benthamiana* using the conditions described for BGMV further suggests that the extraction procedure is not responsible for the absence of BGMV cccDNA. It cannot be ruled out, however, that BGMV cccDNA does exist within infected cells but in quantities that cannot readily be detected.

The identification of possible concatenate structures (multiples of unit-length genome size) containing virus-specific sequences both for TGMV double-stranded DNA[39] and CLV single-stranded DNA[41] (see also Figure 2) suggests a rolling circle mechanism for geminivirus replication.

C. Comparison with Known Replicative Strategies

Table 3 summarizes the replication and transcription strategies of various autonomous virus groups that contain a small DNA genome. As geminiviruses represent a unique group of eukaryotic viruses having a genome of single-stranded DNA (with the reservation discussed below), there is no reason to assume that they will adopt one or the other of these strategies. However, just as a comparison of the features of the otherwise unrelated caulimoviruses with hepatitis B virus (HBV) and retroviruses has recently suggested a replicative process that proceeds via the primed reverse transcription of an RNA intermediate,[66,67] then features of the geminivirus genome in common with other viruses may help elucidate the strategy adopted by this group of viruses. Although relatively little information is available about porcine circovirus (PCV) and its replicative cycle, it is included in Table 3 because it is the sole representative of the animal viruses shown to have a genome of circular, single-stranded DNA,[78] and, as pointed out, bears no relationship to other eukaryotic viruses except the geminiviruses. The size of the PCV genome is exceptionally small when compared with those of other DNA-containing viruses and so it may prove that its genome shows a further analogy with the geminivirus genome in that it is multipartite. Unlike the geminiviruses, however, PCV DNA is encapsidated within single isometric particles.[78]

For those autonomous viruses with a single-stranded DNA genome other than the insect densonucleosis viruses, the primary event on invading a host cell is the synthesis of a second DNA strand complementary to the virion DNA. This function appears to be carried out by purely host-coded factors, although the mechanism varies between viruses. This diversity is exemplified by the bacteriophages cited in Table 3 which require initial RNA primer synthesis on a short stretch of DNA either by RNA polymerase-mediated transcription in the case of M13[83] or recognition and priming of G4 DNA with primase[84] or of ϕX174 DNA with a multienzyme primase-containing "primosome" complex.[85] The primer is subsequently extended in each case using host-coded DNA polymerase III holoenzyme. Densonucleosis virus differs from the other single-stranded DNA viruses of Table 3 in that complementary linear DNA strands are encapsidated in separate particles which upon infection, are uncoated and anneal to form functional double-stranded DNA[86] without the requirement for DNA synthesis. In all three cases cited in Table 3, a region of the bacteriophage single-stranded DNA that can form a stable hairpin secondary structure has been implicated in the host-directed synthesis of double-stranded replicative intermediate.[82,87-89] Viral DNA from both BGMV[7,8] and CLV,[23] when stripped of capsid proteins using either phenol or alkali, has been shown to in infective, demonstrating that virus proteins are not required during the initial infection processes until transcription of the genome is initiated.

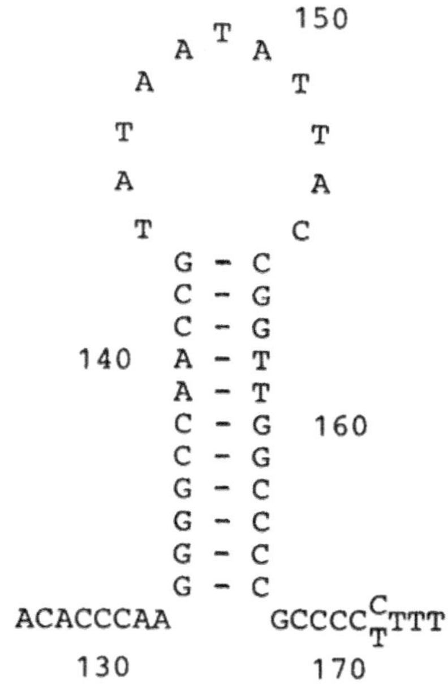

FIGURE 7. Possible hairpin loop structure within the region of the cassava latent virus genome containing sequences common to both DNAs. Nucleotide 171 is either C (DNA 1) or T (DNA 2).

If transcripts are synthesized from the double-stranded DNA form, this implies that complementary strand DNA production relies entirely on host-coded factors. The sequence of the single-stranded DNA of CLV has been screened by computer for potential base-pairing regions. Interestingly, the most stable loop structure found occurs within the region common to DNAs 1 and 2 between nucleotides 133 to 143 and 155 to 165, shown in Figure 7. As it might be expected that both DNAs 1 and 2 interact with identical factors during their conversion to double-stranded intermediates, the region common to both DNAs appears a likely candidate to encompass the replicative origin for this process. The GC-rich stem enclosing an AT-rich loop is reminscent of the primosome recognition site of φX174.[89]

In the case of MSV, several hairpin loop structures are suggested by the sequence.[50a] Two of these are close to a small non-coding region where an 80 nucleotide DNA sequence complementary to the virion DNA strand is located.[89a] The small DNA enables MSV DNA to selfprime *in vitro,* and is possibly the primer for second-strand DNA synthesis in infected cells.

For all single-stranded DNA viruses, the complementary strand DNA within the double-stranded intermediate serves as a template from which the virus strand DNA is copied. The replicative origin for this synthesis may overlap with the complementary strand origin as seen for M13[90] or it may be distinct, as in the case of G4.[91] In addition, synthesis may proceed by an asymetric rolling circle mechanism[92] as for bacteriophages φX174, G4, and M13 or by bidirectional synthesis as employed by SV40.[71,72] As earlier pointed out, the concatenate structures observed in tissue infected with TGMV[39] or CLV[41] suggest a rolling circle mechanism for geminivirus replication. The origin of replication of virus strand DNA synthesis (or synthesis of a pool of progeny replicative form DNA) often appears to occur at a region of destabilization within the double-helical structure. It has been proposed that this destabilization is brought about by the presence of extended inverted-repeat sequences which act by supplying an alternative cruciform structure to the existing double helix, as seen in for example M13,[90] SV40,[93-95] and minute virus of mouse (MVM).[75] With respect to the geminivirus genome, applying the same argument that a process common to both DNA components requires similar if not identical recognition signals suggests the participation of the common region in this second process of DNA replication. Furthermore, the inverted-repeat sequence encompassed in Figure 7 may provide a suitable relaxed region of the double helix to act as a recognition site. The origin of replication of SV40[73,74] and those of both the complementary strand and virus strand DNA of M13[96,97] are to be found within intergenic regions of their respective genomes. The proposal that the origins of replication of both the complementary strand and virus strand DNA of CLV are to be found within the common (intergenic) region of the two DNA components awaits further investigation. It will be of interest to see if other geminiviruses contain not only a region common to both DNA components but also within this region a sequence and/or structure analogous to that depicted in Figure 7. When a suitable transformation system becomes available, it may prove possible to isolate a functional replicative origin within a suitable cloning vector and demonstrate vector DNA replication, mediated by helper virus.

D. Defective DNA Molecules

Analysis of encapsidated DNA from CLV-infected leaf tissue of *N. benthamiana* (Figure 2) has demonstrated that circular single-stranded DNA is produced regardless of whether the inoculum used was native virus DNA or cloned linearized double-stranded DNA. Extended autoradiography of the Southern blots, resulting in overexposure of the major virus-specific DNA reveals, in addition to possible concatenate structures mentioned above, the presence of virus-specific circular single-stranded DNA smaller than genomic size, containing sequences specific to DNA2.[41] While the possibility that DNA 1-specific molecules were also present could not be ruled out at the time, specific probes have now been used to verify that the small DNAs are limited to DNA 2 sequences.[122] Similar virus-specific DNA species have been isolated from TGMV-infected material[39] and it has likewise been suggested that they originate from DNA B (analogous to CLV DNA 2).[48] For both CLV and TGMV this DNA is present when native virus DNA is used as the inoculum but not when cloned DNA is used, suggesting that it represents nonessential defective virus DNA, analogous to that of miniphages,[98] the latter containing variable subgenomic lengths of bacteriophage single-stranded circular DNA. The discrete size of the geminivirus-defective DNA, as judged by polyacrylamide gel electrophoresis (see Figure 2), may reflect packaging constraints imposed by the capsid morphology. As the defective DNA is encapsidated and moves systemically throughout the plant while retaining only approximately half the usual amount of DNA, its cloned counterpart may well have an important role to play in the investigation of insertion of foreign DNA into the geminivirus genome.

VI. TRANSCRIPTION OF THE GENOME

A. Possible Transcription Mechanisms

One of several alternative strategies might be employed for the production of RNA transcripts from the virus DNA genome. In contrast to the prokaryotic system, the scanning mechanism[99] adopted by eukaryotic ribosomes usually infers recognition of the 5′ proximal initiation codon within the eukaryotic message. The majority of the few exceptions to this rule may be explained by "leakiness" of nonfunctional initiation codons, as dictated by their flanking nucleotides.[99,100] Thus, the production of a polycistronic message is not usually seen in eukaryotic systems although exceptions do occur[101] An alternative mechanism adopted to alleviate the problem of eukaryotic ribosome scanning is by synthesis of a polyprotein from a single initiation site. The polyprotein, a product of fusion of a number of genes (reviewed in Reference 100), if subsequently processed by intramolecular cleavage to produce mature gene products (see also Chapter 4). This strategy is clearly not adopted by CLV as an uninterrupted open reading frame is essential. Two related alternatives remain. First, an open reading frame may be preceded by a unique promoter to form a concise transcriptional unit. Alternatively, a single promoter may serve a number of (often overlapping) open reading frames to produce a primary transcript, maturation of which is achieved by controlled splicing events.

B. Bidirectional Transcription of the CLV Genome

The data concerning the open reading frames found within the genome of CLV, as summarized in Figure 5, show that both the virus strand DNA and its complement are probably involved in protein coding. Single-stranded DNA extracted from M13 vectors into which have been cloned parts of the CLV genome, may be used as strand-specific probes. This approach has been used to identify virus-specific polyadenylated RNA originating from both strands of CLV DNAs 1 and 2, confirming the bidirectional transcription strategy.[123] If CLV is representative of the geminiviruses, then the overall transcription strategy of this group more closely resembles that of SV40 than the other viruses listed in Table 3 that have so far been investigated. In contrast to SV40 however, there are no long repeat sequences acting as transcription enhancer elements[102-104] within the CLV genome. SV40 transcription initiation occurs within the intergenic region close to the origin of DNA replication and proceeds along the complementary strands, in opposite directions, to a point approximately halfway round the genome. The primary transcripts are subsequently spliced to give the mature virus messages. Although the 3′ termini of the primary transcripts overlap slightly, the C-termini of the translation products found furthest downstream, VP1 and the T antigen, do not.[73,74] A similar situation may be envisaged for CLV. The regions of the genome in question lie at points approximately diametrically opposite the region common to both DNAs 1 and 2 and are shown in Figure 8 in the form of double-stranded DNA. The 3′ termini of the abutting open reading frames 2 and 6 within DNA 1 and open reading frames 10 and 12 within DNA 2 do not overlap, although the former share two nucleotide pairs in their respective termination codons. The sequences of both DNAs in this region are AT-rich, reminiscent of the analogous region of SV40. Included in Figure 8 are the positions of the polyadenylation signal AATAAA, usually found between 10 and 30 nucleotides upstream from the poly(A) tail.[105] This highly conserved hexanucleotide is thought to be responsible in many eukaryotic systems for determining the 3′ extremity of the primary transcripts by defining either the point of transcription termination or the cleavage site within the transcript (reviewed in reference 106). Conveniently placed polyadenylation signals are present in both strands of DNA 1 but only in the complementary strand of DNA 2. However, as alternative eukaryotic polyadenylation signals have been identified (ATTAAA[107] and AATATA[108]), the region remains an attractive candidate for the termination points of the primary transcripts of CLV DNAs 1 and 2 in both orientations. This being the case, the open reading frames downstream of this region on each DNA strand (open reading frames 7, 8, and 9 for complementary strand DNA 1 and open reading frame 11 for virus strand DNA 2) will not be expressed by this transcriptional mechanism. However, these open reading frames may be transcribed using other as yet uncharacterized promoters or by transcription termination read-through as suggested for the caulimoviruses.[109] The validity of this argument will be substantiated only when identification and fine mapping of the transcripts has been achieved.

The physical segregation of SV40 gene products into early and late genes by placing them on complementary strands provides a mechanism for the control of gene expression throughout the infection process. For example, the T antigen, a product of early transcription, exercises negative control over its own production by binding to the DNA at a position that includes the early promoter. While transcription of the early region is shut off, the adjacent

Table 3
SOME AUTONOMOUSLY REPLICATING VIRUSES WITH A GENOME OF DNA

Virus	Group	Encapsidated DNA: Size (kb)[b]	Encapsidated DNA: Configuration	Replication	Transcription	Ref.
			Eukaryotic			
Cauliflower mosaic virus	Caulimovirus	8.0	ds open circle[c]	Primed reverse transcription[e]	Asymetric	65—67
Hepatitis B virus	—[a]	3.2	ss/ds circle[d]	Primed reverse transcription[e]	Asymetric	68—70
Simian virus 40	Papovavirus	5.2	ds circle	Bidirectional on circular template	Bidirectional	71—74
Minute virus of mouse	Parvovirus	5.1	ss linear	Rolling hairpin on linear (concatenate) template	Asymetric	75,76
Densonucleosis virus	Parvovirus	5.1	ss linear	—[f]	—[f]	77
Porcine circovirus	—[a]	1.8	ss circle	—[f]	—[f]	78
			Prokaryotic			
ϕX174	Isometric phage	5.4	ss circle	Rolling circle	Asymetric	79,80
G4	Isometric phage	5.5	ss circle	Rolling circle	Asymetric	80,81
M13	Filamentous phage	6.4	ss circle	Rolling circle	Asymetric	80,82

[a] Hepatitis B and porcine circovirus have not been classified within a virus group.
[b] The size estimates for all except densonucleosis virus and porcine circovirus are from the deduced nucleotide sequences.
[c] The cauliflower mosaic virus genome shows three triple-stranded interruptions in an otherwise double-stranded (ds) DNA.
[d] The hepatitis B virus genome comprises two complementary linear molecules, one of which is of variable size giving rise to a substantial single-stranded (ss) region.
[e] Not yet verified.
[f] The replication and transcription strategies of densonucleosis virus and porcine circovirus have not yet been investigated.

1210 1220 1230 1240

30,200 → D S I G N ******

DNA 1
GACAGTATTGGCAATT*AATAAA*CATTGAATTTATTTCA
CTGTCATAACCGTTAATTATTTGTAACTTA*AAATAA*AGT

*** Y V N F K I Q ← 15,800

1350 1360 1370 1380

29,300 → S Y V G ***

DNA 2
TCCTATGTGGGATGATTTATTATTTATAAATCTTTATGG
AGGATACACCCTACT*AAATAA*TAAATATTTAGAAATACC

*** L D K H ← 33,700

FIGURE 8. The nucleotide sequence, shown as double-stranded DNA, around the termination codons of the open reading frames of theoretical molecular weight coding capacities of 30,200 (virus strand DNA 1), 15,800 (complementary strand DNA 1), 29,300 (virus strand DNA 2), and 33,700 (complementary strand DNA 2). Amino acids at the C termini of the putative polypeptides are given by the standard one-letter code. Termination codons are denoted *** and possible polyadenylation signals are in italics.

late promoter remains operative[110-112] allowing control of gene expression at the level of transcription. The strong resemblance of the organization of the CLV genome to that of SV40 suggests that CLV genes to be found on complementary DNA strands might be subject to an analogous temporal transcriptional control mechanism. Unlike SV40 however, the additional level of CLV complexity, that of a bipartite genome, might allow further refinement of control of gene expression.

C. Location of Possible Promoters in the CLV Genome

Promoters responsible for the initiation of transcription of eukaryotic genes in most cases have been shown to include a semi-conserved sequence, the TATA box, responsible for defining the origin of transcription possibly by interacting with and so positioning RNA polymerase (reviewed in Reference 106). Whether the sequence per se, the helix stability resulting from the sequence,[113] or an interaction of both factors is responsible for RNA polymerase recognition remains to be seen. The TATA box, defined by the consensus sequence TATA$^{A}_{T}$A$^{A}_{T}$,[114] is usually located between 25 to 30 nucleotides upstream of the transcript start. The position of sequences within the CLV genome that conform to the consensus sequence are indicated in Figure 5. Virus strand DNA 1 contains a single TATA box sequence TATATAA located between nucleotides 251 to 257. As open reading frame 1 lies 35 nucleotides downstream from this signal, the latter is conveniently situated to promote transcription of a message containing a short leader sequence upstream of the initiation codon. Identification of the sequence GGCCAATC upstream of the TATA box (nucleotides 203 to 210), proposed as a signal for promoter modulation,[115,116] provides further support for the presence of a strong promoter located upstream of open reading frame 1. Preliminary mapping of the major virus-specific polyadenylated RNA isolated from CLV-infected *N. benthamiana* has indicated that it encompasses this proposed transcription unit downstream of the TATA box to the putative polyadenylation signal (nucleotides 1221 to 1226) shown in Figure 8,[123] although fine mapping of the termini and identification of splice junctions (if any) await investigation. Open reading frame 2 is located downstream of, and overlaps with open reading 1 and is believed to be involved in coat protein coding[16] (see also Section IV). Whether or not the expression of open reading frame 2 requires transcript splicing remains to be seen. Complementary strand DNA 1 again shows a single TATA box sequence TATATAA at nucleotide 54 (while the numbering of the complementary strand

DNA is identical to the virus strand DNA given in Figure 3, the sequences are read from 5′ to 3′ in an anticlockwise manner) situated 78 nucleotides upstream of open reading frame 3. Transcription from this promoter to the proposed polyadenylation signal shown in Figure 8 covers four overlapping open reading frames which, in order to be expressed, would require either primary transcript splicing or additonal promoters. Transcription of both the virus strand and complementary strand of DNA 2 may be slightly less complex as the open reading frames are not overlapping, but which region of the DNA promotes transcription is open to conjecture. Hopefully, detailed information will soon become available concerning the location of transcripts within the genome and their fine mapping using both nuclease S1 protection experiments[117,118] and direct sequencing techniques.[119] Promoter identification may be facilitated by the use of an in vitro transcription system,[120] recently shown to recognize and correctly initiate transcription from CaMV promoters.[109]

REFERENCES

1. **Mumford, D. L.,** Purification of curly top virus, *Phytopathology*, 64, 136, 1974.
2. **Bock, K. R., Guthrie, E. J., and Woods, R. D.,** Purification of maize streak virus and its relationship to viruses associated with streak diseases of sugarcane and *Panicum maximum*, *Ann. Appl. Biol.*, 77, 289, 1974.
3. **Kim, K. S., Shock, T. L., and Goodman, R. M.,** Infection of *Phaseolus vulgaris* by bean golden mosaic virus: ultrastructural aspects, *Virology*, 89, 22, 1978.
4. **Kim, K. S. and Flores, E. M.,** Nuclear changes associated with euphorbia mosaic virus transmitted by the whitefly, *Phytopathology*, 69, 980, 1979.
5. **Russo, M., Cohen, S., and Martelli, G. P.,** Virus-like particles in tomato plants affected by the yellow leaf curl disease, *J. Gen. Virol.*, 49, 209, 1980.
6. **Thongmeearkom, P., Honda, Y., Saito, Y., and Syamananda, R.,** Nuclear ultrastructural changes and aggregates of viruslike particles in mungbean cells affected by mungbean yellow mosaic disease, *Phytopathology*, 71, 41, 1981.
7. **Goodman, R. M.,** Infectious DNA from a whitefly-transmitted virus of *Phaseolus vulgaris*, *Nature (London)*, 266, 54, 1977.
8. **Goodman, R. M.,** Single-stranded DNA genome in a whitefly-transmitted plant virus, *Virology*, 83, 171, 1977.
9. **Shepherd, R. J. and Lawson, R. H.,** Caulimoviruses, in *Handbook of Plant Virus Infections and Comparative Diagnosis*, Kurstak, E., Ed., Elsevier/North-Holland Biomedical Press, Amsterdam, 1981, chap. 25.
10. **Harrison, B. D., Barker, H., Bock, K. R., Guthrie, E. J., Meredith, G., and Atkinson, M.,** Plant viruses with circular single-stranded DNA, *Nature (London)*, 270, 760, 1977.
11. **Francki, R. I. B., Hatta, T., Boccardo, G., and Randles, J. W.,** The composition of chloris striate mosaic virus, a geminivirus, *Virology*, 101, 233, 1980.
12. **Hamilton, W. D. O., Sanders, R. C., Coutts, R. H. A., and Buck, K. W.,** Characterisation of tomato golden mosaic virus as a geminivirus, *FEMS Microbiol. Lett.*, 11, 263, 1981.
13. **Goodman, R. M.,** Geminiviruses, in *Handbook of Plant Virus Infections and Comparative Diagnosis*, Kurstak, E., Ed., Elsevier/North-Holland Biomedical Press, Amsterdam, 1981, chap. 26.
14. **Goodman, R. M.,** Geminiviruses, *J. Gen. Virol.*, 54, 9, 1981.
15. **Bock, K. R.,** Geminivirus diseases in tropical crops, *Plant Dis.*, 66, 266, 1982.

15a. **Bock, K. R. and Woods, R. D.,** Etiology of African cassava mosaic disease, *Plant Dis.*, 67, 994, 1983.

16. **Stanley, J. and Gay, M. R.,** Nucleotide sequence of cassava latent virus DNA, *Nature (London)*, 301, 260, 1983.
17. **Costa, A. S.,** Whitefly-transmitted plant diseases, *Annu. Rev. Phytopathol.*, 14, 429, 1976.
18. **Jeske, H. and Werz, G.,** Ultrastructural and biochemical investigations on the whitefly transmitted abutilon mosaic virus (AbMV), *Phytophathol. Z.*, 97, 43, 1980.

19. **Goodman, R. M., Shock, T. L., Haber, S., Browning, K. S., and Bowers, G. R.,** The composition of bean golden mosaic virus and its single-stranded DNA genome, *Virology*, 106, 168, 1980.
20. **Costa, A. S.,** Three Whitefly-Transmitted Virus Diseases of Beans in Sao Paulo, Brazil, *Food and Agriculture Organization Plant Protection Bulletin*, 13, 121, 1965.
21. **Reisman, D., Ricciardi, R. P., and Goodman, R. M.,** The size and topology of single-stranded DNA from bean golden mosaic virus, *Virology*, 97, 388, 1979.
22. **Bock, K. R., Guthrie, E. J., and Meredith, G.,** RNA and protein components of maize streak and cassava latent viruses, *Ann. Appl. Biol.*, 85, 305, 1977.
23. **Bock, K. R., Guthrie, E. J., and Meredith, G.,** Distribution, host range, properties and purification of cassava latent virus, a geminivirus, *Ann. Appl. Biol.*, 90, 361, 1978.
24. **Costa, A. S. and Bennett, C. W.,** Whitefly-transmitted mosaic of *Euphorbia prunifolia*, *Phytopathology*, 40, 266, 1950.
25. **Cohen, S., Duffus, J. E., Larsen, R. C., Liu, H. Y., and Flock, R. A.,** Purification, serology and vector relationships of squash leaf curl virus, a whitefly-transmitted geminivirus, *Phytopathology*, 73, 1669, 1983.
26. **Osaki, T. and Inouye, T.,** tobacco leaf curl virus, *CMI/AAB Descriptions of Plant Viruses*, No. 232, 1981.
27. **Uzcategui, R. C. and Lastra, R.,** Transmission and physical properties of the causal agent of mosaico amarillo del tomate (tomato yellow mosaic), *Phytopathology*, 68, 985, 1978.
28. **Thomas, J. E. and Bowyer, J. W.,** Properties of tobacco yellow dwarf and bean summer death viruses, *Phytopathology*, 70, 214, 1980.
29. **Thomas, P. E. and Mink, G. I.,** Beet curly top virus, *CMI/AAB Descriptions of Plant Viruses*, No. 210, 1979.
30. **Francki, R. I. B., Hatta, T., Grylls, N. E., and Grivell, C. J.,** The particle morphology and some other properties of chloris striate mosaic virus, *Ann. Appl. Biol.*, 91, 51, 1979.
31. **Hill, A. V.,** Yellow dwarf of tobacco in Australia. II. Transmission by the jassid *Thamnotettix argentata*, *J. Counc. Sci. Ind. Res. (Australia)*, 14, 181, 1941.
32. **Lindsten, K., Lindsten, B., Abdelmoeti, M., and Juntti, N.,** Purification and some properties of wheat dwarf virus, in *Proc. 3rd Conf. Virus Dis. Gramineae in Europe, Rothamsted, U.K., May 28 to May 30, 1980.*
33. **Hull, R. and Davies, J. W.,** Genetic engineering with plant viruses, and their potential as vectors, *Adv. Virus Res.*, 28, 1, 1983.
34. **Sequeira, J. C. and Harrison, B. D.,** Serological studies on cassava latent virus, *Ann. Appl. Biol.*, 101, 33, 1982.
35. **Stein, V. E., Coutts, R. H. A., and Buck, K. W.,** Serological studies on tomato golden mosaic virus, a geminivirus, *J. Gen. Virol.*, 64, 2493, 1983.
36. **Haber, S., Ikegami, M., Bajet, N. B., and Goodman, R. M.,** Evidence for a divided genome in bean golden mosaic virus, a geminivirus, *Nature (London)*, 289, 324, 1981.
37. **Horiuchi, K. and Zinder, N. D.,** Site-specific cleavage of single-stranded DNA by a *Hemophilus* restriction endonuclease, *Proc. Natl. Acad. Sci. U.S.A.*, 72, 2555, 1975.
38. **Blakesley, P. W. and Wells, R. D.,** Single-stranded DNA from ϕX174 and M13 is cleaved by certain restriction endonucleases, *Nature (London)*, 257, 421, 1975.
39. **Hamilton, W. D. O., Bisaro, D. M., and Buck, K. W.,** Identification of novel DNA forms in tomato golden mosaic virus infected tissue. Evidence for a two component viral genome, *Nucl. Acids Res.*, 10, 4901, 1982.
40. **Bisaro, D. M., Hamilton, W. D. O., Coutts, R. H. A., and Buck, K. W.,** Molecular cloning and characterisation of the two DNA components of tomato golden mosaic virus, *Nucl. Acids Res.*, 10, 4913, 1982.
41. **Stanley, J.,** Infectivity of the cloned geminivirus genome requires sequences from both DNAs, *Nature (London)*, 305, 643, 1983.
42. **Messing, J. and Vieira, J.,** A new pair of M13 vectors for selecting either DNA strand of double-digested restriction fragments, *Gene*, 19, 269, 1982.
43. **Taliansky, M. E., Malyshenko, S. I., Pshennikova, E. S., Kaplan, I. B., Ulanova, E. F., and Atabekov, J. G.,** Plant virus-specific transport function. I. Virus genetic control required for systemic spread, *Virology*, 122, 318, 1982.
44. **Taliansky, M. E., Malyshenko, S. I., Pshennikova, E. S., and Atabekov, J. G.,** Plant virus-specific transport function. II. A factor controlling virus host range, *Virology*, 122, 327, 1982.
45. **Sanger, H. L.,** Functions of the two particles of tobacco rattle virus, *J. Virol.*, 3, 304, 1969.
46. **Goldbach, R., Rezelman, G., and van Kammen, A.,** Independent replication and expression of B-component RNA of cowpea mosaic virus, *Nature (London)*, 286, 297, 1980.
47. **Robinson, D. J., Barker, H., Harrison, B. D., and Mayo, M. A.,** Replication of RNA-1 of tomato black ring virus independently of RNA-2, *J. Gen. Virol.*, 51, 317, 1980.

48. **Hamilton, W. D. O., Bisaro, D. M., Coutts, R. H. A., and Buck, K. W.,** Demonstration of the bipartite nature of the genome of a single-stranded DNA plant virus by infection with the cloned DNA components, *Nucl Acids Res.*, 11, 7387, 1983.
49. **Bock, K. R. and Woods, R. D.,** Etiology of African cassava mosaic disease, *Plant Dis.*, 67, 994, 1983.
50. **Marriott, T. W. and Symons, R. H.,** Characterization of the DNA genomes of Australian geminiviruses, in *Current Communications in Molecular Biology. Plant Infectious Agents,* Robertson, H. D., Howell, S. H., Zaitlin, M., and Malmberg, R. L., Eds., Cold Spring Harbor Laboratories, Cold Spring Harbor, New York, 1983, 59.

50a. **Mullineaux, P. M., Donson, J., Morris-Krsinich, B. A. M., Boulton, M. I., and Davies, J. W.,** The nucleotide sequence of maize streak virus DNA, *EMBO J.*, 3, 3063, 1984.

51. **Maxam, A. M. and Gilbert, W.,** Sequencing end-labelled DNA with base-specific chemical cleavages, *Meth. Enzymol.*, 65, 499, 1980.
52. **Hong, G. F.,** A method for sequencing single-stranded cloned DNA in both directions, *Biosci. Rep.*, 1, 243, 1981.
53. **Messing, J., Crea, R., and Seeburg, P. H.,** A system for shotgun DNA sequencing, *Nucl. Acids Res.*, 9, 309, 1981.
54. **Sanger, F., Coulson, A. R., Barrell, B. G., Smith, A. J. H., and Roe, B. A.,** Cloning in single-stranded bacteriophage as an aid to rapid DNA sequencing, *J. Mol. Biol.*, 143, 161, 1980.
55. **Goelet, P., Lomonossoff, G. P., Butler, P. J. G., Akam, M. E., Gait, M. J., and Karn, J.,** Nucleotide sequence of tobacco mosaic virus RNA, *Proc. Natl. Acad. Sci. U.S.A.*, 79, 5818, 1982.
56. **Lomonossoff, G. P. and Shanks, M.,** The nucleotide sequence of cowpea mosaic virus B RNA, *EMBO J.*, 2, 2253, 1983.
57. **Hohn, T., Richards, K., and Lebeurier, G.,** Cauliflower mosaic virus on its way to becoming a useful plant vector, *Curr. Top. Microbiol. Immunol.*, 96, 193, 1982.
58. **Griffin, B. E.,** Structure and genomic organization of SV40 and polyoma virus, in *DNA Tumor Viruses, Molecular Biology of Tumor Viruses,* Tooze, J., Ed., Cold Spring Harbor Laboratories, Cold Spring Harbor, New York, 1982, chap. 2.
59. **Hull, R., Covey, S. N., Stanley, J., and Davies, J. W.,** The polarity of the cauliflower mosaic virus genome, *Nucl. Acids Res.*, 7, 669, 1979.
60. **Kikuno, R., Toh, H., Hayashida, H., and Miyata, T.,** Sequence similarity between putative gene products of geminiviral DNAs, *Nature (London),* 308, 562, 1984.
61. **Conti, G. G., Vegetti, G., Bassi, M., and Agusta Favali, M.,** Some ultrastructural and cytochemical observations on chinese cabbage leaves infected with cauliflower mosaic virus, *Virology,* 47, 694, 1972.
62. **Shepherd, R. J., Richins, R., and Shalla, T. A.,** Isolation and properties of the inclusion bodies of cauliflower mosaic virus, *Virology,* 102, 389, 1980.
63. **Covey, S. N. and Hull, R.,** Transcription of cauliflower mosaic virus DNA. Detection of transcripts, properties, and location of the gene encoding the virus inclusion body protein, *Virology,* 111, 463, 1981.
64. **Ikegami, M., Haber, S., and Goodman, R. M.,** Isolation and characterization of virus-specific double-stranded DNA from tissues infected by bean golden mosaic virus, *Proc. Natl. Acad. Sci. U.S.A.*, 78, 4102, 1981.
65. **Franck, A., Guilley, H., Jonard, G., Richards, K., and Hirth, L.,** Nucleotide sequence of cauliflower mosaic virus DNA, *Cell,* 21, 285, 1980.
66. **Guilley, H., Richards, K. E., and Jonard, G.,** Observations concerning the discontinuous DNAs of cauliflower mosaic virus, *EMBO J.*, 2, 277, 1983.
67. **Hull, R. and Covey, S. N.,** Does cauliflower mosaic virus replicate by reverse transcription?, *Trends Biochem. Sci.*, 8, 119, 1983.
68. **Galibert, F., Mandart, E., Fitoussi, F., Tiollais, P., and Charnay, P.,** Nucleotide sequence of the hepatitis B virus genome (subtype ayw) cloned in *E. coli, Nature (London),* 281, 646, 1979.
69. **Pasek, M., Goto, T., Gilbert, W., Zink, B., Schaller, H., MacKay, P., Leadbetter, G., and Murray, K.,** Hepatitis B virus genes and their expression in *E. coli, Nature (London),* 282, 575, 1979.
70. **Summers, J. and Mason, W. S.,** Replication of the genome of a hepatitis B-like virus by reverse transcription of an RNA intermediate, *Cell,* 29, 403, 1982.
71. **Fareed, G. C., Garon, C. F., and Salzman, N. P.,** Origin and direction of simian virus 40 deoxyribonucleic acid replication, *J. Virol.*, 10, 484, 1972.
72. **Danna, K. J. and Nathans, D.,** Bidirectional replication of simian virus 40 DNA, *Proc. Natl. Acad. Sci. U.S.A.*, 69, 3097, 1972.
73. **Reddy, V. B., Thimmappaya, B., Dhar, R., Subramanian, K. N., Zain, B. S., Pan, J., Ghosh, P. K., Celma, M. L., and Weissman, S. M.,** The genome of simian virus 40, *Science,* 200, 494, 1978.
74. **Fiers, W., Contreras, R., Haegeman, G., Rogiers, R., van de Voorde, A., van Heuverswyn, H., van Herreweghe, J., Volckaert, G., and Ysebaert, M.,** Complete nucleotide sequence of SV40 DNA, *Nature (London),* 273, 113, 1978.

75. **Astell, C. R., Thomson, M., Merchlinsky, M., and Ward, D. C.**, The complete DNA sequence of minute virus of mice, an autonomous parvovirus, *Nucl. Acids Res.*, 11, 1019, 1983.
76. **Pintel, D., Dadachanji, D., Astell, C. R., and Ward, D. C.**, The genome of minute virus of mice, an autonomous parvovirus, encodes two overlapping transcription units, *Nucl. Acids Res.*, 11, 999, 1983.
77. **Kurstak, E.**, Small DNA densonucleosis virus (DNV), *Adv. Virus. Res.*, 17, 207, 1972.
78. **Tischer, I., Gelderblom, H., Vettermann, W., and Koch, M. A.**, A very small porcine virus with circular single-stranded DNA, *Nature (London)*, 295, 64, 1982.
79. **Sanger, F., Air, G. M., Barrell, B. G., Brown, N. L., Coulson, A. R., Fiddes, J. C., Hutchison, C. A., Slocombe, P. M., and Smith, M.**, Nucleotide sequence of bacteriophage ϕX174 DNA, *Nature (London)*, 265, 687, 1977.
80. **Air, G. M.**, DNA sequencing of viral genomes, in *Comprehensive Virology*, Vol. 13, Fraenkel-Conrat, H. and Wagner, R. R., Eds., Plenum Press, New York, 1979, 205.
81. **Godson, G. N., Barrell, B. G., Staden, R., and Fiddes, J. C.**, Nucleotide sequence of bacteriophage G4 DNA, *Nature (London)*, 276, 236, 1978.
82. **Van Wezenbeek, P. M. G. F., Hulsebos, T. J. M., and Schoenmakers, J. G. G.**, Nucleotide sequence of the filamentous bacteriophage M13 DNA genome: comparison with phage fd, *Gene*, 11, 129, 1980.
83. **Geider, K. and Kornberg, A.**, Conversion of the M13 viral single strand to the double-stranded replicative forms by purified proteins, *J. Biol. Chem.*, 249, 3999, 1974.
84. **Zechel, K., Bouche, J-P., and Kornberg, A.**, Replication of phage G4, *J. Biol. Chem.*, 250, 4684, 1975.
85. **Meyer, R. R., Shlomai, J., Kobori, J., Bates, D. L., Rowen, L., McMacken, R., Ueda, K., and Kornberg, A.**, Enzymatic conversion of single-stranded ϕX174 and G4 circles to duplex forms: discontinuous replication, *Cold Spring Harbor Symp. Quant. Biol.*, 43, 289, 1978.
86. **Kurstak, E., Tijssen, P., and Garzon, S.**, Densonucleosis viruses (parvoviridae), in *The Atlas of Insect and Plant Viruses*, Maramorosch, K., Ed., Academic Press, New York, 1977, chap. 3.
87. **Suggs, S. V. and Ray, D. S.**, Nucleotide sequence of the origin for bacteriophage M13 DNA replication, *Cold Spring Harbor Symp. Quant. Biol.*, 43, 379, 1978.
88. **Sims, J. and Dressler, D.**, Site-specific initiation of a DNA fragment: nucleotide sequence of the bacteriophage G4 negative-strand initiation site, *Proc. Natl. Acad. Sci. U.S.A.*, 75, 3094, 1978.
89. **Shlomai, J. and Kornberg, A.**, An *Escherichia coli* replication protein that recognizes a unique sequence within a hairpin region in ϕX174 DNA, *Proc. Natl. Acad. Sci. U.S.A.*, 77, 799, 1980.

89a. **Donson, J., Morris-Krsinich, B. A. M., Mullineaux, P. M., Boulton, M. I., and Davies, J. W.**, A putative primer for second-strand DNA synthesis of maize streak virus is virion-associated, *EMBO J.*, 3, 3069, 1984.

90. **Meyer, T. F., Geider, K., Kurz, C., and Schaller, H.**, Cleavage site of bacteriophage fd gene II-protein in the origin of viral strand replication, *Nature (London)*, 278, 365, 1979.
91. **Fiddes, J. C., Barrell, B. G., and Godson, G. N.**, Nucleotide sequences of the separate origins of synthesis of bacteriophage G4 viral and complementary DNA strands, *Proc. Natl. Acad. Sci. U.S.A.*, 75, 1081, 1978.
92. **Gilbert, W. and Dressler, D.**, DNA replication: the rolling circle model, *Cold Spring Harbor Symp. Quant. Biol.*, 33, 473, 1968.
93. **Subramanian, K. N., Dhar, R., and Weissman, S. M.**, Nucleotide sequence of a fragment of SV40 DNA that contains the origin of DNA replication and speficies the 5′ ends of 'early' and 'late' viral RNA. I. Mapping of the restriction endonuclease sites within the *Eco* RII-G fragment and strategy employed for its sequence analysis, *J. Biol. Chem.*, 252, 333, 1977.
94. **Subramanian, K. N., Dhar, R., Weissman, S. M., and Ghosh, P. K.**, Nucleotide sequence of a fragment of SV40 DNA that contains the origin of DNA replication and specifies the 5′ ends of 'early' and 'late' viral RNA. II. Sequences of T1 and pancreatic ribonuclease digestion products of RNA transcripts prepared from subfragments of *Eco* RII-G, *J. Biol. Chem.*, 252, 340, 1977.
95. **Subramanian, K. N., Dhar, R., and Weissman, S. M.**, Nucleotide sequence of a fragment of SV40 DNA that contains the origin of DNA replication and specifies the 5′ ends of 'early' and 'late' viral RNA. III. Construction of the total sequence of *Eco* RII-G fragment of SV40 DNA, *J. Biol. Chem.*, 252, 355, 1977.
96. **Tabak, H. F., Griffith, J., Geider, K., Schaller, H., and Kornberg, A.**, Initiation of deoxyribonucleic acid synthesis, *J. Biol. Chem.*, 249, 3049, 1974.
97. **Suggs, S. V. and Ray, D. S.**, Replication of bacteriophage M13. XI. Localization of the origin for M13 single-strand synthesis, *J. Mol. Biol.*, 110, 147, 1977.
98. **Kornberg, A.**, Bacterial DNA viruses and plasmids, in *DNA Replication*, W. H. Freeman and Co., San Francisco, 1980, 494.
99. **Kozak, M.**, Possible role of flanking nucleotides in recognition of the AUG initiator codon by eukaryotic ribosomes, *Nucl. Acids Res.*, 9, 5233, 1981.
100. **Kozak, M.**, Comparison of initiation of protein synthesis in procaryotics, eucaryotics and organelles, *Microbiol. Rev.*, 47, 1, 1983.

101. **Jay, G., Nomura, S., Anderson, C. W., and Khoury, G.,** Identification of the SV40 agnogene product: a DNA binding protein, *Nature (London)*, 291, 346, 1981.
102. **Gruss, P., Dhar, R., and Khoury, G.,** Simian virus 40 tandem repeated sequences as an element of the early promoter, *Proc. Natl. Acad. Sci. U.S.A.*, 78, 943, 1981.
103. **Banerji, J., Rusconi, S., and Schaffner, W.,** Expression of a β-globin gene is enhanced by remote SV40 DNA sequences, *Cell*, 27, 299, 1981.
104. **Moreau, P., Hen, R., Wasylyk, B., Everett, R., Gaub, M. P., and Chambon, P.,** The SV40 72 base pair repeat has a striking effect on gene expression both in SV40 and other chimeric recombinants, *Nucl. Acids Res.*, 9, 6047, 1981.
105. **Proudfoot, N. J. and Brownlee, G. G.,** 3′ non-coding region sequences in eukaryotic messenger RNA, *Nature (London)*, 263, 211, 1976.
106. **Nevins, J. R.,** The pathway of eukaryotic mRNA formation, *Annu. Rev. Biochem.*, 52, 441, 1983.
107. **Jung, A., Sippel, A. E., Grez, M., and Schutz, G.,** Exons encode functional and structural units of chicken lysozyme, *Proc. Natl. Acad. Sci. U.S.A.*, 77, 5759, 1980.
108. **Tosi, M., Young, R. A., Hagenbuchle, O., and Schibler, U.,** Multiple polyadenylation sites in a mouse α-amylase gene, *Nucl. Acids. Res.*, 9, 2313, 1981.
109. **Guilley, H., Dudley, R. K., Jonard, G., Balazs, E., and Richards, K. E.,** Transcription of cauliflower mosaic virus DNA: detection of promoter sequences, and characterization of transcripts, *Cell*, 30, 763, 1982.
110. **Tegtmeyer, P., Schwartz, M., Collins, J. K., and Rundell, K.,** Regulation of tumor antigen synthesis by simian virus 40 gene A, *J. Virol.*, 16, 168, 1975.
111. **Khoury, G. and May, E.,** Regulation of early and late simian virus 40 transcription: overproduction of early viral RNA in the absense of a functional T-antigen, *J. Virol.*, 23, 167, 1977.
112. **Alwine, J. C., Reed, S. I., and Stark, G. R.,** Characterisation of the autoregulation of the simian virus 40 gene A, *J. Virol.*, 24, 22, 1977.
113. **Bensimhon, M., Gabarro-Arpa, J., Ehrlich, R., and Reiss, C.,** Physical characteristics in eucaryotic promoters, *Nucl. Acids Res.*, 11, 4521, 1983.
114. **Breathnach, R. and Chambon, P.,** Organization and expression of eucaryotic split genes coding for proteins, *Annu. Rev. Biochem.*, 50, 349, 1981.
115. **Efstratiadis, A., Posakony, J. W., Maniatis, T., Lawn, R. M., O'Connell, C., Spritz, R. A., DeRiel, J. K., Forget, B. G., Weissman, S. M., Slightom, J. L., Blechl, A. E., Smithies, O., Baralle, F. E., Shoulders, C. C., and Proudfoot, N. J.,** The structure and evolution of the human β-globin gene family, *Cell*, 21, 653, 1980.
116. **Benoist, C., O'Hare, K., Breathnach, R., and Chambon, P.,** The ovalbumin gene-sequence of putative control regions, *Nucl. Acids Res.*, 8, 127, 1980.
117. **Berk, A. J. and Sharp, P. A.,** Sizing and mapping of early adenovirus mRNAs by gel electrophoresis of S1 endonuclease digested hybrids, *Cell*, 12, 721, 1977.
118. **Favaloro, J., Triesman, R., and Kamen, R.,** Transcription maps of polyoma virus-specific RNA: analysis by two-dimensional nuclease S1 gel mapping, *Meth. Enzymol.*, 65, 718, 1980.
119. **Covey, S. N., Lomonossoff, G. P., and Hull, R.,** Characterisation of cauliflower mosaic virus DNA sequences which encode major polyadenylated transcripts, *Nucl. Acids Res.*, 9, 6735, 1981.
120. **Manley, J. L., Fire, A., Cano, A., Sharp, P. A., and Gefter, M. L.,** DNA-dependent transcription of adenovirus genes in a soluble whole-cell extract, *Proc. Natl. Acad. Sci. U.S.A.*, 77, 3855, 1980.
121. **Robinson, D.,** personal communication.
122. **Stanley, J.,** unpublished data.
123. **Townsend, R., Stanley, J., Curson, S. J., and Short, M. N.,** Major polyadenylated transcripts of cassava latent virus and location of the gene encoding coat protein, *EMBO J.*, 4, 33, 1985.
124. **Townsend, R.,** personal communication.

INDEX

A

B

C

D

E

O

P

Q

R

S

T

U

V

W

X

Y

Z

For Product Safety Concerns and Information please contact our EU representative GPSR@taylorandfrancis.com Taylor & Francis Verlag GmbH, Kaufingerstraße 24, 80331 München, Germany

Batch number: 10731074

Printed by Printforce, the Netherlands